ASTRONOMY from the BEGINNING

A history of skywatching from the Upper Palaeolithic to the Renaissance

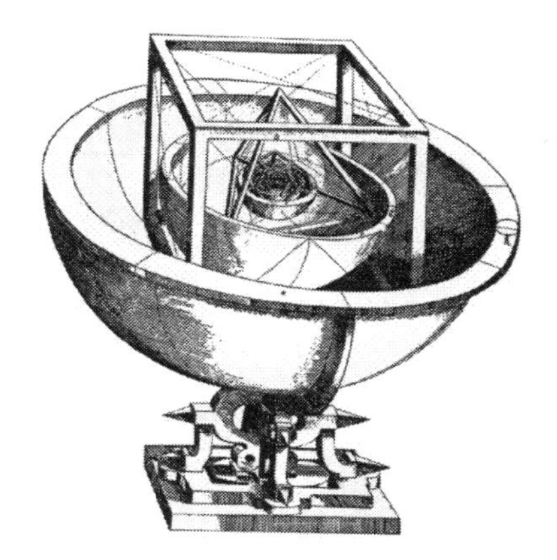

CHRISTOPHER SEDDON

Author of 'Humans: from the Beginning'

~

Astronomy: from the beginning

The incredible story of how we learned about the Universe, from the earliest prehistoric observations to the first telescopes. Early astronomy ranks among the greatest achievements of the human intellect. But how did astronomers of the pre-telescopic era make accurate observations of the Sun, Moon, and planets, and predict their movements? And how can we uncover the ancient knowledge of societies that left few or no written records? The answers are in this fascinating book, which explores the history of astronomy, from prehistoric times to the Renaissance and the birth of modern science. Written by the author of a major guide to prehistory, it is even-handed, accessible, and avoids sensationalism. While aimed at the general reader, it is also fully referenced for students and academics.

Among the topics discussed, you will discover:

★ Why some archaeologists believe that Stonehenge and other Neolithic stone monuments served astronomical functions ranging from lunar and solar markers to 'megalithic observatories', which tracked the movements of Sun and Moon with great precision.

★ What evidence there is that present-day constellations including Taurus and Orion were depicted in Upper Palaeolithic cave art.

★ How a 'void' zone in the star maps of the ancient world – corresponding to stars not visible from the Mediterranean – suggests that many constellations were devised by late Bronze Age sailors as an aid to navigation.

★ How the early Babylonian, Indian, Chinese, and Maya civilisations all developed advanced astronomical methods, for needs ranging from the compilation of reliable calendars to the prediction of eclipses and other ominous celestial phenomena.

★ How ancient Greek astronomers and mathematicians sought to explain as well as predict the movements of celestial bodies; and how their efforts culminated with Ptolemy's pivotal *Almagest*, which set out the 'Theory of Everything' of the ancient world.

★ Why Islamic Golden Age scholars did far more than simply keep alive the ancient world's intellectual tradition during the medieval period; and how they refined it and laid the mathematical foundations for the Copernican revolution.

★ How Copernicus demoted the earth from its place at the centre of the universe; and how Kepler, Galileo, and Newton went on to bring about an essentially modern understanding of the Solar System and the laws that govern it.

CHRISTOPHER SEDDON is the author of *Humans: from the Beginning*, a unique and fully researched guide to our prehistoric past, spanning the vast interval of time from the emergence of the first apes to the rise of the first cities.

~

Copyright notice

GLANVILLE PUBLICATIONS

Published by Glanville Publications, London, UK

ISBN 978-1-9162964-2-8

~

Contents

~

Introduction

The *Concise Oxford Dictionary* provides a suitably concise definition of the word 'astronomy' – "*the scientific study of celestial bodies*". In a present-day academic sense, it includes such fields as cosmology, astrophysics, celestial mechanics, planetary science, and astrometry. All these are underpinned by the physical sciences and mathematics – but the history of astronomy predates any of these disciplines. Ask people what they associate with astronomy and they are likely to reply 'telescopes', but the telescope dates to no earlier than early seventeenth century. The *Concise Oxford Dictionary* definition refers to a tradition of scientific enquiry that dates to ancient Greece in the sixth century BC – but astronomical traditions probably go back much further.

As far back as the Neolithic ('New Stone Age') period, early astronomers were seeking to understand the movements of the Sun and Moon, and mark their rising and setting points with megaliths. Much earlier, possibly as long ago as 40,000 years, Upper Palaeolithic ('Later Old Stone Age') people were recording the phases of the Moon. It is also possible that the enigmatic cave paintings of Ice Age Europe could have incorporated star maps and calendrical data.

As Neolithic societies gave way to the first cities, so astronomers were more likely to have been fulltime specialists. They now tried to understand and predict the movements of the heavenly bodies, and there was a particular emphasis on predicting eclipses. The Babylonians devised complex mathematical models for this purpose. Later, the Greeks not only produced models that represented the motions of the Sun, Moon, and planets with great accuracy; they also sought to understand just how the Solar System worked. The Greeks were probably the first to realise that the Earth is spherical; they also obtained a reasonably accurate result for the diameter. Greek cosmology placed the Earth at the centre of the Solar System where it would remain for the next 1,500 years. India and China also developed a rich astronomical tradition, as did the Arabic scholars of the Islamic Golden Age. Meanwhile, uninfluenced by and completely unaware of developments in the Old World, the Maya of Mesoamerica developed their own distinctive astronomy, underpinned by their unique calendrical systems.

This book is a history of astronomy from the Upper Palaeolithic to Sir Isaac Newton. It begins in Ice Age Europe, where our distant ancestors – people every bit as intelligent and capable as ourselves – quite possibly constructed lunar calendars and star maps. As this is intended as the story of astronomy before the telescope it should, strictly speaking, end with Johannes Kepler. But Kepler's life largely overlapped that of Galileo, and he actually improved upon the latter's telescope. If there is to be a 'natural break' in the narrative, it does not lie between Kepler and Galileo, nor indeed immediately thereafter. Instead, I have chosen to end this book with Newton who, as we shall see, really was standing on the shoulders of giants when he drew together the breakthroughs of the previous century and a half into his epochal *Principia*.

To understand the phenomena studied and recorded by early astronomers, the reader will need to be familiar with the movements of the Sun, Moon, planets, and stars as seen from Earth. All of these go through cycles of varying degrees of complexity that are explained in an appendix, *The Celestial Clockwork*, and I have also included a glossary. Terms in bold type have glossary definitions. The reader might additionally want to make use of planetarium software such as the free open source *Stellarium*, which can show the appearance of the heavens from any point on Earth at any desired time from 100,000 years in the past to 100,000 years in the future.

I am aware that the late Stephen Hawking was once warned that every equation included in his best-selling *A Brief History of Time* would halve the sales, but to do justice to the subject matter of this book a certain amount of mathematics in some of the chapters is unavoidable. However, the mathematics involved is reasonably straightforward (GCSE level and below) and – I would hope – enjoyable.

01
Astronomy in the Upper Palaeolithic

Early modern humans

Modern humans (*Homo sapiens*) emerged in Africa around 300,000 years ago, during the earlier part of the Middle Palaeolithic period[1]. These earliest representatives of our species retained archaic traits including moderately developed browridges and a long, low cranial vault; their cognitive abilities, like those of our cousins the Neanderthals, remains disputed. However, by 200,000 years ago, the modern globular cranial shape, and by implication the modern brain shape, had emerged[2]. Though we lack hard evidence, it is likely that these by now fully anatomically modern people were also behaviourally modern; that is to say, they possessed modern language, and their thought processes were fundamentally no different from ours. By 190,000 years ago, modern humans had ventured beyond Africa[3], reaching China by 100,000 years ago[4], Australia by 65,000 years ago[5], and the New World by 18,500 years ago[6]. As they spread through Eurasia, they encountered and interbred with archaic humans: Neanderthals, Denisovans, and almost certainly others.

Archaic humans first reached Europe over 1,200,000 years ago[7], but the first modern humans did not arrive until 46,000 years ago[8], rapidly displacing the indigenous Neanderthal population[9]. The archaeological record is marked by a dramatic discontinuity that spread from the Balkans across the whole of the continent over the course of 5,000 years[8]. Below the transition point are the largely utilitarian objects of the Neanderthals; above are art objects that are clearly the work of modern minds: ivory sculptures, engraved stone blocks, and bas reliefs. The European Upper Palaeolithic is characterised not just by cave art but by statuettes of animals, human and anthropomorphic figures, representations of both male and female sex organs, and so-called 'Venus' figurines. Although much earlier examples of abstract and graphic art have been found in Africa, there is no fully representational figurative art from that era[10,11]. The period saw four major tool-making industries, which partially overlapped in time: the Aurignacian (43,000-28,000 years ago), the Gravettian (37,500-15,000 years ago), the Solutrean (22,000-15,000 years ago), and the Magdalenian (18,000-12,000 years ago).

Throughout this long period, humans lived as hunter-gatherers. Only at the end of the last Ice Age, 11,600 years ago, did warm, stable climatic conditions make it possible for agricultural economies to emerge in several parts of the world. These in turn gave rise to increased social complexity, from which the first urban societies emerged from around 3500 BC. It should be stressed that hunter-gatherer people were not 'primitive' or 'savages'. They were every bit as intelligent as we are. They lacked our technologies and social complexity, but in all other aspects they were no different to ourselves. They would have had views of the heavens unspoiled by light-pollution, and which must surely have influenced them – but in what way? What can we learn about the astronomical knowledge and belief systems of people who lived tens of millennia ago?

What might the earliest astronomers have observed?

We can begin by asking what astronomical phenomena the first anatomically and behaviourally modern humans might have observed, and what deductions they might have been able to make. Those living in middle latitudes would have been aware that nature followed the regular cycle we call a year, and that in higher latitudes (though not in the tropics) the year divides naturally into four seasons:

1. Summer, when days are long and hot, and deciduous trees are in leaf;
2. Autumn (fall), when days grow shorter and leaves turn brown and fall from the trees;
3. Winter, when days are short and cold, and trees are devoid of leaves;
4. Spring, when days are lengthening, and the leaves return.

It is reasonable to suppose that these people would have noticed a link between the seasons and the behaviour of the Sun: that in summer it rises and sets well to the north of where it rises and sets in winter; and that it climbs much higher in the sky and is above the horizon for longer in summer than in winter. It is also perfectly feasible that they would realise that the Sun moves cyclically between northern and southern rising and setting limits, reversing direction whenever it reaches one of these limits, i.e., that they were aware of the summer and winter **solstices**. It is also possible that they would have noted the times that when the Sun reached the mid-points of these limits, day and night were of equal length (the **equinoxes**), although is we shall see in subsequent chapters, this is less likely.

At night, early skywatchers would have seen the myriad points of light we refer to as stars, many of which are bright enough to form distinctive patterns in the sky, and they would have seen the misty band of light we know as the Milky Way. A casual observer would notice that while the stars move from hour to hour, they remain fixed in relation to one another, and the patterns they form in the sky do not change.

They would soon come to realise that the stars all circle around a fixed point in the sky. For those in Eurasia, that fixed point is the north celestial pole; for their counterparts in Australia, it is the south celestial pole. Any star that is close to either celestial pole (as Polaris currently is to the north celestial pole) will remain motionless in the sky. Some constellations, close to the celestial pole, never set (**circumpolar**). Anybody who familiarised themselves with night skies would soon recognise that non-circumpolar constellations rise earlier each night, and thus different constellations are prominent at different times of the year. A constellation rising in the east soon after sunset in winter will disappear into the western twilight as summer approaches.

The behaviour of the Moon is rather more complex. It is obvious that it changes its appearance from night to night in a regular manner, waxing from a slim crescent in the western skies after dusk to a full disk that rises at sunset, before waning back to a slim crescent seen in the east just before dawn, and then disappearing entirely for a few days. Anybody taking an interest in the phases of the Moon would also be aware that it rises later each night. It is likely that they would notice that:

1. The Moon's rising and setting positions, like those of the Sun, move cyclically between northerly and southerly limits, but over the course of a **lunation** rather than

a year;

2. The full Moon rises very high in winter, but it remains low in the sky in summer.
3. The Moon moves against the starry background, its movements readily perceptible over the course of a few hours;
4. Its movements are always confined to the same narrow band of the night sky;
5. For a given phase at a given time of year, the Moon will always be seen against the same background of stars – for instance, in midwinter, the full Moon will be in or close to the constellation we now call Gemini.

Much less obvious are the opening and closing of the Moon's northerly and southerly rising and setting limits over the course of its **nodal cycle** of 18.61 years, but there is no reason to believe that early skywatchers could not have been aware of them. In higher latitudes, they would notice that in some years the summer full Moon remained even closer to the horizon than usual.

How familiar might our Palaeolithic ancestors have been with the visible planets, and were they aware that their behaviour differed to that of the stars? They would certainly have been aware that on occasions a very bright star-like object (Venus) could be seen in the western skies at sunset, or in the eastern skies before sunrise. This object was sometimes bright enough to cast shadows. Did they realise that the morning and evening objects were one and the same? Did they ever notice that on rare occasions, a second, much fainter object (Mercury) could also be seen in the evening twilight or predawn skies?

We can be sure that that they would have noticed Mars, Jupiter, and Saturn, all of which are very prominent in the night sky. Observation over time will readily show that they do not remain fixed against the stars and move in the same narrow band as the Moon, albeit at a far gentler pace – but were early skywatchers aware of this? Again, there is no reason to believe that they were not.

Lunar and solar eclipses would have certainly been as awe-inspiring to people of these times as they are to us. Lacking our understanding, they would likely have found these phenomena rather more disconcerting. From any point on Earth, lunar eclipses will be seen roughly once a year assuming good weather, but only around 40 percent will be total, and only around half will be visible from start to finish. Nevertheless, at intervals usually no longer than a few years, people would witness what to them must have been a very disturbing phenomenon as a darkness gradually fell over the full Moon, followed by it turning dark red. This phenomenon would be short-lived, usually lasting around an hour but never more than two hours, before the mysterious darkness retreated and the Moon resumed its normal appearance.

Although solar eclipses are more frequent than lunar eclipses, totality affects only a very small region, whereas a lunar eclipse can be seen from anywhere where the Moon is above the horizon. The wait time for a total eclipse of the Sun from any one point on earth is of the order of three or four centuries, hence most people would never see one. However, it is possible that some groups might have had legends of an even more terrifying phenomenon than the 'blood moon' where a darkness that was not a cloud moved across the face of the Sun, turning day into night, and causing the stars to become visible for several minutes.

Partial solar eclipses are visible over a much wider area, and it is possible that among groups there were individuals who claimed to have seen a dark chink appear in the Sun, or even a sizeable 'bite' taken out of one side. If so, it is to be hoped that they did not stare directly at the Sun, or that they noticed the eclipse through light cloud. At some point, it might have been noticed that a solar eclipse could only happen during the few days each month when the Moon is absent from the skies.

Finally, we might want to ask if there are any of the present-day constellations that are so distinctive that the same or very similar groupings were recognised by Upper Palaeolithic peoples (the imperceptible proper movements of the stars through space did not bring them into the configurations we now recognise until around 30,000 years ago). It could be argued that there are at least three:

1. Ursa Major (the Great Bear) containing the seven second-magnitude stars making up the Plough or Big Dipper, all of which are circumpolar in the northern hemisphere and prominent at all times of the year;
2. Taurus (the Bull) containing the compact but extremely noticeable open clusters of the Pleiades (the Seven Sisters) and the Hyades, with the latter combining with the first-magnitude star Aldebaran (α Tauri) to form a prominent V-shape;
3. Orion (the Hunter) containing three bright, equally spaced stars in a row (the Belt), with the naked-eye Orion Nebula to its south, and flanked by the first-magnitude stars Rigel (β Orionis), Betelgeuse (α Orionis), and Bellatrix (γ Orionis).

All can be easily seen on a clear night from central London, which is one of the most heavily light-polluted spots on Earth.

There is absolutely no doubt that Palaeolithic people could have observed and made records of most of the phenomena outlined above. The questions, of course, are did they, and if so for what purposes – and how can we, tens of millennia later, hope to find an answer to these questions?

There are two main lines of enquiry that we can pursue. The first, obviously, is the archaeological record. The Upper Palaeolithic of Europe presents us with a dramatic ensemble of both parietal (cave) and mobiliary (portable) art, and from this it has been suggested that the people of these times recorded astronomical phenomena including constellations and the phases of the Moon. Two questions that might not seem so obvious from a modern perspective are firstly, given that they lacked writing are, did the people of that era have the concept of making records; and secondly did they have the concept of counting or numbers? This brings us to our second line of enquiry, the ethnographic record. By examining ethnographic data on cultures that might have resembled those existing in the Upper Palaeolithic, we might gain an insight into the roles played by astronomy among the latter.

External memory devices

To a present-day amateur astronomer, a notebook is only slightly less important than a telescope. No telescope is required to note and record the phenomena previously outlined, although longer-term phenomena such as the regression of the lunar nodes (opening and closing of lunar standstills) require observations over the course of several decades. The ability to make records of such observations is essential. Although writing did not appear until the Bronze Age, there are other means by which people living in earlier times could have recorded astronomical phenomena. Yet as we shall see, the archaeological evidence for

Upper Palaeolithic notation systems and record-keeping is not widely accepted.

The existence of Upper Palaeolithic record-keeping, if proven, would imply that humans began the recording and storage of data on a medium external to the brain many millennia before the invention of the first writing systems. Cognitive researchers use the term 'external memory device' to describe physical devices specifically designed to record, store, and recover information. Although it is widely accepted that such devices predate the invention of writing, little is known about their origins and early stages of development. Their use implies the use of modern language, without which the symbols used to record data could not be created nor understood. Such systems represent an important stage in the development of both human cognitive abilities and social complexity - an expansion of memory necessitated by the emergence of societies with information requirements too great to be stored and handled by the brains of individuals[12,13].

Numeracy and counting systems

In addition to some form of notation system, Upper Palaeolithic astronomers would have required at least basic numeracy and counting systems. The ability to use numbers is so essential to our everyday lives that we tend to assume it has always been the case: for example, we see no reason to think that an Upper Palaeolithic hunter, on coming across a group of six bison, would not have been able to make a report to back to their fellows that they saw "*six bison by the stream*". In fact, the ability to enumerate or to count either verbally or in writing requires a certain level of social complexity, in addition to language and the use of symbols. Upper Palaeolithic people might have lacked the concept of the number 'six'. Similarly, while they might have recognised that their hands and feet each have the same number of digits, they might not necessarily have had a word for 'five'[14].

Spoken numerals are of two types: atoms and syntagms. An 'atom' in this context is a single morpheme (indivisible unit of language) and syntagms are composites derived from atoms. Their creation involves three components: atoms, a base (which is typically an existing atom used serially to derive larger numerals), and arithmetic operations. For example, in English 2,521 is derived as *two* × *thousand* + *five* × *hundred* + *two* × *ten* + *one*. Here *thousand*, *hundred*, and *ten* are bases and *two*, *five*, and *one* are atoms. In French *quatre-vingts-neuf* (89) is derived as *quatre* × *vingt* + *neuf*. Here, *vingt* is the base and *quatre* and *neuf* are atoms. Bases can themselves be derived from operations and atoms; for example, *hundred* and *thousand* are derived from *ten*. Alternatively, bases can be from unrelated roots; for example, in the Yuchi language spoken by the Native American Yuchi people, the base for 100 means 'road' and the base for 1,000 means 'road large'. In many languages, the word for 'five' is the same as that for 'hand'. The most common bases are quinary (five), decimal (ten), vigesimal (twenty), and combinations of vigesimal and decimal[15].

American linguist Patience Epps, and colleagues[16] surveyed the numeral systems of 193 languages spoken by hunter-gatherer and low-level food-producing groups in North America, South America, Australia, and Africa. They found that hunter-gatherer numeral systems are not always small and simple, nor when compared with systems of neighbouring agricultural groups, are the latter necessarily complex. Factors such as numerical limits (i.e., the largest number that can be expressed), bases (values and multiplicity; a language can use multiple bases), and the transition points from atomic to derived numeral terms tend to correspond primarily to geographical region. Numerical complexity and subsidence type are correlated only weakly, if at all. In Africa and South America, there is a tendency for hunter-gatherer numeral systems to have low overall limits, and correspondingly low uppermost bases, whereas systems wither larger overall limits and higher uppermost bases are more common among agriculturalists. However, restricted numeral systems are also widespread among agriculturalists in South America. No evidence was found for any correlation in North America. Hunter-gatherer systems also tend to be influenced over time by the systems of neighbouring groups, including agriculturalists. Australia (which has no native agriculture) and South America (where most groups are low-level food producers to some extent) both tend toward restricted numeral systems. In Australia, where there was little or no contact with agriculturalist languages before the European arrival, none of the 122 languages surveyed had numeral terms above twenty. Despite the enormous variation among numeral systems in contemporary hunter-gatherer languages, it is possible that such restricted systems were much more common throughout most of human history.

Ethnographic data and ethnological studies

If we accept the existence of notation systems and numeracy in the Upper Palaeolithic, the next question is, were these used to record astronomical phenomena if indeed astronomical phenomena were observed at all? If so, the next question is, why? One line of investigation is to examine ethnographic data on cultures that might have resembled those existing in the Upper Palaeolithic, an approach that is not without its pitfalls[17,18]. Only uncontacted peoples are entirely uninfluenced by modern civilisation, and by definition obtaining ethnographic data about such peoples is problematic. Another issue is the considerable difference between the landscapes, fauna, and flora of the Late Pleistocene and those of the Holocene. Late Pleistocene societies could have faced problems that have no Holocene equivalent. We cannot therefore automatically assume that the social organisation and belief systems of Upper Palaeolithic people have analogues among present-day hunter-gatherers.

With these caveats in mind, Canadian archaeologists Brian Hayden and Suzanne Villeneuve[19] carried out an ethnological study (comparative study of ethnographic data) to assess the plausibility of claims for Upper Palaeolithic astronomy. They began by distinguishing between 'simple' and 'complex' hunter-gatherers and drew up a list of differences between the two types of society.

Complex hunter-gatherers are generally characterised by the following important traits:

1. Relatively high population densities (0.2 - 10.0 people per square kilometre);
2. Seasonal or full sedentism;
3. Private ownership of products and some productive resources by individuals or families;
4. Storage controlled by individuals or families;
5. Significant socioeconomic differences within communities (often reflected in burials);
6. Manufacture and trading of prestige goods based on surpluses;
7. Competitive displays and elaborate feasting activities based on surpluses;
8. Bride prices or dowries;

9. Attempts by elites to control access to the supernatural;
10. Hierarchical and heterarchical socio-political organisations based on economic production;
11. Complex counting systems that extend into the hundreds or thousands.

By contrast, simpler, more egalitarian hunter-gatherer societies are characterised by lower population densities, obligatory sharing, communally owned resources, simple counting systems limited to ten to twenty numbers, minimal socioeconomic inequalities or prestige objects, and negligible economic-based competition or feasting, or marriage payments.

Complex hunter-gatherer societies are known in North America (California, Northwest Coast and Plateau, and Alaska), Siberia, and Japan. Could such societies have existed during the Upper Palaeolithic; for example, in the most productive Eurasian environments such as the Périgord region of southwest France?

Upper Palaeolithic groups in Europe exhibited relatively high population densities, and they were able to establish seasonal or even fully sedentary communities lasting several years in favourable locations. These groups stored large quantities of food, had surpluses of resources in sufficient quantities, and had sufficient time to manufacture prestige goods to trade for shells, amber, and other exotic items from distant locations. The surpluses produced appear to have underpinned socioeconomic inequalities between families. Certain individuals were lavishly buried; for example, the two adolescent children buried with extensive grave goods at Sungir 200 km (120 miles) east of Moscow. Their status was clearly inherited, as they were too young to have earned it in their own right. There is evidence for competitive feasting in the form of prestige eating utensils made from antler and ivory, and thick concentrations of animal bones at some caves. Much Upper Palaeolithic cave art is located deep underground, which thus limits the ritual experiences associated with them to a privileged minority. By contrast, though simple hunter-gatherer groups do sometimes produce images, they are almost never produced in deep parts of caves. Based on this evidence, Hayden and Villeneuve concluded that complex hunter-gatherer societies did exist during the Upper Palaeolithic.

The next task was to survey the ethnographic record for evidence of astronomical activities, with the caveat that ethnographers might not have collected all the relevant data. 82 groups were considered: 7 simple groups and 75 complex groups. This is rather lopsided, but perhaps fewer records were available for the simple groups.

The survey reviewed the following:

1. The degree to which various groups monitored solar or lunar movements, how and why monitoring was carried out, and information on the social context associated with solstices where available;
2. Whether various groups had calendars, the organisation of their calendars, how they were determined, and what celestial monitoring was involved;
3. Whether various groups recognised constellations, and which stars were included in these constellations.

The ethnographic data showed that almost all hunter-gatherer groups, both simple and complex, have a degree of astronomical knowledge. Simple hunter-gatherer groups are broadly aware of the northerly and southerly limits of solar movements, but they do not typically determine the specific days of the solstices, nor do they mark them with special ceremonies.

Of the complex hunter-gatherer groups, 63 out of the 75 monitored solstices. There was no information available regarding the remainder, but there was no evidence to suggest that they explicitly ignored solstices. All the 63 regarded the winter solstice as being important: as the start of the new year, as a time for holding rituals, dances, and feasts, or in some cases starting food restrictions. The summer solstice was generally regarded as being of lesser importance: only 23 groups gave it equal status with the winter solstice. With one possible exception, there was no evidence that any of these groups monitored or attached any importance to the equinoxes.

The solstice observations typically involved specialists making careful and accurate monitoring of solar rising and setting positions with tree, post, or rock alignments viewed from special locations. The role of solstice monitor was often held by shamans or other ritual specialists; for example, 'sun priests', 'old men', 'wise old men', 'calendar experts', 'calendar keepers', or 'official elders'. Their job was often highly political, as it also involved setting a date for ceremonies, rituals, feasting, and dances. Given the competitive nature of these events, it was vital to get the date right. Differences of opinion over the correct date often led to acrimonious disputes, but errors of four or more days were not uncommon. Disputes over timing even form part of the mythology of some groups. But it is likely that the need for accuracy drove refinements in astronomical monitoring.

Both simple and complex hunter-gatherer groups showed an awareness of stars and constellations; they played a part in mythology, were used for navigation at night, and as indicators of the time of year. However, to recognise the ecliptic and to construct a zodiac to monitor solar, lunar, and planetary movements requires far more sophisticated astronomical concepts, and it has been often been suggested to be beyond the reach of all but literate state-level societies such as those of ancient Mesopotamia. It is therefore something of a bold step to propose that such a system existed in the Upper Palaeolithic.

For this part of the survey, Hayden and Villeneuve considered 26 hunter-gatherer groups. Eighteen constellations were mentioned in the literature, but all but four consisted of only small portions of the generally recognised constellation, or indeed of single stars; for example, knowledge of Taurus was restricted to the Pleiades and Aldebaran. Only three zodiacal constellations were referenced (Taurus, Leo, and Gemini) suggesting that this region of the sky was not of particular importance. There was widespread awareness of Orion and Ursa Major, along with the Pleiades, the Milky Way, and Venus, but other constellations were recognised by only three or fewer groups. It would suggest that most 'constellations' were purely local constructs.

Hayden and Villeneuve suggested that hunter-gatherer constellation concepts tend to be quite different from those familiar to present-day astronomers. Individual stars most commonly represent individual people or animals or things in a mythological story, while clusters of stars tend to represent groups of animals or people.

The distinctive Pleiades were recognised by a wide range of hunter-gatherer and agricultural societies throughout the world. They were almost always viewed as a group of individuals: sometimes as a group of women, sometimes chased by another star or cluster representing a man or men; others view them as children. Their importance cross-culturally strongly suggests that they would have been similarly notable in the prehistoric past.

Australia

We can gain a possible insight into Upper Palaeolithic astronomy through the astronomical traditions of Aboriginal Australians. Humans have been living in Australia for at least 65,000 years and probably longer[5]. Throughout that time, they have lived as hunter-gatherers. Barring claims for limited food production in the Murray Valley between Victoria and New South Wales[20], Australia is the only inhabited continent where indigenous agriculture never arose.

Aboriginal Australian culture remained continuous until the arrival of the British in the late eighteenth century, thus it is one of the oldest cultures anywhere in the world. At the time of the British settlement, there were about 300 distinct Aboriginal language groups with nearly 750 dialects. The songs, stories, and belief-systems of each were distinct though most were based on the concept that the world was created by ancestral spirits, whose presence can still be seen on the land and in the sky. According to tradition, these spirits provided a holistic philosophy of how people should live and interact with one other, and with the environment. The Warlpiri people of the Northern Territory use the word *Jukurrpa* to describe the creation period and the philosophy; it is inaccurately referred to as the 'Dreaming' or 'Dreamtime' by English speakers. The arrival of the British colonists did of course have a catastrophic effect on Aboriginal cultures, particularly in southeastern Australia. However, some groups in northern and central Australia still retain most of their precontact language and culture. It has been possible to obtain much information about the astronomical traditions of these groups[21].

Aboriginal cosmology centres on other realms within their cultural landscape to which they can travel in spirit form. These realms are the Skyworld (the heavens) and the Underworld where the Sun Ancestor spends the night. The Skyworld is a reflection of the everyday landscape, with many plants and animals shared between both realms. Many of the planets and constellations were considered to be the Ancestors of living people[22].

The songs, stories, art, and ceremonies of many Aboriginal Australian cultures refer to the Sun, Moon, planets, stars, and the Milky Way. Aboriginal constellations include the Southern Cross, which variously represents the footprint of an emu, a stingray, a possum in a tree, or a red river gum tree. The emu is also represented in many Aboriginal cultures by dark nebulae stretching from Scorpius to the Coalsack Nebula close to the Southern Cross. Orion frequently represents a young man or group of young men, chasing the Pleiades, or paddling a canoe across the sky[21,22].

Aboriginal calendars tend to be more complex than Western calendars, and in the north of Australia they are often based on six seasons. Some Aboriginal groups mark them in terms of the stars which appear during these seasons. For example, the appearance of the Pleiades in the dawn sky in late May marks the start of winter. The appearance or disappearance of stars or constellations also play a role in planning the collection of seasonally available food sources. For example, when the star Spica (α Virginis) sets in the evening sky just after the Sun, the Yolngu know that it is time to gather the tubers of the lotus lily, which is an important food source for them. Similarly, the Boorong people of Victoria make use of the disappearance of Lyra from the evening sky in October, which coincides with when the eggs of the malleefowl (a ground-dwelling bird about the size of a domestic chicken) are laid and ready to be collected. To the Boorong, the constellation of Lyra represents a malleefowl rather than a lyre[23].

In addition to such practical considerations, Aboriginal Australians also sought to understand the motion of celestial bodies and such phenomena as eclipses and tides. For example, the Yolngu people describe the Sun-lady (Walu), who lights a stringybark tree every morning, and then carries it across the sky to her camp in the West. Some coastal believed that when the Sun leaves the Skyworld at the end of the day, it enters the Underworld via the western seas. Conversely, many inland groups believed that the setting Sun enters a cave or a hole in the west similar to those left by a large tree burned in a bush fire, and the next morning the rising Sun emerges from a similar hole in the east. The Yolngu had an interesting explanation for the phases of the Moon. Originally, the Moon-man (Ngalindi) was a fat lazy man (corresponding to the full Moon), but his wives attacked him with axes (resulting in the waning Moon). Ngalindi managed to escape by climbing a tall tree to follow the Sun, but he died from his injuries (the new Moon). After three days, he was reborn, growing round and fat (the waxing Moon), until, after two weeks his wives attacked him again. The cycle continues to repeat every month. The Potaruwutj people of South Australia have a different explanation. To them, the Moon-man (Mitjan) was the male Ancestor of the Quoll (Native Cat) totemic clan. He attempted to steal the wife of another being, but he was driven away. Left wander alone, he was at times well-fed (when the Moon is waxing) but at other times starving (when it is waning)[23,21,22].

At least three groups (the Euahlayi, Yolngu, and Warlpiri) recognise that the Moon is responsible for solar eclipses: all interpreted it as the Sun-woman being hidden by Moon-man as he has sex with her. In the Warlpiri account, the Sun-woman was sent away by the sky spirits because she attempted to seduce the Moon-man. Given the rarity of total solar eclipses in any one location, this explanation implies a remarkable continuity of knowledge passed between generations. People must have seen a solar eclipse, related it to a story passed to them from several generations previously, and then passed on the story for future generations[23,21].

A number of rock carvings in Ku-ring-gai Chase National Park, Sydney depict a man and a woman reaching towards a crescent with its horns pointing downwards. Conventionally, the couple are interpreted as reaching up towards a boomerang, although it is not clear why they should do so; nor is a boomerang typically crescent-shaped. Another possibility is that the crescent represents the Moon, but the horns-down configuration only ever occurs in the morning (waxing crescent) or afternoon (waning crescent) when the Sun is high and the Moon barely visible. The likeliest explanation is that the carvings represent a solar eclipse with the Moon-man partially covering the Sun-woman just before totality[21].

The Warlpiri also attribute a lunar eclipse to the Moon-man being chased by the Sun-woman, who eventually catches up. To realise that an alignment of the Sun and the Moon is responsible for a lunar eclipse is an impressive feat of visualisation. The Yolngu, for their part, noticed the connection between the Moon and the tides, and that spring tides occur when the Moon is new or full. When the tides are high, water fills the Moon as it rises. As the water runs out of the Moon, the tides fall, leaving the Moon empty for three days. Then the tide rises once more, refilling the Moon[23,21].

Also impressive is the Yolngu's apparent understanding of the movements of Venus, which they refer to 'Banumbirr' during its morning apparitions. They believe that Banumbirr is tethered by a faintly visible rope to the mythical island of the dead, Baralku, which lies to the east. This rope allows the living Yolngu to communicate with their ancestors on Baralku, and it also prevents Banumbirr from rising high in the sky. The rope has been interpreted as the zodiacal light (a faint glow caused by sunlight reflected off dust particles in the plane of the Solar System), which is clearly visible in Arnhem Land. The Yolngu stage an elaborate ceremony for communicating with their ancestors, which begins at dusk and climaxes when Banumbirr appears in the predawn skies. The ceremony requires considerable planning, which in turn implies a knowledge of when Venus is going to rise before dawn[23,21].

There is absolutely no reason to suppose that these rich traditions and sophisticated astronomical knowledge have not existed for a very long time, quite possibly for as long as humans have lived in Australia. Equally, there is no reason to suppose that comparable knowledge and traditions did not exist among Upper Palaeolithic people in other parts of the world. As we shall see, there is evidence to suppose that this was indeed the case.

02
Upper Palaeolithic lunar calendars

Lunar and lunisolar calendars

The word 'calendar' derives from the 'calends' (*Kalendae*), which in Roman times was the first day of the month. The word *Kalendae* is in turn derived from the Latin word *calare* ('to call or proclaim'). The Romans originally used a lunar calendar, and the start of each month was proclaimed after the first sighting of the new crescent Moon in the evening sky. This is the same system as is used in the traditional Islamic calendar; other examples of lunar calendars include the Jewish and Chinese calendar, although these mostly no longer rely on the actual sighting of a new Moon.

The earliest calendars are thought to be lunar. The cycle of lunar phases provides a simple, easily observable, means of tracking convenient periods of time. An important consideration is that the **tropical year** (365.242189 days) is not a whole number of **synodic months** (the time for the Moon to go through a full set of phases, i.e., 29.531 days), so lunar calendars typically rely additionally on observation of other astronomical events such as solstices and the insertion of intercalary or 'leap' months as required to keep them in step with the seasons.

Such calendars are known as lunisolar and might have been used in Mesolithic times. At Warren Field in Aberdeenshire, Scotland, an arc formed of twelve pits dating to around 8000 BC was identified from aerial photography in 2004 and was excavated over the next decade. It is thought that the pits were linked to synodic months, and that a marker was used to indicate the current month. Around 8000 BC, as viewed from the central pit, Pit No. 6, the winter solstitial sunrise occurred over a 'notch' in the low hills on the southern horizon; this would have served to periodically reset the calendar[1].

Alexander Marshack and *The Roots of Civilization*

Some forty years before the discovery at Warren Field, the American science journalist and independent researcher Alexander Marshack (1918-2004) claimed that lunar calendars originated far earlier, during the Upper Palaeolithic. He claimed that the notches and lines carved on some portable Upper Palaeolithic artefacts were notation systems for recording the phases of the Moon. He put forward his ideas in several publications, most notably a 1972 monograph entitled *The Roots of Civilization*[2].

Marshack became interested in prehistoric astronomy in 1962 after being commissioned to co-author a book with physicist Robert Jastrow, head of NASA's theoretical science division and former chair of the Lunar Exploration Committee. The book would describe how science had advanced to the point where it was now possible to plan a Moon landing, and also explain some of the technical problems involved.

The project occupied Marshack throughout 1963, during which he met hundreds of American and Russian scientists and engineers. But while trying to write a few pages about the earliest history of astronomy, he became struck by the notion that all the major human advances were supposedly marked by "*a series of 'suddenlies'.*" The sciences, writing, and state-level societies had all appeared 'suddenly' in various parts of the world; before that agriculture had appeared 'suddenly'; and art and decoration had appeared 'suddenly' during the last Ice Age about 40,000-30,000 years ago. Despite lacking any relevant academic qualifications, Marshack became a research associate at the Peabody Museum of Archaeology and Ethnology at Harvard University, Cambridge, MA, with the support of archaeologist Hallam Movius. This gave him access to state and university archaeological collections that would otherwise have been unavailable to him.

We now know that Marshack's scepticism about 'suddenlies' was entirely justified and that all these developments had a lengthy gestation. The origins of agriculture, for example, precede the end of the last Ice Age by more than ten millennia. However, Marshack was writing at a time when it was widely believed that modern human behaviour was a comparatively recent phenomenon that only emerged 40,000-30,000 years ago, well after the emergence of anatomically modern *Homo sapiens*. This view was based on the sudden appearance of cave art in Europe at this time; previously, it was argued, anatomically modern humans might have seemed the real deal, but mentally they had not quite 'got there'; and to our way of thinking were simple-minded and probably lacked our complex, syntactic language.

In a landmark paper, anthropologists Sally McBrearty and Alison Brooks[3] argued that the notion of what biologist Jared Diamond[4] termed the Great Leap Forward was based on an overly Eurocentric reading of the archaeological record. They suggested that the emergence of what we term 'modern human behaviour' was a gradual process that had begun tens of millennia earlier in Africa with the emergence of 'archaic *Homo sapiens*' around 300,000 years ago. Subsequently, evidence from Africa and elsewhere suggested that McBrearty and Brooks were right, and that behavioural modernity emerged well before 40,000 years ago, probably at the same time as full anatomical modernity. However, Marshack[5,6,7,8] was advancing similar arguments more than two decades earlier, and was well ahead of what was then-current thinking.

Marshack considered the then-current theories about origins of agriculture and other developments, and he was rightly sceptical. He took the correct view that agriculture and inventions such as the calendar were products of "*an exceedingly long development*" in human culture. In search of evidence for early calendars, Marshack considered what he termed 'time-factored activities', or activities that require planning and extend over significant spans of time. For example, hunting and agriculture are both time-factored processes. Agriculture requires 'storied' information to anticipate the demands for labour and equipment for the various times of the year. Marshack believed that 'stories' related to subsistence strategies must have existed for thousands of years for agriculture to have come about, as without them the specific needs for a successful harvest could not have been anticipated. Similarly, hunter-gatherers would have required 'storied' information about the seasons in order to plan fishing expeditions to the coast, harvesting fruits and berries, or exploiting annual herd migrations. His view was that the ability of Upper Palaeolithic people to carry out time-factored activities around the changes of the seasons led to the subsequent emergence of agriculture and the rise of complex societies[9,10].

Marshack believed that the earliest evidence for human use of time-factored skills might be found in the mobiliary and cave art of Ice Age Europe. In June 1962, he wrote to French anthropologist André Leroi-Gourhan asking whether the cave paintings depicted seasonal or periodic phenomena: mating or rutting behaviour of herds, migration times, seasonal rains, etc. Leroi-Gourhan's reply was surprisingly equivocal, claiming that there was "*some evidence for the seasonal character of the cave visits*". In fact, there is ample evidence of the type Marshack enquired about: at the 18,000-year-old Lascaux cave complex in the Dordogne, animals are depicted in seasonal coats corresponding to their mating periods: horses at the end of winter and early spring, aurochs in summer, and stags in autumn[11].

In a short paper in the journal *Science*, Marshack[12] described examples of 'lunar notation' from two rock paintings associated with the Mesolithic Azilian culture of Spain, a carved mammoth ivory from Ukraine associated with the Late Upper Palaeolithic Magdalenian culture, and an engraved reindeer bone from the former Czechoslovakia associated with the earlier Upper Palaeolithic Aurignacian culture. The idea that there were any form of notations or tallies in the Upper Palaeolithic was contrary to the prevailing academic opinion, and it is by no means universally accepted to this day.

The term 'notation' was introduced by Marshack, and describes the processes involved in compiling records in a pre-arithmetic context. Anthropologist Judy Robinson[13] defines 'notation' as a two-stage process; firstly, the recognition of a given quantity by an individual, and secondly, the representation of this quantity in a systematic manner. 'Tallying' is a form of non-numerical counting, and it is therefore a notation. Ethnographic and historical evidence suggests that the most common method of recording a quantity or establishing an equivalence between two quantities is by making distinguishable marks on a surface. Thus, records can be kept and subsequently be read by others. The idea that Upper Palaeolithic people kept tallies goes back to the nineteenth century, when French archaeologist Édouard Lartet suggested that notches on the sides of bones were *marques de chasse*, or marks denoting hunting kills.

While notation proposed by Marshack in his 1964 paper does not follow any consistent format between the artefacts (as might be expected for artefacts widely separated both in time and space), he claimed that there is a common theme in that the numbers 29 and 30 are represented, either as a total number of notches, or as part of a repeating pattern. This equates to the synodic month. Marshack believed that Upper Palaeolithic people could not count and had no concept of numerals; for example, three notches did not equate to our concept of the number three. Rather they equated to an interval of time such as the days from the last sighting of the morning crescent Moon to the first sighting of the evening crescent Moon, with a notch being recorded for each night of the Moon's absence.

In *The Roots of Civilization* Marshack[2] continued in a similar vein, and he subjected many artefacts from the Mesolithic and Upper Palaeolithic to a detailed examination. These included a bone tool with a quartz point from Ishango in the former Belgian Congo (now the Democratic Republic of the Congo). The bone was then thought to be Mesolithic, with a date no earlier than 6500 BC, but it is now known to be around 20,000 years old, dating to the later Upper Palaeolithic. Marshack interpreted it as a lunar calendar, recorded over the course of about five and a half months.

He drew similar conclusions about other artefacts, including a 35,000-year-old

Aurignacian engraved bone from Abri Blanchard in the Dordogne, France, and a late Upper Palaeolithic Magdalenian bone baton from Le Placard Cave at Charente, Nouvelle-Aquitaine, France. The Abri Blanchard bone is a small ovoid bone plaque measuring 110 mm (4.33 inches) in length. At one end, there is a 52 mm (1.75 inches) engraved area containing a serpentine pattern of small circular, crescent, and kidney-shaped marks known as punctuations, aligned in gradually changing directions. The marks were made by a technique known as a power-turn, which involves applying the sharp point of a stone tool and rotating it in one direction through an angle of 90 to 130 degrees. This deepens the shallow depression made by the initial application of the tool point. The technique was first recognised in the 1930s, and it has been seen on many Aurignacian artefacts[14]. Marshack interpreted the Abri Blanchard bone as representing the phases of the Moon recorded over two-and-a-quarter months; the Le Placard baton he saw as an extremely close lunar tally, with changes of point, stroke, grouping, or side denoting changes of lunar phase.

In later work, Marshack continued to interpret Upper Palaeolithic artefacts as lunar calendars. These included the La Marche antler from Lussac-les-Châteaux[15] and a small bone plaque from La Grotte de Thaïs, Drôme. The latter, dating to the late Magdalenian period 14,000 years ago, he viewed as a lunar/solar calendar, which recorded solstices as well as lunar phases. Marshack believed that this refinement was the culmination of an evolving tradition that had begun 20,000 years earlier[16].

Marshack[16] suggested that when Neolithic farmers moved into Europe from Southwest Asia, they made use of indigenous hunter-gatherers' non-arithmetical observational astronomical skill and lore, and a seasonal economic and ritual calendar that dated to Upper Palaeolithic times. Megalithic lunar/solar calendars might have their origins in observational lunar/solar calendars such as that represented by the La Grotte de Thaïs bone plaque, themselves the end product of a tradition of non-arithmetical astronomical observation and record-keeping going back to the Aurignacian period.

In addition to his study of artefacts from the Upper Palaeolithic, Marshack considered ethnographic data, in particular a nineteenth century calendar stick originally belonging to Tshi-zun-hau-kau, a chief of the Winnebago of Nebraska. The stick remained in the hands of Winnebago families until the twentieth century, when it passed via a local collector into the collection of the Cranbrook Institute of Science, Michigan. Marshack[9] carried out a detailed examination of the 1.32 m (4 ft 4 in) hickory stick and concluded that it contained two years of lunar notation. He claimed that his analysis documented the notation of a precise, non-arithmetic, observational lunar year of twelve months. There was also evidence for subsidiary months, suggesting the use of an intercalary month every three years to keep the calendar in line with the tropical year. Marshack also suggested that similar to the early Neolithic farmers, the Winnebago had derived their calendrical tradition from the Upper Palaeolithic hunter-gatherers who initially settled the New World after crossing the land bridge that then linked Alaska to Siberia. Marshack believed that there was a widely distributed tradition for the use of calendar sticks by Native Americans, and that this demonstrated that complex day-counts, lunar cycles, and seasonal calendars could be, and were, kept by non-literate societies with relatively simple socio-political organisations. Other researchers have concurred that groups such as the Zuñi and Hopi do indeed keep such calendars[17].

As previously noted, (see Chapter 1) there is an enormous variation among numeral systems in contemporary hunter-gatherer languages, but many of these groups do not live in isolation from agriculturalists. It is therefore possible that more restricted systems were much more common throughout most of human history. In a world where there were no agriculturalists, possibly there were only very restricted numeral systems, as Marshack suggested.

Why calendars?

Brian Hayden and Suzanne Villeneuve noted that the traditional explanation for astronomical calendars is so that ritual elite experts can advise farmers when to plant their crops. Such an explanation would, of course, be inapplicable to Upper Palaeolithic hunter-gatherers, but Hayden and Villeneuve further noted that the explanation has been challenged. For example, South African archaeologists David Lewis-Williams and David Pearce[18] believe that farmers must have realised that the annual cycle of the seasons was driven by nature and not by ritual specialists. Lewis-Williams and Pearce note that people in small-scale present-day societies around the world are quite capable of using non-astronomical cues to know when the change of the seasons is imminent. Hayden and Villeneuve also noted that given variations in the weather from year to year, planning agricultural activities around precise calendrical dates would not have been conducive to producing good harvests.

An alternative explanation, which would be applicable to complex hunter-gatherer groups during the Upper Palaeolithic, is that calendars were used by the elite to plan feasts and the accompanying ceremonials and rituals. Such events require much planning, and preparations can often take a year or more. Major feasts entail much more than just food preparation: they also involve complex debts, obligations, and social relations, and there is an associated competitive element in staging a bigger, better feast than everybody else.

In the absence of refrigerators, the huge amount of food required for a major feast cannot be produced and stored by one household alone; instead, it is necessary to borrow surpluses from other households on the understanding that such debts will be repaid with interest within an agreed timescale. To arrange loans, call in debts all together at a specific time, marshal labour, and avoid dates that conflicted with feasts others were planning were all essential for staging large feasts. All of these involve timing, and for this purpose it seems likely that considerable time and effort was devoted to establishing calendars, monitoring solstices, and tracking **lunations**.

Hayden and Villeneuve conclude that based on this ethnographic data, Upper Palaeolithic groups did use some form of notational system. They believe that it is highly likely that some of these systems resulted from and were used for monitoring solar and lunar movements. The motivation for the accurate monitoring of such celestial phenomena was likely for scheduling major events such as winter ceremonials. Notational systems might have also been to keep track of debts associated with feasts.

The ethnographic record shows that in the majority of recent complex hunter-gatherer societies, elite aggrandizers systematically promote their self-interests above those of the community as a whole, and use competitive feasts, prestige goods, and secret societies to further their aims. Such a pattern might well have begun with the hunter-gatherers of the Upper Palaeolithic. However, there is as yet no archaeological evidence of feasting that can be associated with solstitial or calendrical evidence.

Hayden and Villeneuve noted a detailed study by anthropologists Travis Hudson and Ernest Underhay[19] of the Chumash of southern California. Elite members of these people were specialist astronomers who belonged to a secret society known as the 'antap. They acquired supernatural power and knowledge from celestial beings while on vision quests to the upper world, and their detailed knowledge about the movements of the Sun and other celestial bodies was regarded as esoteric and never shared with commoners. They claimed that only they could understand and influence the celestial system upon which Chumash life depended. If there were such secret societies in the Upper Palaeolithic, it is possible that detailed knowledge of solar and lunar movements formed a part of their esoteric knowledge.

Criticisms of Marshack

Although Marshack's radical ideas won praise[20], academic opinion has been divided, and his methodologies, theoretical perspectives, and ethnographic parallels have been questioned[13]. Hayden and Villeneuve[21] note that the groups Marshack considered in his ethnographic studies included agriculturalists such as the Winnebago, Hopi, and Highland Maya. Such peoples do not provide an appropriate comparison to the purely hunter-gatherer groups of the Upper Palaeolithic.

Hayden and Villeneuve suggest a few of the engraved artefacts Marshack considered do seem to show sequences of counts that could conceivably have been related to day counts of lunar cycles; but other examples such as the La Grotte de Thaïs plaque involve so many line counts that it is difficult to imagine that they represented lunar calendars. They could have been notational counts of something else – possibly tallies of debts and gifts associated with major feasts.

Robinson[13] notes that of the large number of Upper Palaeolithic artefacts Marshack has examined, no two examples fit his lunar scale the same way. Marshack's method is to impose the 29/30-day scale upon the artefacts under consideration – thus, if there are 45 marks, that is one and a half lunations; 57 is two lunations; 73 is two and a half lunations, etc. As the system of counting is non-arithmetic, minor discrepancies can be disregarded, so marks totalling 27 or 32 can be identified as a 'month' with or without the Moon's period of invisibility. The problem is, almost anything can thus be made to fit. Also, it might be expected that the months so recorded would be separated from one another by some form of delimiter such as a line; or that there would be a special indicator for the full Moon – but this is not the case for any of the examples Marshack describes. Robinson also questions whether lunar calendars would need to 'evolve', especially over such a long timescale. Tallies of a phenomenon which produces the same pattern every 29 to 30 days could have been developed in isolation on many occasions.

Even if we accept that the artefacts encode a repeating pattern varying from 27 to 32 days in duration, there might be an explanation other than the lunar calendar hypothesis. Marshack[9] attributes the variation to cloud cover preventing timely observation. But the human menstrual cycle averages 28 days (the similarity to the lunar cycle is pure coincidence, albeit one responsible for the common origin of the words 'moon', 'menstruate', and 'measure' (time)), but it varies from 24 to 38 days. Could this be what was being recorded rather than the phases of the moon? The advantages of knowing when that time of the month is approaching are obvious, and that was probably also the case 35,000 years ago. However, Marshack[10] rejected this explanation.

Marshack has also been challenged on methodological grounds. He failed to use statistical methods to back up his assertions, something modern researchers would consider an essential part of any such studies[22]. He also never proposed a formal definition of a system of notation which could enable such systems to be identified in the archaeological record[23,24]; and he did not give proper consideration to alternatives to his calendrical hypothesis[25].

A strong critic was Italian cognitive researcher Francesco d'Errico. Marshack's lunar interpretation relies on the view that differences between marks reflected them being produced at different times with different tools[26]. D'Errico[27] carried out experimental work with limestone pebbles and demonstrated that it is possible to see with an electron microscope the direction in which the incised lines were made, the order in which superimposed marks were made, and whether all the marks were made by the same tool. But Marshack never attempted to reproduce his supposed calendars experimentally, nor did he present experimentally-verified microscopic evidence that data was recorded by several distinct engraving/notching operations over a relatively long period, or the use of multiple tools. D'Errico's own experimental work showed that morphological changes between two engraved lines or groups of lines, which Marshack attributed to a change of tool, could also result from other causes such as a variation in the tool's orientation, or a microfracture in its path[28]. Marshack[29,15] robustly defended his views, noting that he himself had previously dismissed some of the artefacts d'Errico examined as being notational.

In 2012, an Aurignacian limestone slab was recovered at Abri Blanchard. The slab is engraved with a complex composition combining an aurochs and a large number of aligned punctuations similar to those of the bone plaque analysed by Marshack. Examination has shown that the marks were all made with a single tool in a single sitting[14]. It was suggested that Marshack's lunar calendar theory could be dismissed, but there is no reason to draw such a conclusion. The 2012 slab is clearly ornamental, and not a calendar; but there is no reason to suppose that the same 'power-turn' technique could not have been used for both calendrical and non-calendrical objects.

Marshack's theories cannot be dismissed, but they cannot be proved either. There is nothing intrinsically implausible in the idea that Upper Palaeolithic people made observations of lunar phases, although deciphering the notations might not be as straightforward as Marshack's interpretations imply[30]. The criticisms of Marshack outlined above are valid, and the evidence presented has so far failed to conclusively demonstrate the existence of calendrical records from Ice Age Europe. However, based on the ethnographic evidence, it seems very likely that complex hunter-gatherer societies existed during the Upper Palaeolithic, and that they were aware of lunar and solar phenomena. It also seems likely systems of notation existed for monitoring these phenomena, as well as for other purposes. The search for evidence to support Marshack's claims should continue.

03
Ice age star maps

Upper Palaeolithic cave paintings

The first Upper Palaeolithic cave paintings were discovered at the Cueva de Altamira in Cantabria, Spain, in 1868 by a local man named Modesto Cubillas. Some years later, the find came to the attention of amateur archaeologist Marcelino Sanz de Sautuola, who first visited the cave 1875. In 1880, he published an account of the find and his view that it was prehistoric, only to be met with widespread scepticism. It was not until further discoveries of cave art were made at the start of the twentieth century that the Altamira paintings were accepted as genuine. Since then, around 400 sites have been discovered, mainly in the Franco-Cantabria region of northeast Spain and southwest France. Many of these sites are now UNESCO World Heritage Sites, including Lascaux and Chauvet-Pont d'Arc in southwestern France, and Altamira itself.

Many theories have been proposed as to why the cave art was created. An early theory was that it was *l'art pour l'art*, and that the caves were Upper Palaeolithic art galleries, created and appreciated by people with time on their hand for activities beyond the daily necessities of hunting, gathering, and toolmaking. But much of the art is located in the inaccessible depths of caves. At Lascaux, for example, the Shaft of the Dead Man is, as its name implies, a narrow shaft some 6 m (17 ft) deep, only capable of holding a few people at a time.

A later explanation, popular during the first half of the twentieth century, was that the caves were associated with totemism, which is a belief system where clans are represented by a particular species of animal or plant. The problem is that the caves invariably portray multiple species rather than just the one, as would be expected for totemic cultures. Other proposed explanations were based on hunting magic, which is a belief that an image of an animal or person can influence the subject portrayed. Hunting magic is a belief common in traditional small-scale societies. Making images deep underground, where they cannot be readily seen by others, implies magical purposes. What was important was the making rather than the viewing of the images, to ensure successful hunting[1,2,3].

Structuralism is a philosophical movement that originated early in the twentieth century with the work of the Swiss linguist Ferdinand de Saussure. It was applied to anthropology by the French anthropologist Claude Levi-Strauss, who proposed that the language, kinship systems, and mythology of any human culture can be explained in terms of 'binary opposites' (for example, up/down, life/death, and male/female), and the relationships between them. Subsequently, French archaeologists Andre Leroi-Gourhan and Annette Laming-Emperaire looked for binary opposites in cave art. They saw the images both of animals and abstract designs as either male or female symbols; for example, horses and stags are male symbolisations whereas bison and aurochs are female. Similarly, line or arrow-like figures are male, and broader triangular designs female, in both cases representing genitalia[2].

More recently, South African archaeologist David Lewis-Williams[2] has proposed that Upper Palaeolithic cave art was created as part of a shamanistic belief system. He believes that the universal belief in a multi-tiered cosmos is rooted in the very structure of the human brain, and that in altered states of consciousness, people can experience journeys to realms other than that of our everyday existence. While such altered states are associated with psychotropic substances, they may be induced by other means including intense concentration, chanting, clapping, drumming, prolonged rhythmic movement, and hyperventilation. Sensory deprivation, of the type that may be experienced in the depths of the cave networks, also induces hallucinations.

An astronomical interpretation of cave art

A shamanistic interpretation of the cave art is consistent with the intriguing, but controversial claim put forward by a number of researchers that there is astronomical data encoded in cave paintings. It has been suggested that some Upper Palaeolithic cave paintings can be interpreted as star maps, and that they depict actual star groups as they appeared in the skies of Ice Age Europe.

Pentagonal 'tectiform' images from cave paintings are often identified as tents, but they could represent the constellation Auriga (the Charioteer), which lies adjacent to Taurus. Ethnographic ritual practices provide a possible insight into the significance of these tectiform images. Shamans of the Ojibwa and Algonkian hunter-gatherer groups of the Lake Superior region erect a special 'shaking tent', which they use for travel to a 'Hole in the Sky'. When the bottoms of these structures are covered with vertical sheets of bark, they take on a slightly elongated pentagonal shape. The 'Hole in the Sky' is identified by the Ojibwa as the Pleiades. Shaking tents also feature in the shamanic seances of the Cree of Quebec; and Siberian groups also construct tents for shamanistic journeys to other realms[4].

The belief systems of many present-day hunter-gatherer groups involve shamanistic travel to specific constellations or stars in order to contact the spirits of the Upper World, and it might well have featured in Upper Palaeolithic shamanistic practices. It is possible that Upper Palaeolithic shamans were leading members of secret societies analogous to the 'antap of the Chumash (see Chapter 2); and that their cosmological concepts were encoded in some of the cave art. These might have included the 'stargates' used for journeys to the other realms[4].

The Hall of the Bulls

The Hall of the Bulls is a vaulted rotunda in the Lascaux cave complex, which contains a wrap-around mural depicting aurochs, horses, and stags. The mural is 18,000-19,000 years old, dating to the Late Magdalenian period. Several commentators including Luz Antequera Congregado[5], Frank Edge[6], and Michael Rappenglück[7] have noted that a pattern of six dots above the mural's dominant animal, known to archaeologists as Bull No. 18, resembles the Pleiades; they have also identified a V-shaped set of dots on its face with the Hyades open star cluster; and the bright star Aldebaran (α Tauri) corresponds to the bull's eye. These are all constituents of the present-day zodiacal constellation of Taurus (the Bull)[4]; if the correlation is intentional then the association of Taurus with a bull is very ancient indeed.

A strong proponent of the astronomical theory is German researcher Michael Rappenglück, who has published extensively on the subject. He has made use of planetarium software to mimic the effects of **precession** and reproduce the skies as they would have appeared to observers at various stages of the Upper Palaeolithic between 45,000 and 15,000 years ago. Rappenglück believes that the cosmology of Upper Palaeolithic people recognised their deep links to nature. They associated the terrestrial annual cycles of birth, growth, death, and rebirth with the annual cycle of the night sky. He follows Lewis-Williams in ascribing a shamanistic interpretation to cave art, but he believes that there is an astronomical dimension as well.

In a report on the Hall of the Bulls at Lascaux[7], he suggests that the cave complex was a Palaeolithic planetarium; an example of a world view based on shamanistic and totemic concepts, which were illustrated on the rock walls underground. He notes that from the hill above the cave the view of the night sky would have been excellent, and that it would have been 'surprising' if the changing view of the heavens over the course of a year had not influenced the cosmology of the Lascaux people.

In the cave complex, many rock pictures of animals depict certain seasons. For example, horses, bison, and ibex appear in both winter and summer coats, indicating the change from winter to spring; a rutting stag indicates the coming of autumn; and a pregnant mare indicates the transition from spring to summer. All of this makes it clear to Rappenglück that the representation of seasonality, in particular around the **equinoxes** and **solstices**, were important to the Lascaux people. He believes that the cave complex was a planetarium, a temple, and a repository of cosmological knowledge.

The first indication of an explicitly astronomical depiction is to be found in the Hall of the Bulls, where he concurs with others who have suggested that Bull No. 18 is a depiction of Taurus and the Pleiades. Rappenglück notes that the Pleiades are prominent in many traditions across Eurasia, Australia, and the New World. They are referred to in the mythology of ancient Greeks, Cherokee, Onondaga Iroquois, Japan, the Mongols, and Aboriginal Australians. There are Navajo and Hopi (North America) and Chukchi (Siberia) depictions that resemble the Lascaux depiction. The Pleiades also feature in the logo of the Subaru car company of Japan. The association of the Hyades with a bull goes back at least as far as ancient Mesopotamia.

Rappenglück also sees depictions similar to the spotted face of Bull No. 18 on three bull heads (bucrania) in Room VI B 8 at Çatalhöyük, a large Neolithic settlement on the Konya Plain in central Anatolia, Turkey, dating to 7000 BC. They comprise from seven to fifteen dark-coloured lozenges, each of which contains a white circle in the centre which could represent a group of stars, possibly the Hyades. The three bucrania are accompanied by the head of a ram, and all four are affixed directly above a mural claimed to represent a calendar based upon Aldebaran, the Pleiades, and the bright star Hamal (α Arietis) in the neighbouring constellation of Aries (the Ram). Rappenglück claims that the calendar is related to the annual life cycles of bees and barley[8], and suggests that similar concepts existed around Taurus thousands of years earlier in the Upper Palaeolithic, based on a hunter-gatherer rather than agricultural world view.

To expand upon this idea, it is necessary to know the age of Lascaux and hence the corresponding stage in the precessional cycle. Unfortunately, this is not straightforward: although the cave complex is well within the range of radiocarbon dating, no charcoal was used in any of the paints, and radiocarbon dates have been based upon a few pieces of charcoal found throughout the cave complex. Dating has had to be based upon these and on stylistic comparison between the art and that of other caves. Rappenglück claims that the cave complex dates to the Late Magdalenian and is 15,000-17,000 years old; this is more recent than the 18,000-19,000 years old commonly cited[9]. Jean Clottes[3], a leading authority on Upper Palaeolithic cave art, attributes Lascaux to the Solutrean or Early Magdalenian.

Rappenglück claims that the Pleiades and Taurus signalled the start of spring in the Late Magdalenian. He notes that the Pleiades were close to the **First point of Libra** around 15,300 BC (17,250 years ago), a date he equates to the Hall of the Bulls. At that time, they disappeared into the western skies at sunset on 26 August, a month before the autumn equinox (**heliacal setting**), and they reappeared on 11 October in the eastern evening skies (**heliacal rising**). They are visible for 319 days of the year; around 15,300 BC the mid-point of their visibility would occur just before the spring equinox; i.e., they were opposite the Sun in the sky when the latter was at the **First point of Ares**; at this time Taurus **culminated** at midnight, but its orientation (head down, rump up) is not depicted at Lascaux. Instead, it shows Taurus just before the Pleiades disappears into the sunset in late August.

Taurus and the Pleiades are also represented at the cave of La-Tête-du-Lion at Ardèche in southeastern France. Rappenglück[10] considers a rock panel there he sees as representing Aldebaran, the Pleiades, and the lunar cycle. The cave painting is Solutrean, predating Lascaux. Radiocarbon dating suggests that it is from 20,850 to 22,450 years old. It comprises an aurochs cow, with a stag to the left and two ibex heads below. The animals are all indicative of the second half of the year: the ibexes are in winter coats, and the stags' antlers are fully developed as happens between July and January. The animals are all drawn in red ochre.

The aurochs has a prominent dot and a pattern of seven dots, which Rappenglück interprets as representing Aldebaran and the Pleiades, respectively. Above the aurochs is a serpentine pattern of 21 dots. At the time the cave art was produced, Taurus lay between the points of the summer solstice and the autumn equinox. As at Lascaux, Rappenglück notes the horizontal position of the aurochs and takes this to imply the disappearance of the Pleiades into the sunset; otherwise, Taurus would appear head down, rump up in the sky. At the midpoint of the date range for the cave art, 21,650 years ago, the heliacal setting of the Pleiades coincided with the summer solstice. But six hundred years later (21,050 years ago; near the start of the date range), the heliacal setting of the Pleiades was 21 days before summer solstice; Rappenglück believes that this is what is represented by the 21 dots above the aurochs.

The Taurus theory, or at any rate a part of it, does appear to be feasible. As noted, the constellation is one of the most distinctive in the heavens; and the Pleiades feature in many traditions. The identification of the Hyades in Taurus with a bull goes back at least 4,000 years to ancient Mesopotamia, and it is at least possible that it originated in the Upper Palaeolithic. The relatively accurate placement of dots resembling the Pleiades and Aldebaran with a bull in the correct spatial relationship for Taurus at Lascaux would be an extraordinary coincidence were it not intentional[4]. It therefore seems likely that Taurus is indeed represented in the Hall of the Bulls at Lascaux and at La-Tête-du-Lion. What seems less likely is that there is anything significant about the horizonal depiction of the aurochs; the animals would normally be seen thus and not head down, rump up. The portrayal probably reflects nothing more than a desire to present the aurochs in a naturalistic pose. It is unlikely to be a reference to the time of year, albeit there are clear allusions to the seasons in the Lascaux cave art.

American astronomer Frank Edge[6] goes on to consider the other animals portrayed in the Hall of the Bulls mural at Lascaux. His scheme runs from west (sunset) to east (sunrise); confusingly this is contrary to the scheme adopted by archaeologists, who numbered the animals from left to right as seen on entering the rotunda. Edge associates the constellations of Orion (the Hunter) and Gemini (the Twins) with a second aurochs, Bull No. 13, and that of Leo (the Lion) with a third, Bull No. 9. Uniquely for the mural, these two animals stand head-to-head. Other animals are associated with Canis Minor (the Lesser Dog), Virgo (the Virgin), Libra (the Scales), Scorpius (the Scorpion), and Sagittarius (the Archer). The last three are represented by the mural's last animal (in Edge's scheme), Figure No. 2, which is generally known as the 'Unicorn', despite having two horns. The Unicorn lies on the opposite side of the rotunda to the 'Taurus' Bull; in the skies Scorpius lies opposite Taurus. If we accept Edge's interpretations, then the mural as a whole seems to be a representation of just over half of the prominent star-patterns lying along or close to the **ecliptic**. Today, the mural's lead constellations, Taurus, Orion, and Gemini, are prominent in the winter skies. However due to the effects of precession, in the era of Lascaux these constellations would have been prominent in spring. As summer approached, they would have begun to move down towards the western horizon, and at the summer solstice, they would have stood low down on the horizon just after sunset.

Instead of the three present-day groupings of Taurus, Orion, and Gemini, the Magdalenian people saw a pair of bulls (Bulls 18 and 13), which at this time of the year in Edge's words, "*seemed to walk on the horizon*". The other constellations depicted would have been visible between sunset and sunrise. Edge believes that the stars portrayed in the mural were used in conjunction with the phases of the Moon to predict and keep track of the time of the summer solstice. Following Alexander Marshack[11], he believes people in the Dordogne region were observing the phases of the Moon at least 15,000 years before the Lascaux Caves were painted. He notes that at the time the mural was painted, the summer solstitial full Moon would have appeared low in the sky between Taurus and Gemini.

Though going rather beyond the 'standard' Taurus interpretations, Edge's theory does on the face of it sound plausible. As noted, there is ethnographically documented evidence for hunter-gatherer groups recognising the constellations of Taurus, Orion, and Gemini. However, it should also be recalled that no evidence has been found for any particular interest in the zodiac, so the suggestion that other constellations on or close to the ecliptic were depicted in the Hall of the Bulls is not supported by the ethnographic record for hunter-gatherer societies (see Chapter 1).

The Shaft of the Dead Man

After his report on the Hall of the Bulls at Lascaux, Rappenglück[12] published a report about the enigmatic Shaft of the Dead Man at Lascaux. Often simply referred to as the Shaft, it is deep, cramped, and accessible only by ropes and ladders. Only a few people could have occupied it at a time. On the northern rockface is portrayed a highly stylised man, who has the head of a bird, four-fingered birdlike hands, and a prominent phallus. He is apparently falling backwards while confronting a partially eviscerated bison. Below him, a bird is perched atop a post. With its back to this scene and its tail raised is a rhinoceros, and on the opposite side is a black horse. These images have been interpreted as shamanistic: the death of the bison parallels the metaphorical 'death' of a shaman as he enters the spirit world and his fusion with his spirit-helper, a bird. Birds on sticks are a common motif in many shamanistic cultures worldwide. They are seen as symbolising the interconnection of the realms of a (typically) three-tiered cosmos comprising upper and lower realms in addition to the (middle) realm of everyday existence. Interestingly, the Shaft is characterised by high levels of naturally occurring CO_2, which might have induced altered states of consciousness if people stayed down there long enough.

Like Lewis-Williams, Rappenglück believes that the work has a shamanistic explanation, but he also sees astronomical alignments in the birdman and the bird on a stick. He notes that as viewed from the top of the shaft, the bird on a stick appears inclined rather than vertical, and it is the birdman that appears vertical. He suggests that the composition was intended to be viewed from this vantage point, from where the bird on a stick appears to be inclined at 45.3 degrees to the vertical. The latitude of Lascaux is at 45.1° N; Rappenglück suggests that the bird on a stick is aligned on the north **celestial pole** with the birdman pointing towards to **zenith**. He goes on to suggest that the eyes of the birdman, the bird on a stick, and the bison correspond to Vega (α Lyrae), Deneb (α Cygni), and third magnitude star β Delphini at around midnight at the time of the summer solstice. The grouping is similar to the present-day 'Summer Triangle' **asterism**, where Altair (α Aquilae) replaces β Delphini and forms the apex or a roughly isosceles triangle with Vega and Deneb. Around 14,430 BC (16,380 years ago), the north celestial pole was very close to the moderately bright star δ Cygni in a part of the sky through which the Milky Way runs; this point, roughly midway between Vega and Deneb, is marked by the wrist of the birdman's hand.

This may be a speculation too far. Firstly, it is much simpler to assume that the falling birdman is merely portrayed at an angle midway between horizontal and vertical, i.e., very close to an angle of 45 degrees than to postulate an alignment on the north celestial pole. We have to question whether the Lascaux people could locate the north celestial pole to within a fifth of a degree, and then reproduce the result in the cave.

Secondly, assuming the relative positions of the eyes are reasonably correct, why would the Lascaux people have favoured β Delphini – a relatively faint star in the constellation of Delphinus – over Altair, a much brighter star in the neighbouring constellation of Aquila. Both stars were circumpolar around 14,430 BC, though this is no longer the case. To represent Altair, the bird on a stick would have to move over to the right and squash up the composition, which suggest that artistic rather than astronomical considerations dictated its placement.

The Northern Crown

Rappenglück[13] claimed to have found a representation of the constellation of Corona Borealis (the Northern Crown) in the Cueva de El Castillo in Puente Viesgo, northern Spain. Deep inside this cave is a 5 m (16 ft) panel comprising thirty hand-prints, known as the Frieze of Hands; near the bottom right is a pattern of seven dots arranged in a semi-circle.

Corona Borealis is a small but distinctive constellation of seven stars located near the bright star Arcturus (α Boötis) in the neighbouring constellation of Boötes (the Herdsman). Between 12,000 and 8,000 years ago, the seven stars formed a perfect semi-circle as supposedly depicted in the cave panel; they have since ceased to do so owing to the effects of the **proper motions** of the stars through space. The constellation became circumpolar from northern Spain around 13,000 years ago and was closest to the north celestial pole between 10,000 and 9,500 years ago. Rappenglück suggests that as a distinctive circumpolar constellation, Corona Borealis would have served the same role as the Plough does today, as a celestial clock and an aid to navigation.

The suggestion is plausible, but support for the suggested timescale is weak. Five radiocarbon dates are cited in the report. One gives a range from 12,900 to 11,600 years ago, which partially overlaps with the upper end of the window between 12,000 and 8,000 years ago. However, three others suggest that frieze is from 14,100 to 16,400 years old, and the remaining date suggests that it is at least 19,000 years old. Rappenglück favours the one date that supports his theory, but it might be better to apply **Occam's Razor** and note that the bulk of the radiocarbon dates do not, and that the dots probably do not represent Corona Borealis.

An Aurignacian Orion 'pregnancy calendar'

Geißenklösterle is a cave site near the town of Blaubeuren, southwest Germany. It has yielded large numbers of artefacts dating to the Aurignacian period, including one of the world's earliest-known musical instruments: an ivory flute, which was played at the 2012 edition of the quinquennial Documenta contemporary art exhibition. Other artefacts from the site include a small plaque discovered in 1979 and believed to be 32,500 to 38,000 years old. The plaque measures 38 x 14 mm (1.5 x 0.5 inches), and on it an 'anthropoid' figure is carved in low relief. The figure's posture has been described as 'adorant' (conveying worship), but it has also been described as a therianthropic human-feline about to pounce. Between the legs is an appendage that reaches down to the level of the right heel. On the reverse side of the plaque are carved 86 notches. The figure is similar to two other Central European artefacts of roughly the same age: the Löwenmensch (Lion-man) from Hohlenstein-Stadel, southwest Germany; and the Venus of Galgenberg from Galgenberg, Austria.

Rappenglück[14] interprets the figure as a representation of the constellation of Orion as it appeared 32,000 years ago. It differs slightly from its present-day appearance due to the effects of proper motion; in particular, the faint star θ^2 Orionis lay around three degrees north of its current position. Rappenglück notes that the straddle-legged posture of the figure, with the right foot slightly higher than the left, resembles the lower part of Orion. The narrow waist depicts the Belt and the upraised arms the upper part of the constellation. The appendage corresponds to the Sword of Orion in present-day interpretations. There were of course no swords in Upper Palaeolithic times; Rappenglück assumes that the appendage is the phallic equivalent.

From 44,000 until 33,500 years ago, Orion was not visible in its entirety from Geißenklösterle; but by 32,000 years ago, Betelgeuse (α Orionis) set heliacally 14 days before the spring equinox and reappeared in the evening skies 19 days before the summer solstice. Accordingly, Rappenglück suggests that Orion could have been used to presage the arrival of spring and summer. Betelgeuse remained out of view for 86 days – corresponding to the number of notches on the reverse side of the plaque. One can envisage an Aurignacian observer notching the reverse of the plaque for each night of Betelgeuse's invisibility, until it reappeared in the western skies.

The problem, as with Rappenglück's interpretation of the Frieze of Hands, is that the timescale obtained by radiocarbon dating is a poor fit for the hypothesis. A date of 32,000 years ago is 500 years more recent than the minimum date suggested by the radiocarbon results. Furthermore, even this date range might be an underestimate. The Geißenklösterle ivory flute, noted above, was once thought to be 36,000 years old; but in 2012 a revised date of 42,000-43,000 years old was obtained and it is possible that the ages of other artefacts from the site have also been understated[15]. Any similar upward revision of the plaques' age would of course completely invalidate Rappenglück's hypothesis.

Notwithstanding this difficulty, Rappenglück goes on to propose that the plaque's purpose might have been a pregnancy calendar: One year (365 days) minus 86 days (period of invisibility) is 279 days, or very near to the 40-week average length of the human gestation period. Rappenglück cites Naegele's rule, which is a standard method for calculating the due date for a baby. Named for the nineteenth century German obstetrician Franz Naegele, it works by subtracting three calendar months (around 92 days) from the date of the first day of the last menstrual period, and then adding a year and seven days. Effectively, it is the same as adding a year and subtracting 85 days, or close to Orion's period of invisibility. In other words, a baby conceived at the time Orion returns to the evening skies would be born at around the time it disappears again.

It is entirely right to be sceptical, but one can envisage that the return of Orion would be seen as auspicious; and that babies were conceived during the ensuing festivities. Inevitably, it would be noticed that Orion disappeared again at around the time these babies were born. In time, annual fertility rituals involving sexual activity would take place around the heliacal rising of Orion. If so, it is easy to understand why what we see as the Sword of Orion was viewed by Aurignacian people as something rather different.

Star maps or not?

The strongest case for Upper Palaeolithic star maps is made by the association between Lascaux's Bull No. 18 and Taurus as represented by the Pleiades, Hyades, and Aldebaran. This connection has been noticed by several researchers, and it is probably genuine. It is also likely that Taurus is depicted at La-Tête-du-Lion. Frank Edge's suggestion that Bulls 18 and 13 depict Taurus, Gemini, and Orion at the time of the summer solstitial full Moon also seems plausible.

Michael Rappenglück's otherwise feasible claims for the Cueva de El Castillo Corona Borealis and the Geißenklösterle Orion receive only marginal support from the available radiocarbon dates; in particular, the latter is vulnerable to an upward revision of the date given that other artefacts from the same site have proved to be older than was once believed.

In summary, the case for an Upper Palaeolithic origin for some present-day constellations looks reasonably sound. If so, it is possible that Upper Palaeolithic people recognised other groupings with no present-day equivalent, although there is no strong evidence from the ethnographic record to suggest that this was necessarily the case.

04
Göbekli Tepe and the ancient peril

An early Neolithic temple

Göbekli Tepe ('Potbelly Hill') is an archaeological site a limestone ridge 15 km (9 miles) from the town of Şanlıurfa in southeastern Turkey. Often described as the world's oldest temple, it comprises a series of stone circles that superficially resembles Stonehenge, but it is 7,000 years older and is associated with a Near Eastern culture known as the Pre-Pottery Neolithic A, usually abbreviated to PPNA. The culture and its successor, the PPNB, were first described by the British archaeologist Dame Kathleen Kenyon, one of the most influential archaeologists of the last century.

By 9600 BC, the last Ice Age was over, and a radically new way of life had emerged. Although the origins of agriculture can be traced as far as 22,000 BC, it was not until the onset of the Holocene that the unstable and often cold, arid climate was replaced by warm, wet, and ultimately stable conditions that allowed human groups to adopt a sedentary lifestyle and begin producing their own food. The PPNA people were low-level food producers rather than true agriculturalists. Although they raised crops, there was still an emphasis on hunting rather than herding animals. The PPNA was a relatively short-lived phenomenon, giving way to the fully agricultural PPNB after about a millennium. But the people of this era were far from unsophisticated.

Although Göbekli Tepe has been known since the early 1960s, it was largely ignored until 1994, when it was visited by German archaeologist Klaus Schmidt. He began excavating there the following year, and he continued his work at the site until his untimely death from a heart attack in 2014, aged sixty. At the lowest level of the site, Layer III, Schmidt discovered a series of semi-submerged circular or oval enclosures. Each comprises a dry-stone wall, into which up to twelve T-shaped limestone pillars are set, often joined to one another by stone benches. At the centre of each enclosure are two more pillars, which tend to be larger than the surrounding ones. The pillars range in height from 3 to 5 m (10 to 16 ft), and they weigh up to 10 tonnes. Four enclosures, designated A to D, are currently undergoing excavation, but geomagnetic surveys suggest that at least twenty exist. A radiocarbon date of 9530 BC (11,480 ± 220 years ago) has been obtained from the wall plaster of Enclosure D[1]. It should be noted that the original layout of the Göbekli Tepe enclosures is uncertain. Many of the T pillars are not standing in their original positions and the buildings underwent significant alterations during their lifecycles. Building archaeology studies have shown that many of the T pillars were removed from their original settings and reused elsewhere. Thus, the monuments as they are today are the culmination of multiple phases of building and rebuilding[2].

Many of the pillars are carved with bas-reliefs of animals. These include snakes, wild boar, foxes, lions, aurochs, wild sheep, gazelle, onager, birds, various insects, spiders, and scorpions; where the sex can be determined, they are always male. Different animals are depicted in each enclosure. The images are large, often life-size, and semi-naturalistic in style. Some pillars exhibit pairs of human arms and hands, and they may represent stylised anthropomorphic beings. Possibly, these are gods, shamans, ancestors, or demons. There are also a number of abstract symbols that have been interpreted as pictograms[3,4,5,6,7].

The seeming absence of any houses has led many scholars to believe that Göbekli Tepe was a ritual centre, possibly the first anywhere in the world[4]. One possibility is that the site was used by a number of groups, each of which identified itself with a different totemic animal or animals. These groups performed rituals in their own enclosures, where their particular animals were depicted[7]. Another possibility is that the complex was associated with shamanistic practices[8]. The two explanations are not mutually exclusive, and the role of Göbekli Tepe as a temple is widely accepted. However, in 2017, two engineers at the University of Edinburgh proposed that it was intended to observe incoming meteor showers that supposedly threatened Earth at the end of the last Ice Age.

The fall and rise of catastrophism

Catastrophism is the view that events on Earth such as geological change, the evolution of life, and even human history have been shaped by upheavals of a violent or unusual nature. Until the nineteenth century it was the prevailing view, and it was generally accepted that stories like the Biblical Flood related to actual events. In the late eighteenth century, Scottish geologist James Hutton argued that Earth has been shaped by the same geological processes throughout its history. Changes occurred gradually, and involved volcanic action, deposition of sediment, and erosion by wind and rain, rather than floods and other biblical catastrophes. But Hutton's view attracted little interest in his lifetime, and it was not until the 1830s that they were popularised by fellow Scot Sir Charles Lyell. Nevertheless, Hutton's theoretical approach, for which Lyell coined the term 'uniformitarian', is now considered to be the foundation of modern geology.

This new paradigm remained largely unchallenged for the next 150 years, although as far back as the 1930s astronomers were aware that asteroids can and periodically do approach the Earth, and that impacts must occur from time to time – sometimes with global consequences. In the 1950s, it was proposed that an asteroid impact might have been responsible for the extinction of the dinosaurs[9], and there were suggestions that impacts had been responsible for other mass extinctions. But such views attracted very little interest, and neo-catastrophism was probably set back by what became known as the Velikovsky Affair.

Belarus-born Immanuel Velikovsky moved to the United States shortly before the outbreak of World War II, and after the war he put forward some rather strange views in a book entitled *Worlds in Collision*[10]. He meticulously catalogued accounts of supposed global catastrophes as described in the Bible, the records of the Maya, Aztec, and Inca civilisations, and Greek and Nordic mythology. He worked on the basis that these myths and legends were literally true, and he sought to interpret them as references to catastrophes caused by Earth experiencing near misses first from Venus and then from Mars, occurring around 2000 BC and 800 BC. The book is well-researched as far as the ancient accounts go; but it parts company with the laws of physics more or less from the off. Had Venus genuinely been on a collision course for Earth, it would have taken rather more than an electrical discharge between the two planets to avoid total annihilation; and had its gravitational influence reversed the direction of Earth's rotation, the effects would not have been confined to the walls of Jericho.

Upon its release, *Worlds in Collision* became an immediate best seller in the United States, but the scientific community was rather less impressed. There were even threats by some universities to boycott Velikovsky's publisher Macmillan unless the book was withdrawn, with the consequence that it was transferred to Doubleday. This ridiculous over-reaction only served to lend the book spurious credibility, with the result that it continues to find its way into pseudoscientific speculations to the present day. Often referred to as the Velikovsky Affair, the furore had the unfortunate effect of discouraging rational consideration of catastrophism[11], or of the idea that the myths and records of prehistoric societies and early civilisations might contain important information about astronomical events[12].

The situation finally changed in 1980, when American physicist and Nobel laureate Luis Alvarez, his son Walter, and a number of other collaborators published a paper in which they presented evidence that an asteroid had struck the Earth 66 million years ago. The vital clue was a thin layer of clay laid down between two layers of limestone in rocks near the town of Gubbio, in northern Italy. The clay was located at the boundary between the Cretaceous and Tertiary geological periods: the limestone below the clay contained Cretaceous fossils; that above contained fossils from the Tertiary. It marked the point in geological time when the dinosaurs became extinct. Analysis of the clay revealed anomalous levels of the metal iridium, which is rare in the Earth's crust but relatively abundant in undifferentiated objects such as meteorites. As such, it was interpreted as the signature of a major asteroid impact[13]. The impact triggered firestorms around the world as superheated ejecta from the impact fell back to Earth. The planet was shrouded with clouds of dust and soot, cutting off the light of the Sun. Temperatures plummeted, and plant life died off. It was this 'impact winter' that killed off the dinosaurs. The crater left by the impact was identified in 1990 at Chicxulub, in the Yucatan Peninsula[14], and it is now believed to have been formed 66,038,000 years ago[15]. The proof that the Chicxulub impact event profoundly influenced the history of the Earth re-established catastrophism as a legitimate scientific paradigm.

The British neo-catastrophists

The orthodox view was and still is that the Chicxulub impactor was a random Earth-crossing asteroid that happened to score a bullseye. In 2007, it was even claimed that the origin of the impactor had been traced to a collision that took place in the asteroid belt 160 million years ago[16], although this possibility was later ruled out by NASA. There have also been suggestions that the impactor was a comet rather than an asteroid. This latter suggestion is favoured by a group of British astronomers including Victor Clube, William Napier, Mark Bailey, Duncan Steel, and David Asher, all of whom are proponents of what Steel[12] has termed 'coherent catastrophism'.

In a series of papers published from 1979, Clube and Napier[17,18,19,20,21] proposed that large comets occasionally become destabilised into short-period orbits, where they break up due to outgassing, or to tidal disruption experienced during close approaches to the Sun, or

Jupiter, or other planets. Fragments range from in size from submillimetre dust grains to kilometre-sized objects. While the larger objects will remain in a cluster, smaller meteoroids will spread around the original orbit of the progenitor comet. A very large progenitor will produce secondary objects that will themselves eventually break up, resulting in a complex of related meteoroid streams. If Earth's orbit should intersect such a stream, the planet will experience effects ranging from meteor showers to major impacts.

The more significant effects will occur whenever Earth encounters the denser parts of the stream, close to the progenitor. For example, the comet 55P/Tempel-Tuttle is associated with the Leonid meteor shower, so-called because the meteors appear to radiate from a point (the radiant) in the constellation of Leo (the Lion). The Leonids peak around 18 November each year, and an observer might typically see around 12 meteors an hour at this point. But every 33 years, a far more spectacular display is seen, corresponding to the orbital period of the comet. In 1833, it is believed that around 60,000 meteors per hour were seen over North America[22]. Impressive displays were also seen in 1866, 1933, and 1966.

55P/Tempel-Tuttle is a modest-sized short-period comet, and its effects upon Earth are limited to these periodic celestial firework displays. However, Clube and Napier[21,23] and Steel and Asher[24] have suggested that a much larger comet entered the inner Solar System around 20,000-30,000 years ago. This comet, with an estimated diameter of around 100 km (62 miles), is claimed to be the progenitor of the small short-period comet 2P/Encke, the asteroid 2004 TG_{10}, and the Taurid meteoroid stream. Encke has an orbital period of just 1,204 days, or 3.3 years, shorter than any other known comet. It is named for German astronomer Johann Franz Encke, who computed its orbit in 1819, although it was first seen in 1786. The Taurids are a complex of multiple streams known to be associated with at four related meteor showers, and probably associated with at least four others[25]. The main streams are the Taurids, which comprise a northern and southern component (associated with TG_{10} and Encke respectively) and which appear in late October and early November with many bright fireballs; and the β Taurids, a daytime shower of meteors detectable only by radar which peaks around 30 June each year.

The Taurid Complex is also believed to be associated with the Tunguska event, a massive explosion that occurred at 07:17 local time (00:17 GMT) on 30 June 1908 near the Stony Tunguska River in central Siberia. The explosion is thought to be an airburst of a small comet with a diameter of 100-600 m (330-2,000 ft), with a mass of up to one million tonnes. The detonation occurred at an altitude of 5-7 km (3-4 miles) with an explosive energy of up to 10 megatons; this is over 500 times more powerful than the atomic bombs dropped on Hiroshima and Nagasaki at the end of World War II. The blast felled around 80 million trees over an area of 2,150 sq. km (830 sq. miles). The absence of a crater points to a comet rather than an asteroid, and the direction of travel as reported by eye-witnesses suggests that it originated from the Taurid Complex[26,27].

Fortunately, the region was largely uninhabited, and casualties are said to have been limited to a dozen or so nomadic tribespeople who were slightly injured. But had the comet arrived 4 hours 52 minutes later, the city of St. Petersburg would have been totally destroyed and most of its inhabitants would have been killed or injured. These at the time included one Vladimir Ilyich Ulyanov, better known as Lenin. Just a few hours, and the history of the twentieth century might have been very different. Over the next hour, Helsinki, Stockholm, and Oslo would have been successively imperilled[28].

The Moon has also experienced bombardment from the Taurid Complex. In June 1975, impacts from 100 kg (220 lb) boulders were detected by seismometers left on the Moon by the Apollo astronauts. Less plausibly, it is claimed that a group of monks at Canterbury on 18 June 1178 (Julian style) witnessed an impact that formed the 20 km (12.5 mile) diameter lunar crater Giordano Bruno[29]. The evening crescent Moon is said to have "*throbbed like a wounded snake*", but the phenomenon was not reported anywhere else in the world; nor were there any subsequent reports of the meteor storm that would have resulted from ejecta from the impact reaching Earth[30]. More recently, data from the Japanese SELENE ('Kaguya') lunar orbiter has suggested that Giordano Bruno, though recent in lunar terms, is at least one million years old[31]. It seems likely that what the monks saw was nothing more than an unusual atmospheric disturbance while the Moon was close to the horizon.

At the present time, the Earth's orbit takes it through only low-density regions of the Taurid Complex. However, due to perturbation effects of the planets (in particular Jupiter), the orbits of the meteoroid streams vary cyclically, with the consequence that Earth encounters high-density regions roughly every three thousand years. The coherent catastrophe model proposes that during high-density encounter epochs, Earth experiences not only multiple impacts from objects the size of the Tunguska object and above; but in addition, dust is injected into the stratosphere in sufficient quantities as to cause global cooling. Such episodes are supposed to have occurred during the third millennium BC (see Chapter 5) and during the period AD 400-600. The latter has been linked to the fall of the Roman Empire in the West and the Dark Ages in Europe. The neo-catastrophists warn that the next such epoch is due around AD 3000[12,32,33].

Although it attracted considerable lay interest when it was first proposed, coherent catastrophism has not been widely accepted by astronomers and planetary scientists. The main problem is that there is a good agreement between the terrestrial and lunar cratering records and the currently observed population of near-Earth asteroids, comets, and meteoroids (after making allowances for observational incompleteness). Regardless of whether the craters on the Earth and the Moon were made by coherent bursts of impacts or occurred randomly over millions of years, we can obtain an average cratering rate; this rate turns out to be in line with the predicted rate based on the observed current population of objects in near-Earth space. On the neo-catastrophist picture, the predicted rate (at an 'off-peak' time) would be much less than the observed rate[34,35,36].

On the other hand, there is nothing intrinsically infeasible about comet and asteroid breakups, and they might have affected Earth in the more distant past. It has recently been suggested that dust from the breakup of a large asteroid triggered a relatively brief ice age during the Ordovician period (485.4-443.8 million years ago). This in turn might have been a factor in the Great Ordovician Biodiversification Event, a period of extensive faunal change. Sedimentary rock from this period has been found to contain materials with chemical and isotopic composition consistent with an extraterrestrial source[37].

Claims for a late Ice Age impact

Around 14,500 years ago, the Earth began to emerge from the Last Glacial Maximum, which was the coldest part of the last Ice Age. The warming period, known as the Bølling-Allerød interstadial, lasted until 12,900 years ago, when cold, arid conditions returned. The downturn, which prolonged the Ice Age by another 1,300 years, is known as the Younger Dryas, taking its name from the arctic-alpine flowering plant *Dryas octopetala* that flourished in the northern tundra at that time.

Although the British neo-catastrophist school has long claimed that streams of comet debris play a major role in ice ages, possible evidence linking the Younger Dryas to an impact did not emerge until 2007. A team lead by American researcher Richard Firestone at the Lawrence Berkeley National Laboratory, California claimed that Earth had suffered multiple airbursts and surface impacts from fragments of a comet or asteroid that had previously broken up in space. In North America, the bombardment caused devastating shockwaves and continent-wide forest fires. The net effect was a global 'impact winter', which was responsible for the Younger Dryas downturn.

Evidence for the impacts was claimed in the form of a 12,900-year-old carbon-rich layer or 'black mat'. The layer was identified at around fifty Paleoamerican sites in North America. It is said to contain material consistent with an impact, including magnetic mineral grains, soot, carbon spherules, and so-called nanodiamonds. The latter are minute diamonds formed when carbon particles are subjected to intense heat and pressure by an explosion[38,39,40]. Nanodiamonds, spherules, and other materials associated with impacts and airbursts have also been reported from the Younger Dryas boundary at multiple locations across North America, Mesoamerica, Greenland, Europe, and western Asia[39,41,42,43,44,45]; as has evidence of biomass burning in the form of charcoal and soot in lake, marine, and ice core samples; and at archaeological and cave sites[46,47,48]. Possible impact signatures in the form of anomalous platinum levels have also been reported from Greenland ice core samples[49], from sedimentary sequences from sites across North America[50]. Based on radiocarbon dates obtained from 23 sedimentary sequences, a date of 12,835 to 12,735 years ago has been claimed for the Younger Dryas impact[51].

Spherules and elevated platinum levels have also been found in sediments associated with mammoth and bison skull fragments remains originally recovered in Alaska and the Yukon. It is claimed that the animals died as a result of blast injuries caused by an extraterrestrial impact or airburst. Unfortunately, the remains, which have been in museum collections for decades, lack any useful stratigraphic information[52].

In the most recent development, a 31 km (19 mile) impact crater has been identified beneath Hiawatha Glacier in northwest Greenland. The crater is thought to have been formed by an impact during the Pleistocene, although this only means that it formed within the last 2,588,000 years. More precise dating would obviously be informative, but it will be problematic, as the crater is overlain by up to a kilometre of ice. However, the impactor is believed to have been an asteroid composed primarily of iron. Such an object would not have been associated with the Taurid Complex, nor would it have led to the multiple impacts and airbursts claimed by proponents of the Younger Dryas impact theory[53].

A remarkable coincidence?

In theory, a Late Pleistocene extraterrestrial impact is perfectly feasible, but in practice the timing might be a little suspicious. Other factors were in play at the time, and these could explain the onset of the Younger Dryas. Marking the final stage of the Pleistocene, the Younger Dryas lasted from 12,900 until 11,650 years ago, and both its onset and termination were very abrupt[54,55]. During the preceding Bølling-Allerød warm period, the North American ice sheets had retreated, and the resulting meltwater formed a vast glacial lake known as Lake Agassiz. A widely accepted theory is that Lake Agassiz, which was larger than all the modern Great Lakes put together, discharged a great volume of freshwater into the North Atlantic. This discharge weakened the Atlantic meridional overturning circulation (AMOC), halting the flow of warm seawater from the tropics to higher latitudes. The result was to plunge the Northern Hemisphere back into glacial conditions[56,57]. Effects in the Southern Hemisphere are less certain, though evidence of cooling has been found there also[58].

Lake Agassiz could have drained eastwards via the drainage basin of the St Lawrence River; or southwards via the Mississippi into the Gulf of Mexico; or northwards into the Arctic. During the Bølling-Allerød, the lake overflowed south through the Mississippi basin, and into the Gulf of Mexico; but this overflow ceased during the Younger Dryas. This could be because the meltwater found another outlet, or it could simply be because melting ceased. The eastwards St Lawrence route is not thought to have become deglaciated until near the end of the Younger Dryas[56]. This leaves only the third possibility. In 2010 it was found that beginning 13,000 years ago, Lake Agassiz released a series of freshwater discharges into the Arctic Ocean, through what is now the drainage basin of the Mackenzie River[59].

However, climate modelling research suggests that these freshwater discharges alone cannot fully account for the Younger Dryas cooling, and that other factors must have been involved. Simulations demonstrated that the degree of cooling indicated by climate proxies was best accounted for by a combination of sustained AMOC weakening, changes to atmospheric circulation, and an increase in atmospheric dust of the type that might result from an extraterrestrial impact[60]. The study strengthens the case for an impact; nevertheless, that such an impact should occur at more or less the same time as the freshwater discharge does seem to be a remarkable coincidence.

It should be noted that the evidence presented for the Younger Dryas impact theory is by no means conclusive. Not everybody accepts that the presence of nanodiamonds is robust evidence for extraterrestrial impacts; they also occur in terrestrial deposits that have no association with impact processes[61,62]. Similarly, it is argued that elevated concentrations of supposed impact markers including iridium, magnetic sediments, and magnetic spherules are better attributed to processes common to wetland systems than to extraterrestrial impacts[63]. The 'black mats' do not occur throughout the whole of North America, but rather are located predominantly in the west. They may be algal mats or ancient soils associated with regional increases in moisture[64]. The evidence for the widespread conflagration triggered by the impact has been challenged[65] and it is claimed that such evidence is better attributed to climate change-generated increases in natural wildfires[66]. Small siliceous droplets from North America, previously interpreted as signatures of an impact, have been shown to closely resemble droplets of anthropogenic origin recovered from Late Pleistocene and Early Holocene sites in northern Syria, produced by fires reaching only modest temperatures[67]. The dates tying the supposed impact proxies to the Younger Dryas have also been questioned[68,69]. Unsurprising, such claims are vigorously rejected by proponents of the Younger Dryas impact theory[70].

In what is probably the most significant objection, the notion of a bombardment involving multiple fragments of a comet or asteroid has been challenged. Within months of such a breakup, the fragments would have drifted too far apart to produce the type of bombardment envisaged by Richard Firestone. On average, a comet in a near-Earth orbit has a lifetime of at least a century before it breaks up – so the chances of it breaking up such a short time before a collision are very slim[71].

Overall, I remain somewhat sceptical about the impact theory – but more than a decade after it was first proposed, it is a theory that refuses to go away, and it certainly cannot be dismissed.

Temple or observatory?

If there was indeed a period of increased Taurid activity towards the end of the last Ice Age, how might the night skies have appeared to the peoples of that time? Neo-catastrophist Duncan Steel[12] paints a dramatic picture of what Late Neolithic and Early Bronze Age observers might have seen during a period of supposed heightened Taurid activity 5,000 years ago. Steel suggests that spectacular storms would have been seen in the last week of June every year, reaching a peak each night around midnight. Activity would peak every tenth year: the 3.3-year period of the Taurids means that Earth would only encounter the densest parts of the stream on every third orbit. But these occasions would be accompanied by "*phenomenal meteor storms with associated Tunguska-type explosions and blast waves*".

Engineers Martin B. Sweatman and Dimitrios Tsikritsis[72] suggest that Göbekli Tepe was built to observe such phenomena around the year 9530 BC. They focused on pillar no. 43 in Enclosure D, which is known as the Vulture Stone. The enclosure is similar to the others on the site, consisting of a circular dry-stone wall with eleven T pillars set into the inner surface, with two larger 15-tonne pillars in the centre.

Pillar no. 43 is located in the northwest of the structure. At the top of the pillar is a row of three carvings resembling padlocks and accompanying figures, which Sweatman and Tsikritsis refer to as 'handbags'. Below these are two rows of V-shaped carvings, below which are a series of animals interpreted as follows:

1. A vulture or eagle in flight (Sagittarius the Archer);
2. A circle, located above and to the right of 'Sagittarius', at the visual centre of the composition (the Sun);
3. A flamingo with its legs at an obtuse angle and a 'downward wriggling snake or fish' with a large head (Ophiuchus the Serpent Bearer);
4. Two abstract figures resembling the letters 'H' and 'I';
5. The neck and head of a bird (interpretation uncertain);
6. A scorpion (Scorpius the Scorpion);
7. A duck or goose (Libra the Scales);
8. The head and legs of a wolf (Lupus the Wolf);
9. At the base of the composition is a decapitated man.

Sweatman and Tsikritsis noted that the relative positioning of the constellations depicted is approximately correct. They then used planetarium software to determine when the Sun was in Sagittarius at 'auspicious' times of the year, at different **precessional** epochs. The autumn **equinox** around 4350 BC was too recent; the spring equinox around 18,000 BC was too early; but the summer **solstice** around 10,950 ± 250 BC was closest to the date for Enclosure D of 9530 BC and also very close to the date of the supposed Younger Dryas impact, 12,900 years ago (given as 10,890 BC in the paper, which cites Petaev, et al.[49]). Sweatman and Tsikritsis suggest that the decapitated man alludes to the disastrous aftermath of the impact.

If we accept these conclusions, then the builders of Göbekli Tepe retained a memory of and 'date stamped' a cataclysmic event that occurred 1,360 years before their own time. Alternatively, Sweatman[73] suggests that pillar no. 43 is rather older than the enclosure in which it is set, and that it was originally a free-standing monolith dating to around 10,950 BC. Given that many of the pillars, including no. 43, have been moved from their original setting, this is a possibility. The pillars could be older than the single radiocarbon date we have, which is for the wall plaster of Enclosure D. In either event, the date stamping implies a knowledge of precession, the discovery of which is usually attributed to the Greek astronomer Hipparchus of Nicaea (see Chapter 12), who lived in the second century BC. Like any extraordinary claim, this one demands extraordinary evidence.

To that end, Sweatman and Tsikritsis considered the animals accompanying the 'handbags' at the top of the composition. From left to right, these were seen as a 'downward crawling quadruped giving the impression of a frog' (identified as a bear in a later paper[74]), a charging ibex, and a bending bird. These, they interpreted respectively as the constellations of Virgo (the Virgin), Gemini (the Twins), and Pisces (the Fish) as seen at sunset. The semi-circular disk atop each 'handbag' represents the setting Sun. Around 10,950 BC, the Sun's position at the other 'auspicious' times of the year was Virgo (spring equinox), Gemini (winter solstice), and Pisces (autumn equinox). Sweatman and Tsikritsis suggested that the statistical chances of these constellations being correctly displayed by pure coincidence are less than one in five million, and that any alternative interpretations can to all intents and purposes be disregarded.

The pair then went on to consider pillar no. 18, one of the two central pillars of Enclosure D, which is over 5 m (16 ft) tall. It depicts arms with hands nearly clasped at the front above a belt with a large buckle, decorated with abstract symbols resembling the letters 'H', 'C', and a reversed 'C'. A fox is also depicted, along with a brooch or necklace. The latter consists of a 'punctured' circle above an upturned crescent but below another 'H' symbol. The belt buckle is interpreted as the bow wave of a large object entering Earth's atmosphere; the brooch as an obscuring ash cloud blotting out the Sun; but what does the fox represent?

To answer the question, Sweatman and Tsikritsis considered pillar no. 2 in Enclosure A and pillar no. 38 in Enclosure D. These are the only pillars to depict a sequence of three animals: pillar no. 2 depicts the sequence aurochs, fox, crane, and pillar no. 38 depicts the sequence aurochs, boar, crane. The crane is interpreted as representing the constellation of Pisces, and the aurochs as Capricornus (the Goat). The fox and boar must therefore represent the northern and southern portions of Aquarius (the Water Carrier).

Based on rates of precession of 6 degrees per millennium, in 9530 BC, the radiant of the Taurids would have been located not in Taurus, but around 70 degrees to the east – in Aquarius[75]. Accordingly, Sweatman and Tsikritsis concluded that the fox and boar represent the radiants of the northern and southern Taurids respectively, and that Göbekli Tepe was built as an observatory to monitor activity from the meteoroid complex in case it ever menaced Earth again.

Lascaux revisited

In subsequent work[74,73], Sweatman derived the entire zodiac from Upper Palaeolithic cave art and carved figurines, and from motifs at the Neolithic sites of Çatalhöyük in southern Anatolia and Tell Zeidan in Syria. In addition to the signs he interpreted at Göbekli Tepe, he obtained lion = Cancer (the Crab) and ram = Aries (same as at present) from Çatalhöyük; and rhinoceros = Taurus (the Bull) and horse = Leo (the Lion) from Lascaux Cave. He claimed that this ancient zodiac remained in use with little change from around 40,000 years ago until comparatively recently, and that it did so across a region stretching from Anatolia to France and Spain. Upper Palaeolithic and Neolithic people used animal motifs to represent the position of the Sun in the zodiac at the solstices and equinoxes. They were aware of precession and hence intended these motifs to serve as 'date stamps' for future generations.

Using this methodology, Sweatman interpreted the Shaft of the Dead Man at Lascaux (see Chapter 3) as recording an impact from the Taurid Complex around 15,150 ± 200 BC. The eviscerated bison (or aurochs) was equated to the constellation of Capricornus, which was the radiant for the Taurids at the time in question. The Dead Man, like the headless man at Göbekli Tepe, symbolises the death and destruction wrought by the impact.

Sweatman cites paleoclimate records from Greenland ice cores as evidence of a "*fairly strong climatic fluctuation*" at the time in question, but it is just one of many climatic fluctuations occurring during the Last Glacial Maximum, and he admits that it is hardly convincing evidence. However, he claims that his zodiac is consistent with dates for portable and cave art from the period 40,000 to 17,000 years ago, as well as Göbekli Tepe, Çatalhöyük, and Tell Zeidan. The probability of this occurring by pure chance is "*essentially zero*".

Crazy enough?

Martin Sweatman's theory is certainly intriguing. It recalls a science fiction story that I read as a boy: *Space Agent and the Ancient Peril*, by Angus MacVicar, in which the UN's 'space agent' Jeremy Grant and archaeologist Prof. Spencer Johnson travel back in time to investigate the collapse of the civilisation at Tiahuanaco in the Bolivian Andes, 20,000 years ago. They learn that the Tiahuanacan people have a mythological tradition referring to a rogue planet that on several occasions in the past swung close to the Earth, causing floods and earthquakes. The myth is recorded in carvings on a great megalithic arch known as the Gate of the Sun. The rogue planet is due to make another close approach, but the Tiahuanacan scientific community is largely unconcerned and believes that the effects will be relatively minor. In the event, the rogue planet – which is actually the Moon – is captured into permanent orbit around the Earth, causing global devastation in the process.

The story is based on a fringe theory put forward by Austrian engineer Hanns Hörbiger in the 1920s and popularised by fellow Austrian H. S. Bellamy after World War II. Tiwanaku (to use the modern spelling) actually dates to around AD 800. The carvings on the Gate of the Sun include a figure holding a staff in each hand; this motif occurs frequently in the iconography of pre-Columbian South America, and is thought to represent a weather god.

It is important to realise that the Sweatman theory is *not* pseudoscience, radical though its claims may be. However, it has attracted considerable attention from 'ancient knowledge' enthusiasts and debunkers alike. Consequently, there is a need to examine it with scientific impartiality. As noted above, the evidence for an impact having triggered the Younger Dryas, while compelling, is certainly not conclusive at this stage. Obtaining a date for the Hiawatha Glacier impact crater would certainly be valuable, but this will take time. Even if a date corresponding to the Younger Dryas were to be confirmed, a link to the Taurid Complex seems unlikely. Also as noted, there are problems with the whole coherent catastrophism paradigm. But suppose for the sake of argument that Earth did take a direct hit from the Taurid Complex 12,900 years ago, so triggering the Younger Dryas.

As we have seen, the ethnographic evidence suggests that societies considered to be possible analogues for those existing during the Upper Palaeolithic are generally aware of solstices, but less so of equinoxes. There is also little evidence to support awareness of concepts such as the **ecliptic** or a zodiac – and these are just the basics. It would be necessary for Upper Palaeolithic people to be aware that the Sun will be 'in' a particular constellation at any given time of the year, even though the constellation itself cannot be seen except during a total solar eclipse. It could be argued, of course, that that is exactly how they became aware that there are stars in the sky by day, even though they cannot normally be seen. However, it has been suggested that even the ancient Mesopotamians, thousands of years later around 2300 BC, probably lacked this knowledge[76,77].

Precession shifts the nodes of the ecliptic eastwards by one degree every 71.6 years, or about two lunar diameters in a lifetime. Computerised amateur telescopes need to take the effect into account, otherwise stars begin to drift away from their predicted positions on a timescale of just weeks to months. The accepted view is that the peoples of the pre-telescopic era remained unaware of precession until the emergence of literate, state-level societies with the ability to keep records over the lengthy timescales required for precessional effects to become apparent to naked-eye observers.

Could these effects be discernible to an individual naked eye observer, albeit over the course of many decades, or even a lifetime? The **heliacal risings** and **heliacal settings** of zodiacal groupings such as the Pleiades will fall a day later over the course of a lifetime. But actual observations will depend on weather and atmospheric conditions, and it must also be assumed that Upper Palaeolithic people possessed an accurate calendar. It is possible, though, that if a society maintained a ritual tradition over many centuries around the heliacal rising or setting of (for example) the Pleiades, it might eventually be recognised that the ceremonials were now occurring later in the year than had once been the case. It should be noted, however, that there is no known precedent in the ethnographic record for knowledge of precession. Also, if precession was widely known to Upper Palaeolithic hunter-gatherers and Neolithic farmers, why were the highly sophisticated Mesopotamian astronomers of the sixth century BC apparently unaware of the phenomenon (see Chapter 10)?

The Sweatman theory places considerable emphasis on the use of statistics to suggest that the odds against other explanations are – well – astronomical. I do not claim to be an expert on statistics, but let us suppose that the statistical arguments put forward are valid. The problem is that the theory rests not upon statistics but on the correctness of the interpretation of the Upper Palaeolithic cave paintings, Göbekli Tepe bas-reliefs, and other Neolithic artwork as constellations in the first place.

The archaeologists excavating Göbekli Tepe were critical of the first paper and noted the focus on the "*outstanding (but not exceptional) richly decorated*" pillar no. 43 and consideration of just three others. The emphasis seemed arbitrary, given that there are more than sixty carved limestone pillars. Many of these feature similar bas-reliefs of animals and abstract symbols, and a few are equally complex as pillar no. 43. They also noted that the headless man on pillar no. 43 has a "*clearly emphasised phallus*" which is inconsistent with the association with death and destruction. Instead, they suggested that the image could be rooted in a Pre-Pottery Neolithic mortuary ritual which includes removing the heads of the deceased. The association of the headless man with a vulture could relate to the well-established practice of excarnation by exposing the corpse of the deceased to scavenging animals[2].

In summary, the theory put forward by Martin Sweatman and his collaborators faces a number of quite significant hurdles: ethnographic evidence, evidence from the early historical period, and alternative interpretations of the Göbekli Tepe bas-reliefs. This is in addition to the doubts about the Younger Dryas impact theory and coherent catastrophism. One cannot help but be sceptical of a theory some might describe as crazy – but could it, in the words of Niels Bohr, be "*crazy enough to have a chance of being correct*"?

05
Stonehenge

A brief history of Stonehenge

The transition from hunter-gathering to farming that began in the Near East around 9600 BC reached Europe around 6500 BC. Over the next 2,500 years, farming groups spread westwards, and reached Britain and Ireland around 4000 BC. Although it has long been the view that indigenous Mesolithic Britons adopted agriculture through contact with continental Europe, it is now thought that the dominant process was immigration, probably involving multiple migrations from northern France.

Another view that has changed in the last six decades is that Neolithic Europe was a cultural backwater, inhabited by simple farmers and 'barbarian' tribes. Between 3000 and 2000 BC, the first state-level societies arose in Mesopotamia, Egypt, and the Aegean. Such cultural and technological sophistication that existed in Temperate Europe was assumed to have 'diffused' from the Eastern Mediterranean; a process referred to as *Ex Oriente Lux* ('Light from the East'). But during the 1950s and 1960s, this view fell out of favour as radiocarbon dating showed that many innovations were too old to have been influenced by developments in the East. Stonehenge, for example, turned out to be far older than the Mycenaean civilisation of Greece, which was previously believed to have been an influence on its construction.

The most visible manifestation of Neolithic European innovation was the appearance in Brittany, Britain, and Ireland of megalithic monuments including stone circles, standing stones, and chambered tombs. They vary considerably in size, ranging from single standing stones known only to enthusiasts to world-famous monuments such as Stonehenge, Avebury, and Carnac. The smaller stone circles and stone rows were probably erected and used by local farming communities, but the larger monuments would have been major construction projects. For a long time, it was believed that these required a political hierarchy. For example, in the 1970s, British archaeologist Colin Renfrew[1] suggested that Neolithic Wessex was controlled by a number of chiefdoms, ruled by powerful individuals who could muster the manpower and resources needed to bring large projects to fruition. However, more recently it has been suggested that such hierarchies are not a precondition for complex projects. People might have provided their labour freely, motivated by ritual, tradition, and social customs[2].

The idea that Neolithic and Bronze Age people were highly proficient astronomers has been so widely popularised over the last fifty years or so that it has become almost mainstream. Many laypeople, if asked about Stonehenge, will claim to believe that its purpose was astronomical. Speculation that Stonehenge was intended for astronomical purposes is not new and goes back at least as far as the mid-eighteenth century. In the 1960s, the astronomers C. A. 'Peter' Newham, Gerald Hawkins, and Sir Fred Hoyle all claimed that

Stonehenge incorporated astronomical alignments and was used for predicting eclipses. Their views, along with the wider-ranging conclusions of the Scottish academic Alexander Thom (see Chapter 6) were largely rejected by mainstream astronomers and archaeologists, but they became very popular with 'ancient wisdom' enthusiasts.

Any investigation must take into account that the Stonehenge we see today was the result of around 1,500 years of constructions and modifications. It has long been accepted that the imposing if compact monument that comes into sight as the visitor drives westwards along the A303 from Amesbury did not form part of the original structure. As the site is traditionally understood, Stonehenge was built in three phases, the last of which is subdivided into five subphases. However, work since the beginning of the present century led by archaeologist Mike Parker Pearson, Professor of British Later Prehistory at University College London, has led to a new scheme being proposed, which is adopted here[3].

The original structure, which took shape between 3000 and 2620 BC, is now believed to have been a cremation cemetery. It is often referred to as Old Stonehenge or Stonehenge 1, and it comprised a simple circular bank, about two metres (6 ft 6 in) high, situated within a 110 m (360 ft) diameter ditch enclosure. There is a broad entranceway on the northeast side, and a smaller one to the south. Three stones, designated B, C, and 97, were set up outside the northeast entrance. Inside the outer edge of the enclosure is a ring of 56 pits, each about 1 m (3 ft 3 in) in diameter, which were later filled in. They are known as the Aubrey holes after John Aubrey, the seventeenth-century antiquarian and author of *Brief Lives*, who first noted them. Cremation burials were placed in the ditch, the bank, and the Aubrey holes; and postholes suggest that a timber structure was built within the main enclosure[4,3].

Though this contradicts many astronomical theories, it has recently been suggested that the Aubrey holes originally held bluestones (igneous dolerite and rhyolite) as well as cremated remains. Their dimensions are more consistent with bluestone sockets than they are with postholes, and the compacted and crushed chalk rubble within the holes that have been examined suggests that they once held standing stones[4,3]. These have been provenanced to two quarries in West Wales, Craig Rhos-y-felin and Carn Goedog, confirming the long-suspected Welsh origin. However, the quarries predate Stonehenge 1 by several centuries. It is likely that the bluestones were originally quarried for use in local monuments, and these were later were dismantled and transported to Salisbury Plain. Possibly they were symbols of political unification after a period of conflict, or alternatively communities migrating eastwards brought the stones with them. In either case, it is likely that the bluestones possessed significant ancestral and/or funerary associations, hence the not inconsiderable effort of moving them[5,6].

The migration view would be consistent with the discovery that many of the people whose cremated remains were recovered from the Aubrey holes did not originate from the Wessex region. Strontium isotope analysis of the bone fragments of at 25 individuals suggests that at least ten were non-local. The stable (i.e., non-radioactive) isotopic composition of certain elements such as strontium depends on local geology and a region's distinct 'signature' will thus find its way via the food chain into an individual's bones; this can be used to determine where that individual has spent much of their life. The most plausible origin of the ten non-locals is West Wales, long believed and now confirmed as the origin of the bluestones[7].

Possibly economic circumstances drove a migration from Wales, but why might Stonehenge 1 have been built where it was? Was it a case of 'well, they had to build it somewhere'? The answer is 'no'; it turns out that the choice of site was not arbitrary, nor was Stonehenge 1 the first construction there. Recent work conducted by Wessex Archaeology for English Heritage has shown that Stonehenge was sited near a pair of naturally occurring parallel ridges flanking a deep, corrugated fissure, 30 m (100 ft) wide, that formed as the glaciers retreated at the end of the last Ice Age. This feature just happened to align on the summer **solstitial** sunrise in one direction and the winter solstitial sunset in the other. Postholes dating back to around the eighth or seventh millennium BC suggest that this natural alignment had been noticed in Mesolithic times, long before the erection of the well-known Heel Stone. Thus, the revered bluestones found a new home, on a site that had been regarded as auspicious for many millennia[8].

For completeness, mention must be made of the theory put forward by neo-catastrophist Duncan Steel (see Chapter 4), who proposed that Stonehenge 1 was an early warning system, built in response to Earth facing another episode of peril from the Taurid Complex. Steel noted that the Aubrey holes were just the right size for an observer to sit in, and he suggested that that was in fact their intended purpose. Out of sight of any campfires, meteor watchers could allow their eyes to dark adapt, while being reasonably sheltered from the cold[9].

During the second phase of construction, between 2620 and 2480 BC, the familiar edifice of massive sarsens (a very hard and durable local sandstone) and smaller bluestones began to take shape. Cremation burials continued, but these ended around 2400 BC. The 13.7 metres (45 ft) diameter 'horseshoe' of five sarsen trilithons was set up at the centre of the site, with the Altar Stone within, and the open portion facing northeast. Outside the horseshoe was a double ring of between 50 and 80 bluestones (now the *Q* and *R* holes); and outside this, the sarsen circle of 30 shaped uprights linked by 30 lintels. The four Station stones were set up inside the outer bank. They formed a rectangle, with its long axis forming a right angle to the main axis of the monument. Only two Station stones remain, with earthworks defining the other two. The northeast entrance was enlarged to align on the solstice, and the Heel stone erected. Three large portal stones were set up just inside the entrance (stones *D*, *E*, and *F*), but only *F* – the so-called Slaughter Stone (now fallen) – remains. The *B* and *C* stones were removed[3].

Extensive animal remains have been found at the nearby site of Durrington Walls. The most important animals were pigs, followed by cattle[10]. Analysis of pig tooth wear suggests that many were slaughtered while still immature or subadult, corresponding to their first autumn or winter[11]. Strontium, sulphur, carbon, oxygen, and nitrogen isotope analysis of remains have revealed that the animals were brought from many different locations, with only a few having been raised locally. From exactly where cannot be established from the data. Strontium isotope ratios are consistent with locations in central England, southern and eastern coastal regions, south Devon, Cornwall, Gloucester, and south Wales. It seems likely that Durrington Walls was the site of major feasts, possibly associated with midwinter rituals. Attendees, bringing their own animals, would have travelled distances of anything from 30 to 100 km (18 to 60 miles), or possibly further, to reach the site[12,13,14]. The implications are that the events at Stonehenge drew people from the far corners of Britain, bringing pigs and cattle with them for the feasting. These findings suggest a level of interaction and social complexity in Late Neolithic Britain rather greater than once believed.

There are no burials at Durrington Walls, which served as a camp for the attendees. It was also the location of two more circular monuments – constructed from timber rather than stone. The whole ensemble has been interpreted as a ritual landscape, of which Stonehenge was only a part. Stonehenge was a place of the dead, whereas Durrington Walls was a place of the living[14].

Between 2480 and 1520 BC, there were three further stages of activity, during which a bluestone ring was erected within the central trilithon; subsequently both it and the outer double ring were dismantled and re-built as a bluestone oval inside the trilithon horseshoe and an outer bluestone circle between the trilithon horseshoe and the sarsen circle. The northeast section of the bluestone oval was later removed, creating a horseshoe that matched the shape of the sarsen trilithons. The digging of the 30 *Y* and 29 *Z* holes inside the Aubrey holes (the *X* holes) between 1630 and 1520 BC was the last known structural activity at Stonehenge[3].

Solstitial alignments at Stonehenge

The above would suggest that there were two major incarnations of Stonehenge upon a far older Mesolithic site: as a cremation cemetery and as a major ritual centre. The location appears to have been chosen because of a fortuitous natural alignment on the solstices, but to what extent were astronomical considerations a factor at Stonehenge?

The well-known alignment of the Heel Stone on the summer solstice was first noted by antiquarian William Stukeley in 1740. The alignment is not exact: the summer solstice sunrise, as viewed from the geometrical centre of the monument, occurs just to the left of the Heel Stone. During the third millennium BC, the sun would have risen even further to the left. We do not know whether Neolithic people defined sunrise as the point when the upper limb first appeared, or when the entire solar disk cleared the horizon. But in either event, the Sun would have been well clear of the horizon by the time it sat directly over the Heel stone[15]. One possibility is that it and the now-missing stone 97 straddled the summer solstice[16], although we cannot be certain that the pair were ever present at the same time[2].

The variation is due to cyclical changes in the **obliquity of the ecliptic** (axial tilt). Working on the assumption that the Heel stone alignment was exact when the monument was in use, astronomer Sir Norman Lockyer[17] estimated the construction date to be 1680 BC. Lockyer's definition of 'sunrise' assumed that the upper limb was two arcmins above the horizon. This is a reasonable assumption for 'first light', but obviously we cannot be certain that Neolithic people used the same definition. A more serious problem is that the values he used for changes in the obliquity were incorrect. The obliquity is now known to have been at a maximum around 8700 BC, and it has been decreasing ever since. Consequently, the Sun last rose at the **azimuth** of the Heel stone well thousands of years prior to 8700 BC. This would tend to suggest that the Heel stone was once one of a pair.

A Neolithic computer?

In the 1960s and 70s, astronomers Gerald Hawkins[18,19,20,21] and C. A. 'Peter' Newham[22,23] proposed that Stonehenge was a Neolithic 'computer' or 'observatory', intended to track the movements of the Sun and Moon and predict eclipses. Their views were later supported by astronomer Sir Fred Hoyle[24], who was noted for holding controversial views on a wide range of scientific matters.

Gerald Hawkins studied under radio astronomy pioneer Sir Bernard Lovell at the University of Manchester before moving to the United States, where he became Professor of Astronomy at Boston University. He became a US citizen in 1965. Hawkins used an IBM 7090 mainframe computer to assist with his calculations, which was an unusual step at the time. These room-filling devices were intended for scientific and technological computations rather than for commercial data processing applications, but they were not cheap. In 1960, a typical system cost $2.9 million (equivalent to $18 million today) to purchase outright, or $63,500 a month (equivalent to $500,000 today) to rent. Customers were more likely to be organisations like NASA than individuals, but Hawkins was donated one minute of time on a machine jointly run by Harvard University and the Smithsonian Institution. It should be borne in mind that computers were then regarded as cutting-edge technology rather than mundane household appliances, albeit the most basic present-day laptop is immeasurably more powerful than an IBM 7090. It was probably the 'sexiness' of computers in the public eye in the 1960s that led Hawkins and others to describe Stonehenge as a 'Neolithic computer'.

Hawkins used the Smithsonian-Harvard computer to determine the rising and setting positions of the Sun, Moon, planets, and stars, assuming a date of 1500 BC. For the Sun, he obtained the summer and winter **solstitial limits**; for the Moon, the major and minor **standstill limits**. His computations took into account the apparent altitude of the horizon, and the effects of **refraction** and **parallax**. The next step was to take directions between the Station stones, from the trilithons through the sarsens, and from the centre of the monument to the various outliers. Hawkins reported that to a mean accuracy of one degree, the Sun yielded ten correlations; to a mean accuracy of 1.5 degrees the Moon yielded 14 correlations. The rising limits of midwinter full Moon (i.e., the nearest full Moon to the winter solstice), as viewed from the geometrical centre of the monument, were delimited by the two portal stones *D* and *F* (the Slaughter stone). No significant planetary or stellar alignments were found[18]. In later work, Hawkins claimed there were also solar alignments on the **equinox**, and lunar alignments on the midpoints of the Moon between major and minor standstill. In all, he noted 24 significant alignments out of a maximum of 50 possible pairings; by using a statistical calculation known as Bernoulli's theorem, he calculated that the odds against all arising by chance were around ten million to one[21]. However, some of the alignments involved the *B* and *C* stones, now known to have been removed before the *D*, *E*, and *F* stones were set up (see above).

Hawkins[19,21] noted that when the midwinter full Moon rises over the Heel Stone, it implies that it is at one of the **nodes of its orbit** with the ecliptic and that there is the 'danger' of a lunar or solar eclipse, or both. A lunar eclipse would occur within hours, or a solar eclipse at the next new Moon; however, these eclipses would not necessarily be visible from Stonehenge. Thus, the Heel stone moonrise was nothing more than the warning that an eclipse might be seen. Similarly, when the midwinter full Moon rises over stone *D*, then eclipses are possible around the spring equinox; similarly, when the midwinter full Moon rises over stone *F*, then eclipses are possible around the autumn equinox.

Hawkins[20] also believed that the purpose of the 56 Aubrey holes was to predict eclipses. If a full **nodal cycle** of the Moon is taken to be the nearest full Moon to the winter solstice, then the period is either 18 or 19 years (rather than 18.61 years). The cycle roughly evens out after 19 + 19 + 18 = 56 years; Hawkins noted that 56 is the number of Aubrey holes, and proposed a method by which they could be used to predict these cycles. The Aubrey holes are numbered clockwise 1 to 56, ending with the hole facing the Heel stone. Three white stones (*a*, *b*, *c*) are placed in the Aubrey holes; stone *a* in hole 56 (facing Heel stone), *b* in hole 38 (trailing by 18), *c* in hole 19 (leading by 19). Three black stones (*x*, *y*, *z*) are placed with stone *x* in hole 47, *y* in hole 28 (opposite position 56), and z in hole 10. The stones are each moved once a year anticlockwise at (say) the winter or summer solstice. When a black or a white stone is in hole 56 (corresponding to the Heel stone), then an eclipse will occur in the month of midwinter or midsummer. When a white stone is at hole 51 (corresponding to stone *D*) or hole 5 (corresponding to stone *F*), then eclipses are expected in spring or autumn. The method would hold good for about three hundred years before it began to drift, but it could be 'reset' by advancing the marker stones by one hole. However, it probably would have taken many years of observation to determine an initial setting for this 'computer'.

C. A. Newham[22,23] noted a group of around forty postholes in the causeway beyond the northeast entrance, lying north of the Heel stone. The holes appear to radiate from the centre of the monument and lie in a 10-degree arc. They are arranged into six ranks crossing the causeway. Newham believes that the postholes are where stakes were used as temporary markers to record the point where the midwinter full Moon appeared over the horizon each year over the course of years and possibly decades, in order to obtain the azimuths of the major and minor standstills.

Sir Fred Hoyle[24] also suggested that the Aubrey holes were used to predict eclipses, but his method was different to that proposed by Hawkins. Four markers are used for this method: one for the Sun (*S*), one for the Moon (*M*), and two for the nodes of the Moon's orbit with the ecliptic (*N1* and *N2*). The markers are moved so that the Sun marker *S* completes a revolution every tropical year; the Moon marker *M* every **synodic month** (29.531 days); and the nodal markers *N1* and *N2* every 18.61 years.

1. The Sun marker S is moved one Aubrey hole every 6½ days: i.e., alternately on the morning of the seventh day after an evening move and the evening of the sixth day after a morning move;
2. The nodal markers *N1* and *N2* are positioned on opposite sides of the circle and are moved three Aubrey holes every year;
3. The Moon marker *M* is moved one Aubrey hole in the morning and one in the evening.

It is necessary to set the markers in initial positions and to reset them periodically, using the alignments noted above.

1. The Sun marker *S* can be reset twice a year, at the summer and winter solstices;
2. The Moon marker *M* can be reset twice a month, at full Moon (opposite the Sun marker) and at new Moon (in the same direction as the Sun marker);
3. The nodal markers *N1* and *N2* must be set at right angles to the Moon marker *M* when the Moon is at maximum declination at major standstill.

If the Sun marker *S* and the Moon marker *M* are facing one another (full moon) when at opposite nodes, then there is the possibility of a lunar eclipse. If the Sun marker *S* and the Moon marker *M* are together (new moon) when at either node, then there is the possibility of a solar eclipse.

These astronomical theories attracted considerable interest from the general public. The

Daily Express said of Hawkins' 1965 book *Stonehenge Decoded* that "*If Hawkins is right, and most experts now agree that he must be, then Stonehenge is the Eighth Wonder of the Ancient World*". In fact, most experts were decidedly sceptical. Archaeologist Richard Atkinson[25] dismissed *Stonehenge Decoded* as 'moonshine'. Atkinson made the following criticisms of the book and of Hawkins' interpretations:

1. Hawkins was not an archaeologist, and his understanding of the site's history contained many errors;
2. The plans he had based his calculations on had never been intended to serve as a basis for accurate measurements;
3. The limits of mean accuracy he had allowed in accepting a sightline as significant were 24 times larger than the errors obtained by sighting with a pair of sticks, and even then, many of his claimed alignments fell outside these limits;
4. Crucially, these included the major and minor lunar standstill limits claimed for stones *D* and *F*, upon which much of the astronomical hypothesis rested;
5. He had failed to take into account the effects of erosion on the chalk surface of Salisbury Plain, which has lowered the skyline by about 450 mm (18 inches) since Stonehenge was in use.

Atkinson also pointed out an error in Hawkins' statistical claims. The vanishing small figure was the probability of *exactly* 24 significant alignments out of 50, whereas he should have calculated the probability of obtaining *at least* 24. Archaeologist Clive Ruggles[15] notes that this would improve the odds to one in 12,500 (still small), but that there are actually far more than 50 possible alignments; the number is 182, and even if you eliminate unlikely alignments such as those between two points close together, you get at least 111. The probability of at least 24 successes out of 111 attempts is 0.37. If the three alignments that were more than two degrees out are excluded, the probability of 21 fortuitous successes out of 111 is 0.65, or odds on. By increasing the tolerance to five degrees, two of the three rejected alignments can be included – but the probability that they could have arisen by chance now increases to 0.88, or long odds on.

Ruggles[16] is sceptical as to whether equinoxes were of any interest to Neolithic people. He notes that the days midway between the solstices, the days on which the Sun rises and sets midway between its solstitial limits, or the days on which the Sun rises and sets at opposite points of the compass, all mark the equinoxes – but bear no relation to the modern concept of the Sun being at the points where the ecliptic intersects with the celestial equator. He is also dubious about the lunar standstills, which are not normally observed directly. The Moon's rising or setting at its monthly limit would have to occur on the same day as a standstill limit was reached. All that might be observed is that for some months the Moon would rise and set unusually far north or south; or that it is up for an unusually long time. However, Ruggles[15] concedes that some of Hawkins's alignments, though fortuitous, might have been subsequently noticed and used. Notwithstanding Ruggles' views, there is evidence for possible Neolithic interest in the equinoxes and lunar standstills, which we shall consider in Chapters 6 and 7.

Ruggles[15] challenged the idea of the Aubrey hole eclipse predictor on the basis that it could not have been set up without extensive record keeping. In any case, all it does is identify **eclipse seasons** when an eclipse can occur. This includes penumbral lunar eclipses and partial solar eclipses, which even if visible from Salisbury, would in all probability pass unnoticed. Others have suggested that the Hawkins and Hoyle methods are unnecessarily complicated, requiring detailed knowledge of lunar movements, and that it would have been simpler to have relied on the saros cycle. The ancient Mesopotamians were aware of the saros cycle by 750 BC without any knowledge of the underlying theory[26,27] (see Chapter 9).

The mistake, perhaps, was to consider Stonehenge in isolation. Though an iconic monument with a unique history, Stonehenge is but one of hundreds of Neolithic monuments dotting the landscape of Britain, Ireland, and Brittany. To gain a fuller understanding of possible astronomical knowledge in Neolithic times, it is necessary to consider the broader picture. We turn now to the story of Scottish academic Alexander Thom, who attempted to do just that.

06
Megalithic yards, standardised stone circles, and lunar observatories

The work of Alexander Thom

Alexander Thom (1894-1985) was a Scottish engineer and academic who was Professor of Engineering at Oxford from 1945 to 1962. He had previously spent the war years as the principal science officer at Farnborough, where he designed high-speed wind tunnels. His earlier academic work was referred to for the development of Barnes Wallis's 'bouncing bombs'. Following Thom's retirement from academia, the Engineering department's new tower block was named the Thom Building in his honour. However, Thom is principally known not for his distinguished academic career but his controversial theories about Neolithic stone rings. Often accompanied by his son Archibald and other family members and friends, he spent much of his spare time surveying hundreds of megalithic sites in England, Scotland, Wales, Northern Ireland, and Brittany. Between the 1930s and 1970s, he made over 800 visits to these sites, filling over a hundred notebooks in the process. Many of the sites were located in the Hebridean islands and only accessible by boat: but Thom was an accomplished sailor, who skippered yachts ranging from 25 to 66 ft. He published five books and fifty journal articles – a quite remarkable output in a field that was not his primary occupation[1,2,3].

Thom's main claims were:

1. That Neolithic architects used a standard unit of length which Thom termed the 'Megalithic yard';
2. That the use of this unit combined with a knowledge of Pythagorean geometry enabled Neolithic people to lay out stone rings in accordance with a set of standard designs;
3. That megalithic monuments were used to indicate **azimuths** of astronomical phenomena including the **solstitial** and **equinoctial** rising and setting of the Sun, major and minor lunar **standstill limits**, and the rising and setting of first-magnitude stars;
4. That knowledge of these phenomena enabled Neolithic people to devise a calendar in which the year was divided into eight or possibly sixteen 'epochs' beginning at the spring equinox.

Thom's theories were elaborated in two monographs and numerous papers published from the mid-1950s until shortly before his death. A final work, *Stone Rows and Standing Stones*, was published posthumously in 1990.

Early investigations

In 1955, Thom published a paper, *A Statistical Examination of the Megalithic Sites in Britain*[4], in which he summarised the data he had gathered over the previous twenty years from 250 megalithic sites he had visited in England and Scotland. In this paper, he examined the geometry of the stone circles, the possibility that a standard unit of length was used in their construction, and the possibility that outliers and alignments of two or more stones had an astronomical significance.

Thom concluded that in addition to regular stone circles (Type C), there were two types of flattened circles which he termed Type A and Type B. In a statistically significant number of cases the diameters of stone circles were multiples of 1.66 m (5.43 ft), a length he later termed the 'Megalithic fathom'. In later work[5,6] based upon statistical analysis of the diameters of stone circles in England and Wales, he noted that almost half had a diameter that was an odd number of fathoms. As constructors would have worked in radii rather than diameters, he suggested that the actual standard unit was half a megalithic fathom, for which he obtained a value of 0.829 m (2.72 ft). He termed this unit the 'Megalithic yard' or MY.

Thom then went on to consider possible azimuth markers towards the rising and setting points of the Sun at the solstices and equinoxes, and of first magnitude stars (mag. +1.5 and above). He examined three classes of possible azimuth indication:

1. Outliers from a circle;
2. Alignment comprising a line joining two stones;
3. Alignment comprising a row of three or more stones.

A total of 72 structures at 39 sites were considered. Thom converted the potential azimuths to **declinations**, factoring in the effects of atmospheric **refraction**, and he compiled cumulative probability histograms of his results. These have the advantage that errors caused by factors such as stones shifting over time will be evened out. They also do not presuppose that the targets were astronomical. If Gaussian (statistical) peaks are found at particular declinations, these can be further investigated.

In some cases, it was not obvious in which direction an alignment was supposed to operate. Here, Thom used both directions, accepting that one of them would be spurious. The next step was to plot the declinations of the solstitial and equinoctial sunrise and sunset, and the brighter stars for the period from 2700 to 1300 BC. He factored in the effects of **precession** and changes in the Earth's **obliquity** and using intervals of 200 years.

Thom claimed that his results left "*little doubt that certain of these sites contain a pointer to the rising or setting points of the sun at the solstices and equinoxes*". He suggested that some other sites contained similar indications for the four days mid-way between the solstices and the equinoxes, although he admitted that here the evidence was inconclusive. He also reported indications to some of the brighter stars, but only for the period from 2100 to 1900 BC.

Retirement from Oxford and publication of *Megalithic sites in Britain*

In 1967, five years after his retirement from Oxford, Thom published his first major archaeoastronomy monograph, *Megalithic sites in Britain*[7]. This book re-examined Megalithic yards, stone circle classification, astronomy, and calendars. By now, Thom had visited 450 sites, and had surveyed 300 of them. He extended his database of azimuth indications to 261 at 145 sites.

He proposed that Megalithic yards were in use in Iberia, where it became the basis of the *vara* ('rod' or 'pole'), a traditional Spanish unit of length that was later used in the New World. Various definitions of the *vara* of typically around 0.838 m (33 inches) were in use at different times and places. Thom proposed that standard rods were in use throughout Neolithic Britain and Brittany, which were presumably sent out from a central headquarters. He was unable to say whether this base was located in Britain or in continental Europe. He re-estimated the value of the Megalithic yard to be 2.720 ± 0.003 feet, i.e., the same value as in his earlier papers, but to an additional decimal place of precision. The uncertainty is 0.1 percent, making it known with more precision than any other unit from antiquity. Thom suggested that the Megalithic yard was sometimes subdivided into halves or quarters, but never thirds. He would later refine the estimate to 0.8297 m (2.722 ft). In addition to the Megalithic yard, there was also a Megalithic rod of 2½ Megalithic yard and a Megalithic inch of 0.817 statute inches (20.725 mm) or 1/40 of a Megalithic yard. The Megalithic inch was based on analysis of the so-called cup-and-ring marks found on many monuments[8,9]. These marks are usually interpreted as purely decorative[10], and they consist of small concave depressions incised into the stone surface and often surrounded by concentric circles.

The stone ring classification scheme was extended to include Type D flattened circles, ellipses, and Type I and Type II egg-shaped rings. Thom suggested that the construction of these various types of stone ring entailed a knowledge of Pythagorean triangles[7]. This scheme would be later further extended by Thom and his son Archibald to include modified Type B flattened circles and Type III egg-shapes[11]. Thom believed that megalithic architects had a considerable knowledge of practical geometry and attained a proficiency in measurements matched only by a present-day surveyor. They focused on geometrical figures which had as many dimensions as possible arranged to be integral multiples of their units of length, and they avoided lengths that were based on irrational numbers. These various types probably arose from attempts to obtain diameters and circumferences that were integral numbers of Megalithic yards[9].

The book further considered stellar, solar, and lunar azimuths, for which Thom claimed that indications were set up with great precision. To obtain the required accuracy, he suggested that Neolithic people used a backsight and a foresight which line up on the astronomical phenomenon of interest. The backsight could be a stone, a hole in a stone, a gap between two stones, or a staff at the centre of a circle. The foresight could be a distant stone or a pole at the centre of a circle. The ideal foresight would use a distant mountain peak, a distant notch in the horizon, or a feature out to sea. For such natural foresights, an indicator could be used to distinguish it from other features, but it only needed to provide enough accuracy to avoid confusion. For solar and lunar indicators, precision was improved by preferentially observing the upper and lower limbs of the Sun and Moon.

Azimuth indicators served three uses: time indication, calendrical purposes, and studying the movements of the Moon. Time indication relied on a meridian indication to determine the moment of upper and (for **circumpolar** stars) lower transits of the meridian. There are many sites with definite indications of a north-south line. Obviously, which stars are transiting at a given time depends on the time of year.

Thom enlarged on his earlier idea of what he now termed a 'Megalithic calendar' in which the year was divided into eight or possibly sixteen 'epochs' of roughly equal length (22-24 days), beginning at the spring equinox. Thom suggests that the day of the spring equinox was found by locating (through trial and error) the point where the rises Sun on a day in spring, and then at the same point on a day half a year later in autumn. Having determined the day of the spring equinox, Neolithic people might simply have counted the days, inserting an intercalary day as necessary. Alternatively, they could have set up markers to show the rising points of the Sun for the start of each epoch. Such a system would eliminate the need for intercalary days. The sixteen epochs vary in length between 22 and 24 days because the Earth's orbit is elliptical: epochs corresponding to the time of year when the Earth is close to **perihelion** will be shorter than those when it is further away from the Sun in accordance with Kepler's Second law of planetary motion (see Appendix). Thom claimed that his histograms provided evidence that such a calendar was in fact in use. It explains many otherwise puzzling azimuth lines.

Focus on 'lunar observatories' and higher precision indications

The 1967 monograph only briefly mentions lunar observations, but these were the subject of much of Thom's subsequent work. Publications included his second major work, *Megalithic lunar observatories*[8], and many papers concerning possible lunar azimuth indicators. Some of these were general, for example, Thom[12], Thom & Thom[13,14,15]; others focused on indicators at specific sites or places. These included Stonehenge[16], Argyllshire[17], Orkney[18,19], and Carnac, Brittany[20,21,22,23,24,25]. Thom claimed that the principal function of many stone monuments in England, Scotland, and France was to observe the rising and setting of the Moon at its major and minor standstill points.

The major standstill has a greater effect on the Moon's behaviour at higher latitudes such as in Scotland. At 55° N, the midwinter full Moon rises before sunset and is still up at sunrise. In Neolithic times, the Earth's obliquity was greater than it is now, and at major standstill the midwinter full Moon was almost circumpolar as seen from the Shetlands.

Much of this later work focused on high-precision indications using distant foresights. Thom suggested that these were typically peaks or notches in rows of distant hills or mountains (which are common in Scotland) that are grazed by the rising or setting Sun or Moon as it approaches its maximum or minimum declination (solstices, lunar standstills). A stake was used to mark the backsight each day at the point where the Sun's upper limb could just be seen sliding along the slope of the hills or appearing and disappearing in a notch. Possibly, a group of observers stood at regular distances apart so one or more of them would catch the phenomenon. The process was repeated each day, with a new stake being used to mark the viewing position. Once the Sun or Moon reversed direction and moved back from its extreme declination, the trail of stakes would also reverse. The stake representing the northerly or southerly limit would be replaced with a permanent marker or backsight.

Thom noted that his histograms displayed twin peaks around lunar and solar azimuths, which he claimed demonstrated the preferential use of the upper or lower limbs of the Sun or the Moon for indicating solstitial or standstill events. He claimed that the precision obtained by Neolithic observers was such that it could resolve northerly and southerly limits of sunrises, sunsets, moonrises, and moonsets to a single day; and resolve lunar movements to the extent that a 9 arcmin 'wobble' could be discerned in its movements. At such a level of precision, it was necessary to factor in **parallax** effects due to observations being made from the surface rather than the centre of the Earth. Parallax effects vary over the course of a month, being greatest when the Moon is close to **perigee**.

The Moon's orbital inclination oscillates around a mean due to the Sun pulling it towards the plane of the ecliptic. The periodicity of this wobble is half an **eclipse year** which is defined as the time for the Sun to make successive passages through the same **node of the Moon's orbit**. The eclipse year is 346.620076 days (see Appendix). In Europe, the wobble was first noticed by Tycho Brahe (see Chapter 17), though Arabian astronomers (see Chapter 14) were aware of it earlier. The effect is at a maximum of ± 9 arcmin from the mean when the Sun, Moon, and Earth all line up – which the only time an eclipse is possible. Thus, the maximum 'wobble' coincides with an **eclipse season**; i.e., one of the two periods in the eclipse year when there is the 'danger' of an eclipse.

The main problem with setting up an indication on an extreme lunar or solar declination is that the solstices or standstills only rarely coincide with sunrise, sunset, moonrise, or moonset. In the case of the Moon, this can lead to very significant inaccuracies. Accordingly, Thom suggested that Neolithic observers must have means of extrapolating from the stake positions obtained by two or three nights of observation to derive the position for the maximum declination for that lunation. Thom describes two geometrical methods by which a present-day astronomer would tackle the problem: the triangle method and the sector method. At several places in Caithness, there are fanlike arrangements of stones. Thom believes that their purpose was for extrapolation using the sector method.

Reaction to Thom's claims

Unlike Gerald Hawkins, Thom's work was based not on a single site but was backed up by the data from hundreds of monuments subjected to a rigorous statistical analysis. Archaeologist Richard Atkinson, while scathing of Hawkins' claims (see Chapter 5), was impressed by Thom's meticulous fieldwork and analysis[26]. Conversely, to Thom's annoyance, his theories were enthusiastically embraced by 'ancient wisdom' enthusiasts and other members of the pseudoscientific fringe[3]. Although most mainstream archaeologists were sceptical, they took the view that Thom's theories were at least deserving of serious consideration[27].

Recent work at Stonehenge has provided evidence that rituals and other activities brought people there from across the whole of Britain (see Chapter 5). The scale of this intercommunity mobility demonstrates a level of interaction and social complexity in Late Neolithic Britain that was not appreciated in Thom's day[28]. The existence of standard units of length and standardised forms of stone circle are at least plausible.

Megalithic yards, long feet, and chalk drums

The existence of the Megalithic yard did not gain widespread acceptance. Statistician P. R.

Freeman[27] applied Bayesian statistics to Thom's data. Bayesian inference is named for the eighteenth-century mathematician Rev. Thomas Bayes. It is a powerful but computational-intensive statistical method that has been brought to the fore by the increased 'number-crunching' abilities of modern computers. Freeman concluded that only the Scottish data provided any support for the Megalithic yard, and even with that there were two other values that fitted the data equally well. Archaeologist Clive Ruggles[10] noted that "*...statistical reassessment of Thom's data both from classical and Bayesian viewpoints reached the conclusion that the evidence in favour of the Megalithic yard is at best marginal, and even if it does exist the uncertainty in our knowledge of its value is only of the order of centimetres, far poorer than the 1 mm precision claimed by Thom. In other words, the evidence presented by Thom could be adequately explained by, say, monuments being set out by pacing, with the 'unit' reflecting an average length of pace.*"

However, others were more sympathetic to Thom's claims. Mathematician Alan Davis[29] analysed cup-and-ring markings from sites at Ilkley, Yorkshire and in Northumberland, and concluded that there was tentative evidence for measurements of one, three, and five Megalithic inches. The problem is that five Megalithic inches is close to the width of a human hand and one Megalithic inch could be a mean finger-width. Davis noted that it would be difficult if not impossible to distinguish between the two on the basis of the data. He stated that his work was no more than a preliminary attempt to build on the pioneering work of Thom. Also supportive was archaeologist Margaret Ponting, who suggests that a bone artefact found at a site 12 km (8 miles) from the Hebridean site of Callanish was a Megalithic rule. Markings on it corresponded closely to quarters of a Megalithic inch and when these were averaged out, they gave a value of 20.416 mm for the Megalithic inch; this is within 2 percent of 20.725 mm value claimed by Thom[30].

More recently, the case for standard units of length in Neolithic times has been made by archaeologist Mike Parker Pearson, who has suggested that a unit of length he termed a 'long foot' was in use at Stonehenge. The long foot is 1.056 statute feet or 0.32187 metres. Parker Pearson claims that some features at Stonehenge have diameters that are multiples of 30 'long feet' of 1.056 statute feet (0.32187 metres); for example, the Aubrey holes are 270 long feet in diameter and the exterior bank is 300 long feet in diameter. However, at nearby Durrington Walls a 'short foot' of 0.96 statute feet or 0.293 metres was in use. Ten long feet equals eleven short feet. Parker Pearson notes that the 11:10 ratio is found in some historically documented systems of linear measurement, including the British statute system in which the chain, furlong, and mile are multiples of 11 statute feet. (1 chain = 66 ft (11 x 6); 1 furlong = 660 ft (11 x 60); 1 mile = 5280 ft (11 x 480)). A wheel with a diameter of seven 'short' units has a circumference of approximately 20 'long' units; this is because of the well-known approximation of 22/7 to the mathematical constant π (the ratio of the circumference of a circle to its diameter). The effect of the 11:10 conversion ratio between the measurement systems reduces the diameter length of 22 'short' units to 20 'long' units. Thus, the 'long' system could have been used for laying out the dimensions of monuments using a wheel whose diameter was calibrated in multiples of the 'short' unit[31].

Parker Pearson later claimed that the enigmatic 'chalk drums' of Folkton and Lavant were Neolithic measuring devices[32,33]. The three Folkton drums are carved chalk cylinders that were found in a Neolithic round barrow in 1889 in Folkton, North Yorkshire. They are not identical, progressively increasing in size with height/diameters of 86/103.2 mm (3.39/4.06 inches), 105/126 mm (4.13/4.96 inches) and 117/145.2 mm (4.60/5.72 inches). Until recently, it was assumed that they were unique and purely decorative objects. However, in 1993 a fourth drum was discovered at a Neolithic site in Lavant, West Sussex. The Lavant drum is undecorated, and measures 105/115.0 mm (4.13/4.53 inches). The ages of the drums are not known, but context suggests that they date to around 3000 BC and are thus broadly contemporary with Stonehenge and the spread of Grooved Ware pottery in Britain and Ireland.

The circumference of smallest Folkton drum (Folkton III) is 0.3242 m (1 ft 0.764 in), which is almost exactly one long foot. The circumferences of larger drums (Lavant, Folkton II and Folkton I in order of size) are close to integer subdivisions of 10 long feet. A distance of 10 long feet can be marked out by ten rotations of Folkton III, nine rotations of Lavant, eight rotations of Folkton II, and seven rotations of Folkton I. Parker Pearson believes that this sizing was deliberate, and that the drums form a graduated set intended to be used as standards of linear measurement. Chalk is not a particularly durable material for repeated use as a measure, so it is likely that the drums themselves were replicas of wooden drums that were actually used for making the measurements, or perhaps reference standards for making the wooden drums. In use, cords would have been wrapped around the drums to obtain the required lengths, which would then have been used to lay out Neolithic monuments.

Unfortunately, only a few circular Neolithic monuments have been excavated in sufficient detail to provide diameter measurements that are sufficiently accurate to test the long foot hypothesis. However, the very large monuments of the Ring of Brodgar in Orkney and the Great Circle at Newgrange, County Meath in the Republic of Ireland appear to have similar diameters to the ditch at Stonehenge. It is therefore possible that they conform to the same measurement standard. Parker Pearson believes that the standard measure could have been part of a Neolithic cosmological understanding associated with circular monument building and the adoption of Grooved Ware pottery.

It remains to be seen if further evidence will support the existence of the long foot. It could be that while Thom was wrong about the Megalithic yard, his belief that Neolithic people in Britain and Ireland employed standard units of length will nevertheless be vindicated.

Stone ring classification

Thom's geometrical scheme for stone ring classification has also met with mixed reactions. Civil engineer Ronald Curtis[34] independently surveyed Hebridean stone rings and found a good match with Thom's geometrical scheme. On the other hand, archaeologist Thaddeus Cowan[35] has been more sceptical. He suggests that ring builders were actually attempting to lay out circular rings, and that the effects Thom believed to be intentional were nothing more than artefacts of the methods used, which would not normally produce exact circles.

To people unfamiliar with a compass, it was probably less obvious than it is to us that a circle is a figure of equal radius. Conversely, it would have been apparent that it is a figure of equal width. The width is bounded by well-defined edges and can thus be seen, whereas the centre end of the radius must usually be inferred. The circle is the most obvious example of a figure of equal width, but it is also possible to have cornered (i.e., non-circular) figures that are of equal width. The seven-sided British fifty and twenty pence coins are examples of such figures. They are technically known as Reuleaux polygons and are made up of an odd number of arcs of equal length. However, it is not necessary for a figure of equal width to be composed of arcs of equal length; indeed, the figure need not even be symmetrical. The only requirement is that it must have an odd number of arcs.

If we assume that the megalith builders were attempting to construct circles, then their methods might have been based around the diameter rather than the radius; they might have used arcs of radii equal to the intended diameter of the stone circle they were attempting to lay out. Producing a figure by such means is known as diameter construction. Two planners would keep taut a rope of a length corresponding to the intended diameter of the stone circle. While the first remained stationary, the second paced out a short clockwise arc of 60 degrees or less. The second then remained stationary, while the first paced out a similar clockwise arc. This was then repeated until a cornered, non-circular, closed figure of equal width was defined; for example, three arcs of 60 degrees will form a Reuleaux triangle.

Cowan notes that there are several stone rings that can be best fitted to a Reuleaux pentagon. He also suggests that cornered equal-width figures could also have been used as a base to construct larger figures with rounded corners, and that such methods were employed in laying out rings of the more complex types described by Thom. The geometrical methods that Thom claimed were used to delineate these rings require only minor changes to make them commensurate with Cowan's diameter construction methods. Cowan suggests that Thom was probably unaware of the concept of equal-width figures, or he would have referred to it in his analyses.

Astronomical and calendrical theories

Anthropologist Euan MacKie[36] is broadly sympathetic to Thom's astronomical and calendrical theories. He asks, did structures with simple alignment on solar events for ceremonial reasons evolve into 'long alignments' capable of making useful calendrical observations? If so, it might well indicate the presence in late Neolithic Britain of full time 'wise men' including trained celestial observers – not unlike and possibly ancestral to the Druids.

The existence of such long alignments would imply a desire for an exact calendar, which would go well beyond the needs of Neolithic farmers. The use of backsights and foresights to achieve great precision represents an intellectual leap from the basic idea that the approximate time of the year can be determined from watching the cyclical changes of the sun's position against a hilly horizon and observing how these correspond to the season. It would then have taken many years to find suitable sites and, having done so, establish the exact position of the required backsights. Such an undertaking implies the existence of full-time specialists. We should note that these specialists would have required conceptual knowledge that is not normally thought to have existed prior to Babylonian times, particularly regarding the unequal lengths of the seasons (which is due to the Earth's elliptical orbit around the Sun).

Developments might have included defining the exact length of the year; this would lead to a reasonably accurate solar calendar and probably to interest in defining the four cardinal points - south by the position of the noonday sun and north by the circling of the stars around the north **celestial pole**. As soon as the equinoxes were defined by dividing the year between the solstices, the importance of east and west would be appreciated.

Astronomers could have determined the length of the year by setting a backsight for an arbitrary day. They would note that the year could be divided into two unequal halves. For example, a sunset alignment on 10 April will repeat on 1 September, dividing the year into 144 days through the summer to the autumnal repeat; and 221 or 222 days through autumn and winter to the spring repeat. The date of the summer solstice is obtained by dividing 144 days by 2 and adding to 10 April (21 June); the date of the winter solstice is obtained by dividing 222 days by 2 and adding to 1 September (21 December). The spring equinox lies midway between the winter and summer solstices; the autumn equinox lies midway between the summer and winter solstices. It would soon have been realised that the year is just over 365 days long and that an extra day must be added every fourth year to keep the sunset alignment in step with the calendar.

Once this level of calendrical knowledge had been achieved, MacKie believes that structures could have been aligned on calendrically important astronomical phenomena such as solstices, or towards a cardinal point. Such structures might not necessarily have served a particular calendrical function and could instead have been orientated for ceremonial or religious reasons. An example is Newgrange, a passage grave in County Meath, Republic of Ireland, where a solar alignment was incorporated into the monument. Above the entrance is a roof-box that allows the rising sun to reach the back of the chamber for a few days on either side of the winter solstice. While it is possible that the alignment is coincidental, the roof-box serves no other obvious function. Notably, the roof box was left open after the entrance to the tomb was blocked off, suggesting that although the living could no longer enter, the midwinter sun was still permitted to do so. Newgrange was not a dedicated observatory, but powerful astronomical symbolism was intentionally built into its design[10], and its construction must have relied on a pre-existing calendar. The existence of such large, elaborate structures suggests a class of priests or wise men, and a society more complex than that usually envisaged for the Neolithic. It seems unlikely that farmers and herders would have gone to all the trouble of building them.

MacKie studied a Neolithic site at Brainport Bay, Loch Fyne in Argyllshire, which was never surveyed by Thom. Previous work suggested that it was an important site for sun worship. There are two main areas: the main linear site on low ground next to the bay, and the now fallen Oak Bank stone. At the southwest end of the main linear site is a cairn and an earthen bank. The various parts of the latter form a 'rifle barrel sighting device' comprising a small standing stone seen through a notch in the main outcrop, which points at the summer solstice sunrise.

MacKie suggests that the Oak Bank stone is a backsight to two notches on the horizon; one of these corresponds to the equinoctial sunset, and the other to the winter solstice. Cup-marks on the fallen stone served as indicators to the horizon notches. But the winter solstice marker is not accurate enough to pick out midwinter to the nearest day because the horizon is only 1.8 km (1.1 miles) away; moving 180 mm (6 inches) between one evening and the next would put it out by 24 hours. The line-up might have been a 'reminder' instrument, intended to do no more than give warning that the shortest day of the year was approaching. MacKie supposes by the time it was set up, an accurate calendar must have already been in existence.

Like Thom, MacKie also believes in a 'Megalithic calendar'; a solar calendar consisting of eighths, sixteenths, and possibly even 32ths. He suggests that this ancient solar calendar is the origin of English and Scottish 'quarter days'. These are days fixed by custom to mark quarters of the year; they are used to fix beginnings and endings of tenancy agreements and due dates for rents. The English quarter days correspond reasonably closely to the solstices and the equinoxes: they are Lady Day (25 March), Midsummer Day (24 June), Michaelmas (29 September), and Christmas Day (25 December). The Scottish quarter days differ, corresponding closely to the points of the solar year that lie midway between the solstices and the equinoxes: they are Candlemas (2 February), Whitsunday (15 May), Lammas (1 August), and Martinmas (11 November).

The 'megalithic' solar calendar comprises sixteen 'epochs' beginning on the following days: *22 March*, 14 April, *7 May*, 31 May, *23 June*, 16 July, *8 August*, 30 August, *21 September*, 13 October, *4 November*, 27 November, *20 December*, 11 January, *4 February*, and 27 February. MacKie suggests that the italicised dates correspond closely to the English and Scottish quarter days. In fact, the correspondence is not that close, and dividing the year into eighths could have been reinvented on many occasions. The division of a year around solstices and equinoxes is pretty obvious, as is its further subdivision into eighths. For example, the four Gaelic seasonal festivals occur roughly at the mid-points between solstices and equinoxes: Imbolc on 1 February, Beltane on 1 May, Lughnasadh on 1 August, and Samhain on 1 November. These festivals are very ancient, but not necessarily prehistoric, in their origins.

In conclusion, MacKie believes that the double alignment on Oak Bank gives powerful independent support for the hypothesis that the main linear site was designed for the ceremonial observation of the summer solstice. The main site must originally have been a chance discovery in prehistoric times. The main outcrop is natural, as is one side of the rock notch. It is also due to chance that there is a clear view from it up Loch Fyne to two distant peaks close to the summer solstitial sunrise position. This 'natural observatory' would have seen miraculous and endowed the site with a religious aura which persisted for 3,000 years until the coming of Christianity. Ceremonial activity there continued until as late as the tenth century AD.

Archaeologist Aubrey Burl[37], best known for his studies of megalithic monuments, is also sympathetic to Thom. He notes that only a few advances in stone circle research have been made by mainstream archaeologists. Other than Stonehenge, megalithic monuments have been largely ignored by the archaeological community. As late as 1976, they were being attributed to continental Beaker folk, and were thought to date to around 1700 BC rather than 3200 BC. Hence there has been very little mainstream theorising about their likely role in Neolithic society.

Burl notes that Cumbria hosts over fifty stone circles, some of which are quite large at 30 m (100 ft) or more in diameter and include Long Meg and her Daughters, Penrith, and Castlerigg, Keswick. The larger stone circles in the region were located near axe-making sites, and hence Burl believes that their primary role was as axe-trading centres. However, they also likely served as a focus for communities ranging in size from single families to sizeable populations. They were meeting places for ceremonials as well as trading.

Burl claims that calendrical orientations and features marking the cardinal points of north, south, east, and west are to be found at many Cumbrian rings. These alignments are typically not very precise: if potential sightlines at any one site were considered in isolation, they could be dismissed as accidental. However, the same alignments recur in ring after ring throughout Cumbria, suggesting that they were intentional. But how were they fixed?

It might be expected that Neolithic people must at least have used elementary systems of enumeration for counting livestock, etc. Burl believes that semi-numeracy is suggested from an analysis of numbers of circumference stones (i.e., not counting outliers or central stones) in rings, where the original number is known with reasonable certainty. There are preferred numbers, which do not appear to have been a function of ring diameter; rings with the same number of stones show a considerable variation in diameter[38]. Burl believes that in Cumbria, counting systems were based upon either 6 or 3. For base 3, the number seven would be calculated as 3 + 3 + 1; for base 6, simply 6 +1. Such systems would not have worked efficiently for numbers above 20, unless some form of tallying was used. There is no archaeological evidence in Britain for tally sticks from this period.

Thus, Neolithic people would not have been able to derive the dates of the equinoxes from the solstices by counting the days. Instead, they could have halved the distance between markers aligned on the midwinter and midsummer sunsets. The presence of hills and valleys on the skyline would invariably mean that the centre point did not correspond exactly to the equinox, explaining the imprecision of many of the alignments.

Equinoctial alignments are consistent with the idea that autumnal gatherings with outsiders took place when work at the axe factories finished at the end of each summer. Such gatherings would have been of great importance as they involved the presence of strangers. Burl sees a parallel with stone axe trade of Aboriginal Australians where gift exchange and periods of non-hostility required sanctuaries and rituals at accepted times of the year. Conversely, the winter solstitial alignments at the more remote Cumbrian circles might have been for local rituals, to which outsiders were not admitted. North-south alignments are important to many peoples around the world; there is a notion of the 'immutability of the north', which is based on the observation that the Sun, Moon, and stars revolve around the north celestial pole, which does not itself move.

In summary, Burl believes that a stone circle was a place for exchanging axes and gifts, for holding annual gatherings, and to where the bodies of the dead were brought for burial. Most importantly, though, it was a place of great cosmological significance, and the most sacred of places to its local population. Burl notes that this interpretation has come not from mainstream archaeology but from the work of Alexander Thom and his predecessors such as John Aubrey.

Clive Ruggles reassesses Thom's astronomical claims

Beginning in the 1970s, Clive Ruggles[39,40,10,41] carried out a major reassessment of Thom's astronomical and calendrical work. He grouped the evidence into four stages of ever greater precision, which he labelled Levels 1 to 4. The Level 1 data came from Thom's 1955 and 1967 analyses, and Ruggles suggested that the claimed precision was half a degree, or roughly one solar or lunar diameter. The Level 2 data related to additional analysis Thom had carried out on around thirty solstitial declinations and forty lunar major and minor standstill limits where upper and lower solar and lunar limbs had been preferentially observed; the inferred precision was within ten arcmin, or a third of a lunar or solar diameter. The Level 3 data related to Thom's 1969 and 1971 analyses of lunar movements, where the focus had been on natural foresight on distant horizons; these were precise to within three arcmin, or a tenth of a lunar diameter. Finally, the Level 4 data related to work published by Thom and his son Archibald[14,15], in which 44 sight-lines including 17 from eight new sites had been painstaking examined, with effects of parallax and refraction being calculated for the time of the year and day when readings would have been taken rather than assume mean parallax and refraction corrections. Thom noted that refraction might have been different to that of the present day due to climate change since Neolithic times. A precision of 1 arcmin was claimed, or a thirtieth of a lunar diameter.

Ruggles first considered the Level 1 data and replotted Thom's cumulative probability histograms. He noted that there are peaks at some declinations and a complete avoidance of others. This would not be expected if the indicators had nothing to do with astronomy. There are peaks at the solstices and to some extent around the four lunar limits, especially in the south. But the histogram also showed peaks that do not correspond to astronomically significant positions of the Sun or Moon, including one at a declination of +32° which is well outside the range of movement of both. Thom's suggestion that these relate to bright stars is problematic; due to precession, stellar declinations change significantly on a timescale of a few centuries. By choosing a suitable date, it is not difficult to fit a stellar explanation to any alignment. Ruggles also found several methodological issues with Thom's work:

1. Both the types of megalithic monuments and the types of azimuth indications were extremely varied: monuments include stone circles, short rows of standing stones, pairs of standing stones, longer rows of small stones, and cairns; indicators include sighting along stones in a row, along flat face of a single stone, from centre of circle to an outlier, between centres of two circles, and between two stones on opposite sides of a circle. This was the case even with sites supposedly offering very high levels of precision. If astronomical and calendrical practices throughout Britain at that time were standardised as Thom claimed, then such a variety seems surprising, especially at the highest-precision lunar observatories.
2. Had Thom selected the sites used fairly? The 1955 work contained all the available data, but the 1967 book comprised only those which were surveyed and contributed to the material of the book. For each site used, there were a further two in an unpublished site list that were not included. Ruggles was concerned that only sites with promising alignments were surveyed.
3. Had Thom selected potential indications fairly? Again, in the 1955 work, rigid selection criteria were applied: centre to outlier, two slabs in line, or a row of three or more stones. But the 1967 work was more subjective, with some potential indications being omitted. Was the decision to include or exclude a potential indication influenced by astronomical possibilities? Around 20 percent of the indications use prominent notches and peaks as natural foresights. The problem is that at many sites there are large numbers of equally prominent horizon features, and astronomical possibilities could have influenced the decision to select a specific feature.

Ruggles then turned to the higher precision data, beginning with Level 2. He immediately regarded the twin solar peak as suspicious because the Sun's glare would make sighting on its lower limb very difficult. In the case of the lunar data, Thom had selected 38 lines at 34 sites from the 1967 dataset lines with a listed declination within 0.8 degrees of a mean major or minor limit. But of these, only 23 seemed trustworthy. The issues with the other 15 included foresights that could not be seen from the backsights; foresights listed as stones that could not be found at all; and outliers of doubtful authenticity. Others were archaeologically doubtful, possibly marking modern parish boundaries. When these dubious lines were removed from consideration, the twin lunar peaks disappeared leaving no evidence for preferential observation of upper or lower limbs.

The Level 3 data fared no better under Ruggles' analysis. Thom claimed the moon's 9.4 arcmin wobble could be detected, and that this shows up at 6 to 9 arcmin, and 15 to 17 arcmin, with a very large peak at 24 to 25 arcmin. The semi-diameter of Moon is 16 arcmin, explaining central peak; but many putative foresights mark positions 25 arcmin away from mean lunar limits. But of 40 horizon features, 21 were not indicated at all, or the indication was off by some degrees. Five more foresights could not be seen from the backsight due to intervention of local ground, and one was non-existent. Thus only 13 out of 40 represented indicated horizon features. When these were plotted, little remained of the peaks at 25, 16, and 9 arcmin from the mean. As with the Level 3 data, only 14 of the Level 4 sightlines actually indicate foresights, and some of these were archaeologically dubious. The remainder represent a diverse collection of indicators, with strong evidence that selection of foresights was influenced by astronomical possibilities. In both cases, as noted above, Thom took into account that moonrise or moonset is unlikely to coincide with the timing of a major or minor standstill limit: his solution was to extrapolate between risings and settings actually observed; but Ruggles claimed that the errors in this method would be greater than the wobble itself.

In summary, Ruggles believes that the supposed trends in the Levels 2 to 4 data can be explained by selection effects and the many parameters that can be adjusted to provide a close fit between the high precision lunar theory and the data. At Level 3 and above, there are in any case enormous practical difficulties involved in observing and marking the Moon's motions to the precision claimed. Changes in atmospheric conditions from day to day are greater than Thom believed, rendering observations to a precision greater than around 6 arcmin infeasible. Variations in parallax depending on whether a standstill limit occurs at perigee or at apogee are another factor. As noted above, Thom took into account that moonrise or moonset is unlikely to coincide with the timing of a major or minor standstill limit: his solution was to extrapolate between risings and settings actually observed; but Ruggles claimed that the errors in this method would be greater than the wobble itself. Ruggles therefore concluded that lunar observations to very high degrees of precision were not undertaken.

While Thom's Level 1 data could not be dismissed, Ruggles felt he could not simply revisit it because of concerns about selection criteria. He accordingly conducted his own survey between 1975 and 1981 under severe methodological constraints. The fieldwork was conducted over eight months in 1979 and 1981, and the results were published in 1984. The emphasis was on methodological rigour. The objective was to survey and analyse a large number of alignments, independently of Thom. Like Thom, the researchers focused on standing stones rather than tombs. All the monuments in well-demarcated geographical areas were investigated – not just the ones selected by Thom.

The area chosen was the Hebrides, which was divided into ten regions. The monuments were identified from five sources. A site was deemed to be a single entity if each feature was within 300 m of at least one of the others; unless divided by a sea channel or natural rise. An initial list was compiled of 322 sites that had been reported by at least one source. However, 133 were excluded from further consideration for various non-astronomical reasons; for example, being of dubious authenticity, being chambered tombs, being inaccessible, or simply not being found. Thus, 189 sites remained. Inter-site indications among these were also considered. Ruggles started from a null hypothesis that any alignments are purely chance.

Azimuth and declination data showed no obvious trends, appearing much as might be expected by chance. The clustering around certain declinations seen in Thom's Level 1 data was not apparent. Statistical methods were then applied to see how many alignments might be expected by chance versus those that were in the data. Overall trends were found at three levels of precision.

At the lowest level of precision, declinations between +15° to -15° were seemingly avoided; but this could have reflected a preference for structures to be orientated N-S, NW-SE, or NE-SW, rather than E-W, or anything astronomical.

At the second level, there was a preference for southern declinations between -31° and -19°, and for northern declinations above +27 degrees. The declination range above +27° corresponds to a part of the horizon further north than the Sun or Moon ever rose or set. But the range from -31° to -19° corresponds to within a degree or so of the range of the southerly limits of the Moon's monthly movements at different points in the 18.61-year **nodal cycle**. This might be expected if the structures were aligned on the southerly rising or setting points of the moon over a short period of a year or two, rather than with any intent to track the 18.61-year cycle.

At the third and highest level, there was marginal evidence for six declinations: -30°, -25°, -22.5°, +18°, +27°, and +33°. Three of these (-30°, +18°, +27°) could have indicated an interest in lunar standstill limits but if so, why only three out of the four limits? A declination of -25° could have referred to the winter solstice. A declination of +33° is well outside the range of solar or lunar movements.

There was no evidence for the calendrical 'epoch' dates, or even for an interest in the summer solstice. Also, there was no evidence for precision greater than one degree. Even preferences for other declinations at precision of one to two degrees is marginal and statistically insignificant. Accordingly, Ruggles concluded that his results gave no support for Thom's Level 1 conclusions, or to the idea that Neolithic monuments incorporated alignments that were more precise to anything much better than a degree.

Thom's theories in a modern context

A reassessment of Thom's theories in the light of social complexity of late Neolithic Britain as it is now understood can begin from the premise that there is nothing intrinsically infeasible about them. It would therefore be no surprise if further evidence emerges in support of Neolithic standard units of length, even if Thom's Megalithic yard has been supplanted by long and short feet. The case for standard off the shelf 'types' of stone circle is less convincing. Cowan's diameter construction methods would seem to provide a likelier explanation for the various stone ring types seen.

What, then, of Thom's astronomical hypotheses? Have they disappeared, as astrophysicist Ray Norris[42] puts it, "*in a puff of statistics*"? There is no doubt that Neolithic people were aware of the solstices; indeed, the evidence from Stonehenge (see Chapter 5) suggests an awareness going back to Mesolithic times, if not earlier. On the basis of Burl's work in Cumbria, it also seems highly likely that they were aware of the equinoxes and possibly the calendrical midpoints. But how much further can we go?

An exact calendar of the type advocated by Thom and MacKie implies a degree of numeracy beyond that proposed by Burl. However, Burl[38] also notes that in the present-day ethnographic record, counting systems invariably precede systems of measurement. If Megalithic yards or long and short feet were in widespread use, numeracy in Neolithic Britain might have been greater than Burl believed. The existence of a solar Megalithic calendar cannot be ruled out, but it remains unproven.

As we have seen, Ruggles' reappraisal gave no support to Thom's lunar hypothesis. Quite apart from the practical difficulties outlined by Ruggles, it is difficult to believe that it would have even occurred to Neolithic people to look for small irregularities in the Moon's rising and setting positions amounting to a third of its apparent diameter, much less a thirtieth. It is likely that there was an interest in extremes of lunar movements, but high-precision observations of a type commensurate with eclipse predictions seem improbable.

It is reasonably certain that many megalithic monuments encoded astronomical data pertaining to the movements of the Sun and to a lesser extent the Moon. Their purpose was probably ritual and possibly calendrical; however, it seems less likely that any of them served as lunar observatories for obtaining the high-precision data necessary to predict eclipses.

Controversial though Alexander Thom's theories have been, he must be remembered as one of the founding fathers of archaeoastronomy, as well as for his significant contributions to non-astronomical megalithic studies. That he combined this with a highly successful career as an engineer and academic is all the more remarkable.

07
Drawing down the Moon

The recumbent stone circles of northeastern Scotland

As we have seen, Alexander Thom's 'megalithic lunar observatories' failed to gain acceptance with mainstream archaeologists. A problem highlighted by Clive Ruggles[1] is that there was a great deal of variation in both the types of megalithic monuments and the types of **azimuth** indications supposedly used for high-precision lunar observations. What was needed for statistically significant study were well-defined groups of similar sites confined to a given area, sufficient in number to provide a reasonable database, and with a design such that one direction is clearly of special significance.

In the Aberdeenshire region of northeastern Scotland, there are large numbers of so-called recumbent stone circles[2,3,4,5,6]. Certain general features in their design and placement seem to reflect a ritual tradition that was adhered to over a wide area. They are distinguished by and named for the presence of a single large stone placed on its side and flanked by two tall uprights. About a hundred have been documented, half of which are in a reasonable state of preservation. Only 71 are currently considered to be reliable examples, but it is probable that many more have been dismantled over the centuries. They are confined to a region of some 50 × 80 km (30 × 50 miles), but even in areas of the highest concentration, they are several kilometres apart, occupying small territories of around 10 sq. km. (3.8 sq. miles). The available resources of each could have served the subsistence farming requirements of no more than twenty to thirty individuals, and some might have represented the domains of single-family units of ten to fifteen individuals. No prestigious artefacts have been found at any of these sites, suggesting that the recumbent stone circles were built and used by small, egalitarian societies that shared a common belief system. The majority of the sites probably date to the Scottish Early Bronze Age from 2500 to 1750 BC (roughly contemporary with the sarsen phase of Stonehenge), though an earlier date within the Late Neolithic around 3000 BC has also been proposed.

The 'recumbent' stone is usually by far the largest stone in the circle, and the focal point of the structure. The stones in the remainder of the circle are often graded in height, rising towards the 'flanker' stones and with the smallest stone directly opposite the recumbent. The recumbents and flankers are positioned with great care, but the circle stones appear to be positioned rather casually. Quite often, the recumbent lies askew to the circumference of the circle. This led some to conclude that the circle stones were added after the recumbent was in place, but it is now thought to be the other way around. Overall, the circles are not large, typically measuring from 18 to 24 metres (60 to 80 ft) in diameter. None are more grandiose than any of the others.

The recumbents and flankers are often of different colour and material to the other stones in the circle. For example, at the site of East Aquhorthies, near Inverurie, the recumbent is red granite, the flanker stones are light grey granite, and the remaining stones are pinkish porphyry. It was necessary, therefore, to obtain the recumbent stone from some distance away, although locally available stone was used for the rest of the monument. It would have been a major undertaking to transport the recumbents, which could weigh as much as 50 tonnes. Though small groups could have erected the circle stones, the large recumbents would have required the co-operation of neighbouring groups.

Small, bowl-shaped depressions known as cup-marks are sometimes ground into the sides or tops of the recumbent, flankers, or stones immediately adjacent to these, but not elsewhere. Small scatters of white quartz are frequently found in the vicinity of the recumbent, but again not elsewhere. Frequently, the stone circle encloses a ring cairn: a low, open-circular bank of small stones around an open central court in which small quantities of cremated human bones are found. In some cases, the central court is as small as 3 m (10 ft) across; in others the cairn is a narrow bank surrounding a large space occupying most of the space within the recumbent stone circle.

For many years, studies of recumbent stone circles were hampered by a lack of reliable excavated data[4], but more recently archaeologist Richard Bradley has excavated the recumbent stone circles at Tomnaverie, Cothiemuir Wood, and Aikey Brae. Bradley[7] found that at Tomnaverie the ring cairn predates the stone circle, with a radiocarbon date of 2580-2220 BC; the radiocarbon date for the stone circle is 2300-1700 BC. The first use of the site was for cremation, and residue of burnt soil, charcoal, and fragments of human bone formed a low mound about 3 m (10 ft) in diameter. The ground had been levelled to provide a platform for the ring cairn. The surface of the platform included from seven to thirteen radial divisions pointing towards the centre. They were constructed from large stones and located mainly in the eastern half of the ring with an emphasis towards the northeast direction. No fewer than eight of the lines point to where later stones were set, suggesting that the entire construction sequence from platform and cairn to the final circle was planned at the outset. These findings overturned a long-standing view that the recumbent stone circles were built to demarcate an open space that was used for ceremonies.

The circles were always placed in conspicuous settings, on hillsides in terraced locations with long clear views of the horizon; sometimes the sites were artificially levelled. When viewed from the centre of the circle, the low recumbent stone and its tall flankers directs the viewer's attention to the portion of the horizon so framed. The recumbent and flankers were placed on the SSW of the circle. As viewed from the centre, they are invariably orientated between SSE and WSW, i.e., the alignments all lie within the same quarter of the available horizon from 157°30' to 247°30'. This preferred orientation was noted as long ago as the sixteenth century. Suggestions of an astronomical connection go back to the early part of the last century. The obvious interpretation is that celestial bodies were observed on or above the stretch of horizon framed by the recumbent and its flanker stones.

A lunar connection

The first systematic analysis of the recumbent stone circle orientations was conducted by Aubrey Burl[8], who considered fifty sites with known azimuths. These had been surveyed by a number of individuals including Sir Norman Lockyer in 1908, Alexander Thom, and Burl himself. The azimuths, defined from the centre of each ring to the centre of its recumbent stone, ranged from 155° to 235°. Based on this data, Burl concluded that the focus was on the Moon; not necessarily when it was rising or setting, but when it was up in the sky. As seen from higher latitudes such as in Scotland, the midsummer full Moon stays close to the horizon, and this is especially pronounced in years of a **major standstill** when it rises to no more than five degrees above the horizon. It is highly likely that such a dramatic sight would have been of interest to Neolithic people.

For the latitude of the recumbent stone circles (57° N), assuming a horizon altitude of zero degrees, the major standstill Moon rises at azimuth 155° and sets at azimuth 255°; the **minor standstill** Moon rises at azimuth 127° and sets at azimuth 233°. Forty-two of the fifty azimuths (84 percent) lie between 155° and 205° (broadly south), distributed evenly between the southerly extremes of Moon's rising and setting points at the major standstill. Of the remaining eight, all but one lie between 229° and 235° (broadly SW): close to southernmost moonset at the minor standstill. Burl also suggested that the cup-marks on some monuments indicated specific events, such as **solstices** and the maximum and minimum southerly setting limits of the Moon.

However, subsequent work by Clive Ruggles in conjunction with Burl[1,3,4] suggested that matters were rather less clear cut. A list of 99 sites was drawn up from existing lists, of which 64 were investigated. Of the remainder, some were in too poor a condition for further investigation, a few were in fields under crop or otherwise inaccessible, and some could not be found at all. This time, the horizon altitude was taken into account, and sightlines were considered not just from the centre of the monument to the centre of the recumbent but also a line perpendicular to the long axis of the recumbent. As the recumbent is not always tangential to the ring, the two measurements could differ by up to 20 degrees. But rather than clarifying Burl's seemingly neat conclusions, these considerations confused matters.

Of the 64, 38 of the centre line azimuths fell between 157° and 236° (broadly SSW with a range of 79 degrees); and 47 of the perpendicular azimuths fell between 147° and 237° (broadly SSW with a range of 90 degrees). Even taking into account the 17 outliers, the probability that these ranges would occur by chance is vanishingly small. But in both cases, the groupings around major and minor lunar standstill were more scattered than reported by Burl, with several azimuths falling between the two.

Ruggles and Burl concluded that what was of interest was the midsummer full Moon. As it appeared to skim low over the recumbent stone, it would shed light into the stone circle, and might have provided the backdrop for rituals. In response to Bradley's findings, Ruggles[5] later conceded that they undermine this view. However, the possibility that the orientation of the monument reflected astronomical considerations was still valid. Bradley[7] also suggested that at Tomnaverie the viewing position was not the centre of the circle, but between the two stones opposite the recumbent. This would considerably reduce the area of sky framed by the recumbent and its flankers. In the case of Tomnaverie, the azimuth range falls from 32° to just 14°[9].

Given that a recumbent stone circle could have been constructed at any point in the Moon's 18.61-year **nodal cycle**, this would result in a spread of alignments similar to that observed. If it were important for the Moon to pass above the recumbent rather than rise there, it would explain the bias (SSW) towards westerly over easterly orientations. A reference to the white light of the Moon might be suggested by the scatters of white quartz found around the recumbent stones at many sites. However, Ruggles[4,5] also noted that but for the smaller number of sites facing towards WSW, the data could be interpreted in terms of the winter Sun, low in the sky, rather than the Moon.

Assuming that the focus was on the Moon, ignorance of the nodal cycle might have had disconcerting consequences. If a recumbent stone circle happened to have been set up at a time when the nodal cycle was close to minor standstill, in subsequent years the midsummer full Moon would have set ever further to the left as viewed from the centre of the monument. As the major standstill drew closer, the Moon would eventually fail to pass over the recumbent at all, setting while still approaching from the left.

In addition to the astronomical hypothesis, Ruggles and Burl considered two other possibilities: an interest in features on the ground, and absolute directions. Both possibilities were rejected. The geographical area spanned by the sites is too large for orientations upon particular types of feature to have been a possibility and the local topography over this area is too varied to give rise to an overall orientation trend. More likely is that a particular direction was considered sacred for some reason, possibly because it was the direction from which ancestors came. The problem here is that consistently determining this direction over a wide area could not have been achieved without using astronomical indications. Other possibilities such as the direction of prevailing wind were also rejected; wind directions in this part of Scotland are highly variable.

Clava cairns and axial stone circles

The Clava cairns of northern Inverness-shire and axial stone circles of County Cork and County Kerry are thought to be typologically connected to the recumbent stone circles, and Clive Ruggles evaluated them for possible alignments on the midsummer Moon. The Clava cairns are a group of around thirty cairns named for Balnuaran of Clava, near Inverness, where the first examples were found. There are two types: passage tombs (earth-covered burial chambers entered by narrow, stone-lined passage) and ring cairns. In common with the recumbent stone circles, they are invariably orientated upon the SSW quadrant. Another shared feature is that they are usually surrounded by a stone circle with stones of graded height, with the tallest stones towards to SSW. Burl[2] suggested that they represented a predecessor tradition to the recumbent stone circles, but they are now thought to date to around 2200-2000 BC, making them broadly contemporary. However, the structural similarities are evident, and a connection between the two types of monument seems likely[4].

Burl[10] obtained **declinations** that suggested that the passage tombs fell into three groups: rising or setting of the midsummer full Moon at major and minor standstill limits, and the setting sun at the winter solstice. However, Ruggles[4] felt that a simpler explanation of the data is that the monuments were all aligned on the midsummer full Moon at various points in the nodal cycle, as with the recumbent stone circles.

The axial stone circles of County Cork and County Kerry, in the southwest of the Republic of Ireland, also have many similarities to the recumbent stone circles. Again, there is a recumbent ('axial') stone, but it stands alone, without flankers. Instead, there are two portals on the opposite side of the ring. The monuments tend to be symmetrical about an axis through the axial; the ring consists of paired stones with a tendency for height increasing towards the portal. As with the recumbent stone circles, there is a preferred orientation, but it is slightly different, being concentrated roughly on the quadrant between due south and due west. Both the axial stone and the rings themselves tend to be smaller than their Scottish counterparts; many have just five stones in total. Radiocarbon dates from the monument at Cashelkeelty suggest that it was constructed in the Late Bronze Age, dating to around 1250-800 BC, which would suggest that some and possibly all the axial stone circles are surprisingly late for megalithic monuments. Despite the significant geographical and temporal separation – 900 km (560 miles) and over a millennium – it is possible that the axial stone circle tradition is in some way derived from the earlier Scottish tradition[4].

Ruggles investigated the astronomical potential of the axial stone circles in 1994. A starting list of 48 monuments was drawn up from sources, with those consisting of just five stones being ignores. Visits were made to the 31 sites where the axial stone and/or both portal stones were still present. Ruggles found that unlike the recumbent stone circles, there was no preference for distant horizons behind either the axial stone or the portals, nor for conspicuous hill summits. No consistent relationship with any astronomical body or event could be found, other than at the well-preserved circle at Drombeg. This is aligned on both a conspicuous hilltop notch and winter solstice sunset – but the alignment is not precise, nor is it repeated elsewhere. Therefore, it is impossible to demonstrate that the alignment was intentional[11,4].

Beyond a preference for the south-west quadrant, there is little that can be said about the orientation of the axial stone circles. If they do indeed derive from the recumbent stone circle tradition, then perhaps various subtleties of symbolic association were lost. This might explain why the Irish axial stone circles are smaller, and include examples with only five stones, suggesting a dying tradition. Against this, they seem to be more symmetrical and precisely laid out than the Scottish circles. There is clearly no simple interpretation of the axial stone circles[4].

The short stone rows of western Scotland and southwest Ireland

Clive Ruggles also investigated a second type of monument, the Bronze Age short stone rows that are found in western Scotland[12,13,4,14,15,16,17] and in southwest Ireland[18,19,4,15]. Several hundred rows consisting of three to six stones, typically under 10 m (33 ft) and rarely under 25 m (82 ft) in length, are found in Britain, Ireland, and France, particularly in Argyll, the Inner Hebrides, County Cork, and County Kerry, often in remote coastal locations. Although very different to the recumbent stone circles, their large numbers offer similar scope for investigating astronomical alignments.

Short stone rows are thought to represent the tail end of a tradition that began around 2800 BC with avenues attached to stone circles, such as the West Kennet Avenue at Avebury, Wiltshire. Over time, these became shorter, reducing from up to six stones and eventually to just a pair (albeit the number of stones now existing might not always reflect the number originally present). The few radiocarbon dates that are available suggest that three-stone and four-stone rows continued to be built through to the Late Bronze Age. In the Republic of Ireland, hundreds of rows with up to six stones were constructed between 1700 and 1100

BC. The Scottish short stone rows might be later, dating to around 1250 to 900 BC.

These monuments represent relatively modest construction projects, but their function is unknown. They are not domestic sites, nor was their purpose defensive. They are typically described as ritual or ceremonial monuments, but nobody really knows what they represent nor their intended purpose. They were probably 'family' monuments, easily erected by a small number of people. It is uncommon to find a stone weighing more than 3 or 4 tonnes, and such megaliths could easily have been set upright by a gang of no more than twenty or so workers, with families possibly combining when a new site was proposed. It has been suggested that they marked the boundaries of territories, or that they were markers used by a broader community of nomadic herders. They do not enclose space, as might be expected of a ritual centre used repeatedly by a group of people within their own territory. Possibly the act of constructing them was in itself the important ritual and ceremonial act, and the rows were not used for subsequent rituals. In present-day Melanesia, megalithic monuments are erected in the context of male initiation ceremonies. The initiates are expected to bring about the construction of the monuments, and to provide a feast, but there is no subsequent use of such monuments. But the most obvious possibility is that the short stone rows were pointing at something.

The short stone rows of western Scotland attracted the attention of archaeoastronomers from the late 1970s onwards, but Ruggles chose to focus initially on the monuments in southwest Ireland. These had been largely ignored, and Ruggles felt they provided the opportunity for a fresh start. Accordingly, he carried out a program of fieldwork there between 1991 and 1993. The starting point was lists and inventories comprising a total of 79 rows of between three and six stones. Of these, 21 were not visited as fewer than three stones were known to be still standing; four could not be located, and one was found to have been dismantled and moved. Horizon surveys were carried out at 49 of the remaining 53 rows.

It was found that there was a strong preference for a northeast-southwest axis; only seven rows fell outside this range. A problem for interpreting this result is that unlike the recumbent stone circles, the short stone rows are two-directional, with no obviously preferred direction of indication. In some cases, the stone height is graded, with the tallest stone at one end. However, this is not always the case; there are monuments where there are tall stones at each end; and in many cases the tops of stones have broken off, making it very difficult to establish the original height graduation. Even where a direction of indication can be established, this was as often to the northeast as it is to the southwest. What was notable is that there was a sharp increase in the proportion of distant horizons in the supposedly indicated direction, suggesting that it was indeed the preferred direction.

If the indications were astronomical, the southwest-aligned monuments would indicate setting objects, and the northeast-aligned monuments rising ones. The south-westerly alignments were found to give reasonable agreement to the southern major and minor lunar standstill limits; and the north-easterly alignments to the northerly limits, though perhaps slightly less so. Where prominent hilltops were indicated, it was found that azimuths of these from the monuments gave a better agreement with the standstill limits than did those obtained directly from the monuments. The case for a relationship with the Moon seems strong, but it is unexplained why some monuments are seemingly aligned on the rising midwinter full Moon, others on the setting midsummer full Moon, but none on the rising full Moon in summer, or the setting full Moon in winter.

Ruggles than returned to the short stone rows of western Scotland. He started with a list compiled from various sources and backed up by data from National Monuments Record of Scotland, and he applied a length limit of 25 m (82 ft). Although the list covered the whole of Scotland, the study area was eventually whittled down to seventeen short stone rows and stone pairs in Argyll and northern Mull. A tight grouping was found within a degree or two of declination -30°, with a second, broader grouping around declination -23°. Notably, monuments not orientated close to -30° are always located close to another monument that does indicate -30°; sometimes within a few hundred metres or less.

The primary indications suggest an orientation on the rising or setting of the Moon at its southerly major standstill limit. But these extreme southern declinations are very rare; the Moon might be seen rising or setting there on no more than two or three consecutive summers every eighteen or nineteen years. Ruggles asks, were people so interested in these extreme movements that they waited many years for them to occur? He speculated that the monuments were erected for ceremonials intended to stop the Moon going any further south and disappearing completely. The extreme declination might have been obtained by constructing a monument aligned on any midsummer full Moon. If the Moon rose or set further south in subsequent years, then another monument was built nearby.

The case for a connection between the short stone rows and the Moon's southerly major standstill limit is reasonably strong, but it is not without issue. If there was such interest from a ritual point of view in the extreme southerly movements of the Moon, it seems odd that monument builders made a seemingly arbitrary choice between moonrise and moonset, with both being indicated in roughly equal numbers.

A different viewpoint

A picture that once seemed clear has become increasingly confused in the light of more complete statistical analyses and a better archaeological understanding of the recumbent stone circles. Should alternatives to the lunar theory be explored? Archaeologist Liz Henty[9] believes so.

Rather than reduce Neolithic monuments to cold statistics, she took an 'immersive' approach. She visited Tomnaverie at several different times of the year and considered what people at the site might have seen and experienced at these times. She ruled out the Burl/Ruggles midsummer moon theory, because at this time of the year, sunset is very late at night, and the sky never gets very dark. By contrast, in winter sunset occurs in the early afternoon and the hours of darkness are very long. "*The sepulchral function of the circle with its earlier funeral pyres could have been associated with the setting of the sun or the moon in the winter which metaphorically symbolises death before the spring renewal.*" With this in mind, Henty decided to focus on the winter months.

She then considered the view from the centre of the ring and noted that from this vantage point the upper surface of the recumbent lies below the horizon, demonstrating that this could not have been the viewing point intended by the builders, as Burl and Ruggles had assumed. Only when the recumbent and flankers are viewed from between the two stones (stones 8 and 9) on the opposite side of the ring do they appear outlined against the sky. That the direction of interest was the southwest was also definite. Although it is possible to stand directly behind the recumbent at Tomnaverie, and look towards the northeast, this is not possible at many sites, where the recumbents are taller than a human.

The next step was to use planetarium software to search for significant phenomena that might have occurred above the recumbent during the winter months at around the time the site was in use (due to the effects of **precession** and changes in **obliquity**, these would not be the same as the present day). From between stones 8 and 9, the field of view is from 233° for the east flanker to 247° for the west flanker, with the centre of the recumbent at 240°. Henty assumed a date of 2580 BC.

The Sun only set over Tomnaverie's recumbent in October and January, i.e., not at the winter solstice; and the winter full Moon would have been 45 degrees above the recumbent. Henty did not consider either to have been the focus of the recumbent stone circles. She went on to consider a largely neglected phenomenon, the lunar crossover, which occurs briefly when the Sun and the Moon at a given phase are at the same declination. Thus, in winter, the Sun is low in the sky and the full Moon is high and in summer these are reversed; but around the spring and autumn **equinoxes**, the Sun and full Moon cross over, so on one day their rising and setting azimuths are the same, or very nearly so. The same phenomenon occurs for other phases of the Moon at other times of the year: in particular, for the evening (waxing) and morning (waning) crescent phases, where the crossovers occur around the solstices. Archaeologists Fabio Silva and Fernando Pimenta[20] note that actual crossovers will only rarely be observed, but the north-south reversal of rising or setting positions of the Sun and Moon will empirically confirm that it has occurred. Crossovers are easier to determine than solstitial full Moons, as for the latter it is necessary to first determine the solstice and then to equate this to the nearest full Moon. As such, it is possible that equinoctial full Moon crossovers served as a basis for a seasonal calendar composed of two periods of roughly six or seven **synodic months** each. Interestingly, the evening crescent crossover moonset peaks at a declination close to the Moon's minor standstill setting limit. Silva and Pimenta suggest that this is a possible alternative interpretation of supposed alignments on the Moon's minor standstill limit.

Henty used the planetarium software to investigate crossovers for the winter solstitial evening crescent Moon. She found that more than two-thirds of the moonsets occurred to the left of the area delimited by the recumbent and flankers, but all of these were within the range of the radial division identified by Bradley at azimuth 212° in the southwest of the circle. There are three cup-marks, in the southwest of the monument, including two on the recumbent stone. Neither of the latter match the lunar minor standstill, but they are in range of the winter evening crescent moonset. A radial division at azimuth 272°, to the north of the recumbent and flankers, is close to the declination of the equinoctial full Moon. The declination of the third cup-mark is also in the range of the equinoctial full Moon.

Accordingly, Henty believes that crossovers were used for calendrical purposes at Tomnaverie, as per Silva and Pimenta's views, but only during the earliest use of the site. Later, ritual needs became more focussed on events in the southwestern skies. What might these have been? Henty suggested that the frequent use of red stones hints at an interest in red stars such as Betelgeuse (α Orionis) and Aldebaran (α Tauri), and white stones an interest in white stars such as Sirius (α Canis Majoris). The first two set to the right of the recumbent, and the latter to the left; but all three set within the radial divisions. The three stars comprising the Belt of Orion set almost horizontally over the left-hand side of the recumbent, and Bellatrix (γ Orionis) set over the right-hand side of the recumbent. On the night of the winter solstice, these stars would set within an hour of each other beginning at 23:00 hrs. They would have created a spectacular display as they moved across the window formed by the recumbent and its flankers.

As noted in Chapter 3, interest in Orion might go back to Upper Palaeolithic times, and a possible alignment on the setting of the Belt has also been claimed for the Neolithic monument complex at Thornborough, North Yorkshire[21]. However, from the likely viewing positions, recumbent and flanker arrangements each typically frame only about a sixth of the 90 degrees of horizon spanned by the group as a whole. Therefore, only a small number of the recumbent stone circles would have captured the setting of the Belt of Orion. It is possible that a variety of stellar phenomena were of interest to the builders of these monuments, in addition to lunar and/or solar phenomena, and that these interests changed over time. The evidence from Tomnaverie suggests that a lengthy and complex history, and it is possible that what was of interest changed even over the lifetime of individual monuments. It is becoming clear that no single astronomical phenomenon can account for the preferred alignment of the recumbent stone circles and other monuments considered in this chapter, and that our understanding of their construction and their purpose is far from complete.

08
Spherical Earth

Eratosthenes's experiment

George and Ira Gershwin wrote some of the most memorable songs of the last century, but this particular line highlights a still-common misconception:

They all laughed at Christopher Columbus
When he said the world was round

In reality, when Columbus' small fleet sailed from Palos de la Frontera in 1492, it had been the best part of two millennia since any Western scholar had believed that the world was flat (although the Chinese retained a flat-Earth view for much longer; see Chapter 15). The ancient Greeks were not only aware that the Earth was spherical, but around 240 BC they made an estimate of the circumference and obtained a surprisingly accurate result. Eratosthenes of Cyrene (274-194 BC) was a Greek polymath who held the post of Chief Librarian at the Library of Alexandria. He had heard that in Syrene (now Aswan; not the same place as Cyrene), the noontime Sun on the day of the summer **solstice** lit up a well, casting no shadow on the side and implying that it was directly overhead. At Alexandria, however, the Sun did not quite reach the **zenith** and therefore did cast a shadow. By measuring the length of the shadow cast by a gnomon (vertical rod) of known height, the angular distance of the Sun from the zenith in Alexandria at noon could be determined. Of course, Eratosthenes could not simply look at his watch to see when it was noon, so he would have relied on the shadow being at its minimum length at noon.

Eratosthenes found that the angular distance of the Sun from the zenith was a fiftieth of a whole circle (i.e., 360/50 = 7°12'): the distance from Alexandria to Syrene was therefore a fiftieth of the circumference of the Earth. He then used a value of 5,000 *stadia* for the distance between the two cities to obtain a value of 250,000 *stadia* for the Earth's circumference. The *stadion* (from which we get the word 'stadium') was the length of a Greek running race that featured in the Panhellenic and Olympic Games. Unfortunately, we do not know for certain the exact value of the *stadion* Eratosthenes used, as different values were in use in different parts of the Greek world. However, it is likely that he used the Attic *stadion* comprising 600 Attic feet of 308.3 mm each (12.14 inches; i.e., slightly longer than a statute foot)[1], for a distance of 925 km (574 miles) from Alexandria to Syrene. This would make the circumference of the Earth 46,250 km (28,750 miles), rather higher than the accepted value of 40,075 km (24,901 miles); but if the actual distance of 843 km (524 miles) is used, the answer comes out at a more accurate 42,150 km (26,200 miles).

Eratosthenes made the following assumptions: firstly, the Sun is so distant that rays of light reaching Alexandria and Syrene are effectively parallel; secondly that Syrene is located on the Tropic of Cancer (the latitude where the Sun is directly overhead on the summer solstice); and thirdly that Alexandria is at the same longitude as Syrene. His first assumption was correct, but the other two were not entirely accurate. Syrene, at 24°05' N 32°54' E, does not lie exactly on the Tropic of Cancer. In Eratosthenes's day, the Tropic of Cancer was located at approximately latitude 23°43' N, 22 arcmin south of Syrene. Due to changes in the Earth's **obliquity**, the Tropic is currently located at latitude 23°26'12.7" N. Eratosthenes might have relied on travellers' accounts that at noon on the longest day of the year the Sun cast no shadow in Syrene[2].

Syrene also lies three degrees further east than Alexandria, which is located at 31°12' N 29°55' E; however, this would not have produced significant errors. In an era before standardised time-zones, 'noon' was simply defined as the time at which the Sun reached the local **meridian** and it would have occurred in Syrene twelve minutes before it did in Alexandria.

The distance from the zenith of 360/50° measured by Eratosthenes was slightly low: the correct value is almost exactly 360/48 = 7°30'[3], although a difference of 18 arcmin (four percent) is just about within the limits of error that could be expected by using the gnomon method[2].

Essentially, Eratosthenes's experiment entailed simultaneous measurements of the sun's **altitude** at two separate locations on the same meridian. The experiment was simplified by choosing the solstitial sun at noon, and a second location that he either believed or approximated to be due south and on the Tropic of Cancer. Today we could easily replicate the experiment with two observers equipped with clinometers at John O' Groats and Weston Super Mare, which are very close to sharing a common meridian and are 810 km (503 miles) apart.

The experiment demonstrates how fundamental data can sometimes be obtained from a subtle but easy to measure phenomenon, and it is possible to go further. If θ_s is the altitude of the Sun on the day of the summer solstice, and θ_w is the altitude of the Sun on the day of the winter solstice, then:

Latitude of observer $l = 90 - (\theta_s + \theta_w)/2$

Obliquity of ecliptic $e = (\theta_s - \theta_w)/2$

This simple instrument can thus be used to make four important determinations: the direction of due north, the days of the summer and winter solstices, the observer's latitude, and the obliquity of the ecliptic. The gnomon was undoubtedly invented independently on many occasions. The earliest known use of the gnomon was in China around 2300 (see Chapter 15). According to Herodotus, the Greeks learned about them from the Babylonians in the fifth century BC. However, the gnomon does suffer from the problem that the edges of shadows cast by extended objects such as the Sun are indistinct, making it difficult to locate the tip of the shadow. There is also a fundamental problem with determining the solstices by this method. When close to the solstices, the Sun's rate of change of **declination** is very slow, so for a week either side of the summer or winter solstice the noontime altitude and hence the length of a shadow hardly changes.

In the *Almagest*, Ptolemy recorded a value for $2e$ of between 47⅔° and 47¾° (47°40' - 47°45'); the correct value in his day was 47°21'. He notes that this value agrees well with that obtained by Eratosthenes. This is the only evidence we have that Eratosthenes measured the obliquity, because his own work has been lost[3].

It is uncertain as to how Eratosthenes obtained his distance from Alexandria to Syrene. It is usually claimed that he used a land survey obtained by surveyors pacing out the distance, but as much of the as-the-crow-flies route lies across desert, this has been doubted. It has been suggested that he unknowing used a map that was itself based upon earlier astronomical measurements, which would have rendered his result circular in more ways than one.

In his major work *Geographica*, the Greek geographer Strabo (63 BC-AD 24) provides a description of the Nile from Meroe at a latitude of 17° N to the Delta, documenting the river's many twists and turns as it makes its way north to the Mediterranean. It has been claimed that the distances between each twist and turn are based on a large unit that can be repeated halved (or at least approximately so) and that this unit was a fraction of the Earth's circumference – possibly $^1/_{48}$ (which is close to the distance from Syrene to the Mediterranean). This suggests that the latitudes of each point where the Nile changes course were determined by astronomical means; subsequently a pre-existing estimate of the Earth's circumference was used to convert the angular distances into *stadia*[4,3].

It was suggested that this estimate of the Earth's circumference was obtained by determining how far out at sea a lighthouse flame of known height remains visible. The method is perfectly sound, but atmospheric **refraction** would make the lighthouse remain visible for some time after it had actually fallen below the horizon. Failure to take this effect into account would have overstated Earth's circumference and hence the distance from Alexandria to Syrene[4]. I am somewhat sceptical about this theory: as head of the most important library in the ancient world, Eratosthenes would surely have been aware of the provenance of any maps he used, and he would have been unlikely to rely on data from unknown sources.

Viewing the curvature of the Earth

In Eratosthenes's day, it was common knowledge that the Earth is spherical, but only a few centuries earlier, the prevalent view was that it is flat. Early Egyptian, Mesopotamian, Hebrew, and Homeric Greek cosmologies are all based around a flat Earth. The ancient Hebrews believed that the Earth is a disk supported by pillars and surrounded by a primal Ocean. Below it is an underworld known as Sheol and above it, again supported by pillars, is the Firmament of Heaven, a solid dome separating the Sun, Moon, stars, and planets from the Ocean of Heaven[5,6].

The shift to the modern view of a spherical Earth might have begun with the Pre-Socratic Greek philosophers, though this is uncertain. The discovery has been attributed to Parmenides (512-400 BC) or even Hesiod (c.650 BC). More plausibly it was Pythagoras (*c.*570-495 BC) or at any rate the Pythagorean school first put forward the idea of a spherical Earth; but when Socrates (470-399 BC) was in his youth, the matter was still apparently being debated. Plato (428-348 BC) was a firm advocate of the spherical Earth, as was Aristotle (384-322 BC), and by the end of the fifth century BC no Greek writer of repute believed in a flat Earth[7,8]. There is, however, no clear account of how the change in perception happened.

The first question we must ask, is it possible that under suitable conditions, the Earth's curvature can be perceived with the naked eye? Strictly speaking, what we mean here is a curved horizon rather than the curvature of the Earth. It is, nonetheless, an artefact of a spherical Earth. If Bronze Age people were on occasions able to see a curved horizon, it might have given them a strong hint that they were living on a sphere, not a plate.

It is often claimed that the curvature of the Earth can be seen from an aircraft, a mountain, or even a tall building. The idea is that if you sight the horizon looking out to sea over a level straight edge such a ruler from approximately a metre, you should be able to see a convex meniscus. Unfortunately, you need to be very high up for this method to work. Standing on a clifftop looking out to sea you will see nothing (I have tried). There are many photographs on the internet purporting to show a clear curvature from such a vantage point, but it is invariably an artefact of barrel distortion by the lens.

The curvature is clearly visible at 60,000 feet from high-altitude aircraft such as Concorde, and astronomer David Lynch[9] has reported that it is discernible from an ordinary airliner at 35,000 feet under ideal conditions. He speculates that it might also be visible from very high mountains, although he was unable to see it from Mauna Kea (elevation 13,796 ft; 4205 m) and Haleakala (elevation 10, 223 ft; 3116 m), despite a relatively unobstructed horizon in several directions. Lynch notes that the curvature might be measurable in photographs taken from as low as 20,000 ft (6,000 m), a height which is exceeded by many mountains worldwide, but the air is too thin to breathe unaided. He believes that nobody recognised the curvature with their own eyes until the high-altitude balloon flights of Auguste Piccard and others in the 1930s. We can therefore rule out any possibility that Bronze Age people could have seen the Earth's curvature for themselves.

Ships on the horizon

The classic example of a proof that the Earth must be round is the appearance or disappearance of a ship over the horizon. As it sails away from the land, it will gradually disappear; first the hull, then the superstructure, and finally (for a sailing ship), the masts and sails will sink below the horizon. Conversely, the upper parts of an incoming vessel will be seen before the hull comes into view. The same effect can be seen for tall buildings viewed from across a large body of water; for example, the Chicago skyline from across Lake Michigan or Malmo's 190 m (623 ft) Turning Torso building from across the Øresund. The effect can clearly be seen with a telescope or a pair of binoculars, and the internet abounds with photographs and videos taken with high-zoom cameras.

Bronze Age traders were voyaging across the Mediterranean two millennia before the time of Socrates and Plato. The Minoan civilisation of Crete had substantial contacts with Mesopotamia and Egypt, and the Minoan elite imported large quantities of luxury items from these regions including Egyptian scarabs, faience, ivory, lapis lazuli, Levantine seals, and semi-precious stones[10]. While the ships of Minoan traders often kept close to land, prevailing winds meant that it was far easier for them to sail directly from Crete to Egypt, only coasting on the return voyage[11]. The question then, is why did the Minoans not notice the gradual disappearance of their ships below the horizon?

There are two reasons why this might have been harder than is customarily assumed. The first, obviously, is that there were no telescopes, binoculars, or high-zoom cameras in the Bronze Age. The second is that Bronze Age ships were far smaller than even coastal freighters and tankers of the present day. If the evidence of the fourteenth century BC

Uluburun shipwreck is anything to go by, a typical merchant vessel of that time was only around 15-16 m (50 ft) in length, about the size of a present-day Moody 54 sailing yacht.

Would a keen-eyed observer have been able to see that the hull of an outward-bound Bronze Age ship vanished while its sails remained visible? Even if they could, would they have been able to do so with enough consistency to realise that it was a distinct phenomenon and not just an artefact of sea, weather, or lighting conditions?

Suppose that an individual of average height (eye level 1.7 m or 5ft 7 in) was standing on the shore, watching a ship of comparable size to the Uluburun merchant vessel standing out to sea:

Length of hull:	15 m approx. (50 ft).
Height of hull above waterline:	2 m approx. (6 ft 6 in).
Hoist of mainsail:	15 m approx. (50 ft).
Width of mainsail:	15 m approx. (50 ft).

Using a computer program that takes eye height and target distance, and calculates target hidden height, we find that the hull would have just disappeared as the ship reaches 10 km (6.2 miles) from the shore; for a second observer, sited on a clifftop at a height of 10 m (32 ft), the whole of the ship would still be visible. At a distance of 10 km, the sail would subtend 15/10,000 = 0.0015 radians or 5 arcmin. This is about the same apparent size as a UK five pence coin or a US or Canadian dime viewed from 12 metres (40 ft). But the hull would subtend an angle of just 2/10,000 = 0.0002 radians or 0.7 arcmin above the waterline, or less than the maximum apparent diameter of Venus. The theoretical maximum resolving power of the human eye is about 0.4 arcmin, but most people cannot resolve better than about 1.0 arcmin. Even if our observer possessed the sufficient visual acuity, it would have been quite difficult to distinguish the hull from the sea except under the most favourable conditions of clarity and calmness.

By the time that the ship was 15 km (9.3 miles) out, roughly half the sail would have disappeared as viewed from shore level (although the top half of the hull would still be visible from the clifftop). At this point, the sail would be subtending only 15/15,000 = 0.0010 radians or 3.3 arcmin. Would the shore-level observer have been able to pick out enough detail against the background of an open sea to realise that the sail was disappearing over the horizon? Would they have thought that the ship was doing anything other than vanishing into the distance?

By the fifth century BC, ships had become larger than Bronze Age merchantmen: the present-day Hellenic Navy Ship *Olympias* is a full-sized replica of the triremes that came into service early in that century and saw action at the Battle of Salamis in 480 BC. The ship is 36.9 m (121 ft 1 in) in length, 2½ times the length of the Uluburun merchant vessel. Such ships were probably large enough to be visibly 'hull down' on the horizon, but they were not suitable for long sea voyages and ventured out of sight of land only occasionally. The largest shipwreck known from the fifth century BC is the remains of a merchant vessel found off the coast of the Aegean island of Alonnisos, which measures at least 25 m (82 ft) in length and 10 m (32 ft) in beam. However, the ship dates to the last quarter of the century, by which time, as noted above, the spherical Earth model was becoming widely accepted[12].

The converse of a ship dropping below the horizon as seen from the land is the disappearance of the land from the view of those on board. Again, though, would Bronze Age mariners conclude other than that the land was simply too far away to see? On a clear day, the Taurus Mountains of southern Turkey can be seen from far out to sea – but with the naked eye it is difficult to see that the shore is no longer visible. The lighthouse flame experiment noted above would provide a strong clue, but without prior knowledge that the Earth is round, its disappearance could again be put down to distance.

Overall, I am not convinced that casual observations of the disappearance of either ships or shore-based landmarks can explain the discovery that the Earth is not flat. How, then, might Greek scholars of the fifth century BC have deduced that they were living on the surface of a sphere?

How did the Greeks learn that the Earth is round?

The night skies provide several clues that the Earth cannot be flat. The most often cited is the curved shadow of the Earth as it moves across the face of the Moon during a lunar eclipse. It would long have been known that a lunar eclipse only ever happens when the Moon is full; and that a full Moon happens when the Moon is on the opposite side of the sky to the Sun, its entire Earth-facing surface illuminated by the sun. A lunar eclipse must therefore be caused by Earth getting in the way of the Sun and casting a shadow across the surface of the Moon.

This shadow – a shadow of Earth – always appears circular. While a flat plate could cast a circular shadow, it would not always do so. The shadow would depend on the angle of the plate with respect to the Sun, and it would typically appear elliptical. But, regardless of where the Moon is in the sky when an eclipse occurs, the Earth's shadow is circular. This can only be explained by a spherical Earth, which casts a circular shadow from all angles.

The phases of the Moon also provide a strong hint. When the Moon sets just after sunset, it is seen as a crescent with the illuminated side facing the western horizon; a half Moon (either waxing or waning) is always to be found 90 degrees away from the Sun; a full Moon always rises at sunset; finally, when the Moon rises just before sunrise, it is seen as a crescent with the illuminated side facing the eastern horizon. The phases of the Moon can easily be simulated by illuminating a ball with a torch in a darkened room and observing it from different angles. The ancient Greeks could have carried out the same exercise, substituting the torch for an oil lamp or candle. A spherical Moon does not necessarily imply a spherical Earth, but it is inconsistent with a flat Earth.

However, the paradigm shift required to go from the Hebrew cosmology to a model that can explain the above might have been too great. It seems far likelier that this evidence might have been nothing more than confirmation of a by then-emerging view that the Earth was a sphere floating in space. Was there an intermediate step, half-way between the Hebrew model and a spherical Earth? I believe that there probably was.

Even though the light-pollution of the twenty-first century, the night skies in Athens at 38° N appear significantly different from those in Riga on the Baltic at 57° N. From the latitude of Athens, the pole star Polaris is 19° closer to the horizon; the constellation of Ursa Major, ever-present in the skies above Riga, is barely **circumpolar** and scrapes along the northern horizon at lower **culmination**. There are also more subtle differences, which will be apparent to an experienced skywatcher; for example, the Belt of Orion, as it rises in Riga, is inclined at around twenty degrees from the vertical, but in Athens it is very nearly perpendicular to the horizon.

The European Bronze Age was characterised by an increase in long-distance interactions and trade[13]. Amber from the Baltic region was highly prized, and found its way through trade to the Mediterranean region. Thousands of beads from Mycenaean tombs dating to around the middle of the second millennium BC have been identified as Baltic amber. There is no substantial source of amber within a thousand miles of Greece, and there is no doubt that it arrived there through trade[14]. Baltic amber is also believed to have made its way via Mycenaean settlers in Lebanon to sites in Israel and Jordan[15]. Traders moving in either direction would surely have been aware of the changes in the night sky as they moved along the so-called Amber Road between the Baltic and the Mediterranean.

Similarly, the Egyptians had established trading contacts with a region in East Africa known as Punt from at least the Fifth Dynasty of the Old Kingdom (2494-2345 BC). Trade missions were sent there to obtain gold, aromatic resins, African blackwood, ebony, and slaves. Trade continued until the time of the Twentieth Dynasty ruler Ramesses III (reigned 1184-1153 BC). The location of Punt is uncertain, but it is now thought to have been in southern Sudan or Eritrea. Egyptian traders travelled either by sea from Quseir or Mersa Gawasis, or down the Nile to Kurgus at the fifth cataract and then overland to Punt[16]. As noted above, the Egyptians were also trading with the Minoans in Crete. Eritrea lies at 15° N, some 16 degrees of latitude south of the Nile Delta, and 20 degrees of latitude south of southern Crete. Again, traders would have seen significant changes in the night skies.

We should note that the changes that would have been seen then are not the same as those that are to be seen now. Between 2400 and 1500 BC, there was no bright star at the north **celestial pole**. The second magnitude star Kochab (β Ursae Minoris), slightly fainter than Polaris, was the closest, but it never got closer than eight degrees away from the pole. Around 1400 BC, Ursa Major remained circumpolar as far south as Luxor. However, the bright stars Arcturus and Vega, and the prominent constellations of Cassiopeia and Corona Borealis, were circumpolar from Riga but not from Athens. Meanwhile, the Southern Cross, which had all but disappeared from the Mediterranean skies, remained clearly visible from Luxor.

Some of the people travelling between northern Europe and the Mediterranean during the Bronze Age, or from Egypt to the land of Punt, might have noticed the night skies looked different to those of their homelands, and asked themselves how these differences might be explained. They might have begun query the orthodox cosmological model and consider a revised version in which the Earth disk is replaced by a hemisphere interior to the Firmament of Heaven dome.

It would have been a relatively straightforward mental leap to go from a hemisphere to a complete sphere, concentric with the Firmament of Heaven, with Sheol presumably relocated to the centre of the spherical Earth. This step might have been taken when travellers began to realise that the stars did not begin to 'run out', even as far south as the land of Punt. In summary, I would suggest that by 1400 BC, travellers – or at least those who took an interest in such matters – were aware or at least strongly suspected that Heaven and the Earth were not as they had been taught.

If so, why did it take another millennium for such knowledge to become mainstream? It must be recognised that long-distance traders were not ritual or religious specialists; nor were they were not part of the urban intellectual elite; and they were not navigators of the

Mediterranean. The latter, who would have been the most likely to have noticed differences in the night skies, were making voyages upon a sea whose principal axis lies east-west rather than north-south. Their long-distance voyaging took place around the 36th parallel. Even a voyage from Crete to Egypt spans only four degrees of latitude.

Any traveller deducing the true nature of the Earth was in a milieu where their insight would at best have been of passing interest. Most likely, their views would have attracted the ridicule and disbelief of their peers. They might also have felt it unwise to put the suggestion to a priest or ritual specialist. Contrary to the Gershwin brothers' suggestion, the priests might not have laughed. Indeed, they might have been distinctly unamused and reacted much as their Italian counterparts would do, thousands of years later, to Galileo's 'heretical' ideas (see Chapter 17).

Let us consider the pre-requisites for a society in which the strands of evidence outlined above might have led to a general acceptance that the Earth is spherical:

1. A society in which trade and other long-distance interactions occur, where travellers have opportunities to observe night skies in distant lands.
2. A literate society with writing technology capable of keeping records of astronomical phenomena.
3. An intellectual climate in which rational investigations of astronomical phenomena are likely to occur.

As we have just noted, the first criterion was certainly met by around 1400 BC. The second criterion was also met, because astronomical record-keeping in Mesopotamia goes back to before 1600 BC (see Chapter 10). Astronomy was also used by the Egyptians for calendrical and timekeeping purposes. The astronomers of the Old Babylonian Empire and Egyptian Old Kingdom could surely have deduced that the Earth is spherical. Yet apparently, they did not. Knowledge of the spherical Earth remained confined to the possible speculations of a few traders with unusually enquiring minds. What was evidently lacking was the intellectual climate in which the paradigm shift could occur.

At this point, we should examine a possible reason why belief in a flat Earth might have been so entrenched. South African cognitive archaeologist David Lewis-Williams[17] has suggested that it might be an intrinsic feature of the neural architecture of the human brain. He has noted that a central tenet of many religions is the existence of a three-tiered cosmos with realms located 'above' and 'below' that of our every-day experience. The Abrahamic tradition of Heaven and Hell is only one example of such a cosmology; the Hebrew tradition is another. Lewis-Williams suggests that the widespread belief in the existence of these other realms arises from visions and hallucinations experienced in altered states of consciousness as may be induced by meditation, psychotropic substances, and various ritual practices. All human brains are wired up the same way, and so all will experience broadly the same visions and hallucinations. The specifics of how they are interpreted varies from culture to culture, but they share the same basic aspects of a Heaven, Earth, and Underworld.

Could this have hampered attempts to interpret celestial phenomena that could not be explained by the standard flat Earth model? If so, it took the rationalism of the sixth century BC Pre-Socratic philosophers to break an ancient, hard-wired mindset. It is not clear who made the breakthrough, even assuming it was just one person. It seems entirely plausible that these Greek scholars learned of a controversial and generally ridiculed speculation that had been circulating for centuries amongst long-distance traders and travellers. They realised that this model was consistent with the curved shadow seen on the Moon's surface during a lunar eclipse; and that if the Moon were also spherical, its phases could be explained. Much remained to be done, but a crucial step towards understanding the Solar System had been taken.

09
The origin of the ancient constellations

Gradualist or uniformist?

Ever since 1922, the sky has been divided into the 88 constellations currently recognised by the International Astronomical Union. Some years later, the Belgian astronomer Eugène Delporte drew up the boundaries between them: a series of horizontal and vertical lines between specific **right ascensions** and **declinations** for the **epoch** B 1875.0. The 88 are a mixture of ancient and modern: as we have seen (Chapter 3), some constellations might have been recognised in Upper Palaeolithic times; conversely, there are some that were established as recently as the mid-eighteenth century.

In the second century AD, the Graeco-Roman astronomer Claudius Ptolemy (*c.*AD 100-170) listed 48 mainly northern and zodiacal constellations in his astronomical treatise the *Almagest* (see Chapter 12), although the huge Argo Navis (Argo the Ship) was eventually broken up into Carina (the Keel), Vela (the Sails), Puppis (the Poop Deck), and Pyxis (the Compass Box) to make a total of 51. The remaining 37 were added between 1590 and 1752 as European navigator-astronomers explored the southern hemisphere and charted the regions of the sky that had always been below the horizon of the ancient Greeks.

Just when and where the original 48 ancient Greek constellations originated remains uncertain. Researchers examining the problem have adopted two methodological approaches. The first, mainstream, view is a gradualist approach. On this model, the star-maps of the ancient world were assembled from several unrelated sources, some of which were very early. The problem for researchers is to identify the origins of these sources and to determine the circumstances in which each of the celestial figures was placed in the sky. The second view, less widely accepted, is the uniformist model, which proposes that all or most of the ancient constellations were intentionally 'designed' at the same time. Proponents of this model have repeatedly sought to identify the date when and the latitude where the constellations were established[1].

Babylonian origin of the zodiac

The best-understood origins are those of the twelve constellations of the zodiac, and four 'para-zodiacal' constellations that lie in close proximity. It is generally accepted that they originated in ancient Mesopotamia where there was a long history of astronomical practices (see Chapter 10). British astronomer John Rogers[2] suggests that in Mesopotamia, constellations developed into two distinct traditions, 'divine' and 'rustic', which both developed in stages between 3200 and 500 BC for different purposes. The divine tradition was intended for religious purposes and it depicted heraldic animals and divine figures. It included the twelve zodiacal constellations and the associated animals Serpens the Serpent, Corvus the Crow, Aquila the Eagle, and Piscis Austrinus the Southern Fish. The rustic tradition was an annual farming calendar and it depicted rustic workers and animals. Although many constellations belonged to both traditions, only the zodiacal and associated constellations from the divine tradition formed part of the ancient Greek tradition.

Mesopotamian artwork including ceramics and cylinder seals commonly featured lions, bulls, scorpions, and water carriers, which later featured as the zodiacal constellations of Leo, Taurus, Scorpius, and Aquarius. The latter originally represented Enki (Ea), the god of freshwater, male fertility, and knowledge. Although it is not known when these constellations were defined, Rogers believes that it coincides with the first urban societies, from around 3200 BC. Between 4400 and 2200 BC, the cardinal points of the **ecliptic** (i.e., the equinoctial and solstitial points) lay within the boundaries of Taurus (spring equinox), Leo (summer solstice), Scorpius (autumn equinox), and Aquarius (winter solstice); and the first-magnitude stars Regulus (α Leonis), Aldebaran (α Tauri), and Antares (α Scorpii) all lay very close to the cardinal points around 2800 BC. To people interested in the movements of the Moon and the planets, the cardinal points would have been of great importance. Notably, these four constellations remained unchanged in all later Babylonian star-lists.

From 2300 BC onwards, after the conquest of Sumer (present-day Iraq and Kuwait) by Sargon of Akkad, Sumerian and Akkadian traditions were combined into a standard iconography. The Sumerian gods were predominately associated with fertility and animal husbandry, whereas the Akkadian gods were mainly astronomical objects representing the Sun, Moon, and stars. Many cylinder seals depicted divine figures including those that were included in the zodiac. These depictions might not have been astronomical in intent, but they portrayed gods who were also represented as constellations.

The first recorded star system, the *Three Stars Each* tablets, date from around 1100 BC. The sky was divided into three 'Ways': the northern Way of Enil, the equatorial Way of Anu, and the southern Way of Enka. The boundaries lay at 17° N and 17° S, so that the Sun spent three consecutive months in each Way. The *Three Stars Each* assigned a star from each Way to each month of the year. Some of the 'stars' are indeed single stars; others are constellations. The earliest such catalogue lists the 3 × 12 stars by month, with notes on their relative positions, rising and setting times, and agricultural and/or mythological significance of the **heliacal risings** and **settings**.

The MUL.APIN, dating to around 1000 BC was the second formal compendium of Babylonian astronomy. It takes its name from the first constellation of the year, MUL ('star') and APIN ('The Plough'), a constellation identified with the present-day Triangulum plus the star γ Andromedae rather than the present-day Plough. It comprised the star catalogues produced up to that time, which were the direct descendants of the *Three Stars Each* lists. It included all the stars from the earlier work, but it had been revised on the basis of more accurate observations around 1000 BC, and it was much more extensive and accurate. The information recorded includes heliacal risings and settings; pairs of constellations that rise and set simultaneously; and pairs that are simultaneously at the horizon and the **zenith**.

There are 18 'regions through which the Moon passes in the course of a month' (i.e., the zodiac). The list begins with the Pleiades and includes most of the zodiacal constellations, though Pisces, Aries, and Virgo were substituted for their rustic alternatives of the Field, Hired Labourer, and Furrow. They are not strictly organised into the twelve signs, and other some constellations intrude. Thus, the Babylonian zodiac of twelve constellations had not yet been consolidated. The list was headed by the Pleiades, which lay close to the **First point of Aries** around 2240 BC. The MUL.APIN thus references the much older Sumerian tradition. The MUL.APIN records **circumpolar** constellations for the first time, and more 'rustic' constellations than *Three Stars Each*. The latter might not actually be new and could simply have been recorded for the first time. The twelve zodiacal and four 'para-zodiacal' constellations reached the Greeks around 500 BC. Other than these, the only Babylonian constellations to survive on the modern sky map are Orion, Perseus, and Andromeda, but all are human figures that could have been independently invented. Another possibility is the Mad Dog (Lupus). On the other hand, the rustic constellations might have been transmitted to the Bedouin Arabs and persisted in their star-maps until the first millennium AD. However, the transmission of no more than twenty Babylonian constellations to the Greeks leaves the bulk of the 48 listed by Ptolemy unaccounted for.

The early origins of Ursa Major

Some of the 48 ancient constellations are thought to have very early origins. As discussed in Chapter 3, it is likely that Taurus and possibly other constellations were depicted in the cave art of Upper Palaeolithic Europe. Another constellation with probably very early origins is Ursa Major (the Great Bear). The familiar Plough is only part of this constellation, but it is a highly conspicuous grouping nonetheless, and one of which there is a widespread awareness in the ethnographic record (see Chapter 1). What is perhaps surprising is that there is also a widespread identification of this constellation with bears or bear myths throughout Eurasia and North America[3]. Surprising, because the grouping of the seven most prominent stars looks nothing like a bear. The resemblance to a ladle or a plough is obvious; to a bear much less so. The implication is that the connection of ursine folk-law with Ursa Major has a common origin. Humans reached the New World before 18,500 years ago[4]; but analysis of DNA from ancient human remains in Alaska and Siberia suggests that Native American and North Eurasian populations have been genetically distinct for 25,000-20,000 years[5].

American astronomer Bradley Schaefer suggests that the possibility that the connection is a result of recent European 'contamination' is unlikely because there are three different version of the myth, none of which are likely to have originated with European settlers or missionaries. Furthermore, common specific details in the myths are consistent between Eurasia and North America. If one of the later migrations (Aleut-Eskimo, Na-Dene, or Saqqaq) brought the myths to the New World, then their very widespread distribution in North America is difficult to explain[6]. Bears do occur in European Upper Palaeolithic cave art, albeit less frequently than species that were hunted as food[7,8], suggesting that they might have been of ritual significance. Spanish researcher Luz Antequera Congregado[9] has claimed that Ursa Major is depicted at Altamira, albeit as a female bison rather than a bear.

However, fellow American astronomer Roslyn Frank[1] is sceptical about a common origin. She notes that veneration of bears as ancestors or relatives of humans, and accompanying rituals, have survived across a large northern geographical zone inhabited by hunter-gatherer groups. She believes that it is more probable that the establishment of an ursine ancestral figure, a bear hunt, or the figure of a bearlike hunter as the constellation of Ursa Major, and its integration into a wider belief system could be a natural consequence of the subsistence strategies and lifestyles of hunter-gatherer groups living in higher latitudes. In other words, the constellation could have arisen independently on multiple occasions.

Frank[1,10] believes that the earliest constellations were associated with a pan-European cycle of orally transmitted folktales known collectively as the 'Bear Son Tales'. Many of the tales are shamanistic, featuring the half-human half-bear Hartzkume. That the tales are so widespread in European folk tradition suggests that they are very ancient. From them, Frank claims, came the constellations of Ursa Major, Ursa Minor (the two Sky Bears), and Draco. The latter possibly represents the Serpent-Snake Shaman, who in many versions of the tales is seated at the foot of a fruit tree, the Sky Tree. Hartzkume himself might have been placed in the sky as the constellation of Engonasin (the Kneeling-One), later associated with the Greek superhero Heracles (Hercules). The 'Hartzkume group' might also include the constellations of Eridanus, Lepus, Canis Major, Aquila, Centaurus, and Lupus[10]. Around 4000 BC, the latter two, along with the Southern Cross, would have been visible from Europe. Regardless of any shared cultural significance of bears, the problem is the idea that multiple groups would all associate same, rather un-bearlike group of stars with a bear.

Touching the void

It is likely that some – though not many – of non-zodiacal constellations were of Greek origin; for example, the constellations of Perseus, Andromeda, Cassiopeia, Cepheus, Pegasus, and Cetus were probably invented as a complete set in order to illustrate the legend of Perseus (albeit some of these constellations might have been repurposed from Babylonian groupings)[11]. But what of the remainder, which number around twenty constellations? A controversial but long-running hypothesis is that many of the non-zodiacal constellations were intentionally 'designed' as an aid to navigation at sea.

The earliest literary references to non-zodiacal constellations are to be found in the works of Homer and Hesiod. Homer was writing around 710 and 760 BC, as Greece emerged from its Dark Age[12]. Hesiod refers to the **achronycal rising** of Arcturus occurring sixty days after the winter solstice, suggesting that he was writing around 650 BC[13]. In the *Odyssey*, Odysseus finally sets sail from the island home of the nymph Calypso, having been detained by her for seven years. He navigates by Ursa Major, the Pleiades, Orion, and Boötes; Homer notes that Ursa Major "*never bathes in Ocean's stream*" (i.e., it is circumpolar). Hesiod's poem, *Works and Days*, is a farmer's almanac based around a star calendar, but featuring only a few constellations. Roslyn Frank[1] believes that it is unlikely that these prominent star groups, familiar to all cultures, would represent the extent of the astronomical knowledge of Greek seafarers at that time. Possibly they acquired their knowledge of the other non-zodiacal constellations through contacts with other, older seafaring people.

John Rogers[11] is of the same opinion. He suggests that the bulk of the northern non-zodiacal constellations were invented by pre-Greek Mediterranean seafarers for navigational purposes. He terms these people the Navigators, and notes that there is no documentary evidence for who they were. If many of the non-zodiacal constellations were indeed all established in accordance with a predetermined plan, how might we identify the constellation makers?

One possible method would be to consider a 'zone of avoidance' on early star maps. Not all the stars are represented in the constellations of the ancient world – the region around the south celestial pole is a 'void'. There is nothing mysterious about this – not until you are close to the Equator will you be able to see the most southerly stars (see Appendix). Nobody from the Eastern Mediterranean had ever been far enough south to see these stars, hence they were not included in the early star maps. In theory at least, the void would be roughly circular; its radius would correspond to the latitude of the constellation makers; and its centre would correspond to the south celestial pole. Due to the effects of **precession**, the latter would not be the present-day location of the south celestial pole, but that in the epoch when the constellations were defined. It would thus be possible to determine the latitude and time period at which the constellation makers had lived. It would, however, be necessary to assume that all the constellations, or at least those bordering the southern void, were defined at roughly the same time and place.

This 'void zone' argument is actually an old idea, first proposed by British astronomers Richard Procter and Richard Allen in the nineteenth century. In the early part of the last century, two more British astronomers considered the problem. Edward Walter Maunder[14,15] deduced that the constellations were 'designed' in around 2700 BC at a latitude of between 36° N and 40° N. Some years later, Irish astronomer Andrew Claude de la Cherois Crommelin[16] obtained a radius of 36 degrees for the zone of avoidance (corresponding to a latitude of 36° N for the constellation makers) and a date of 2460 BC. Schaefer[6] has dubbed Maunder and later proponents of this argument the 'voidists'.

The Sphere of Eudoxus

The void zone argument was taken up again in the 1960s by another British astronomer, Michael Ovenden[17], who published his conclusions in an influential paper in *The Philosophical Journal*. He followed Maunder and Crommelin in the starting assumption that the ancient constellations had all been established at the same time and place, and for a specific purpose. He suggested that if the distribution of the constellations is ordered at all, it would be with respect to either the ecliptic or the **celestial equator**. If the latter were the case, then constellations would be 'drawn' orthogonally on the sky and their axes of orientation would indicate the north celestial pole of that epoch; for example, a human torso would be depicted on a north-south axis.

Ovenden noted that to the north of the zodiac is a ring of constellations including Auriga, Perseus, Hercules, and Boötes, but they have no consistent relationship with the 'pole' of the ecliptic (which lies in Draco). Therefore, the relationship, if any, would be to the north **celestial pole** at the epoch these constellations were defined. Was there ever a time when they were all equidistant from the north celestial pole? In the 1960s, it was not simply a matter of starting up the planetarium software on your laptop. Ovenden considered each constellation to delimit a rectangular block of sky, then laboriously estimated the location of the two points where a great circle bisecting the appropriate sides of the rectangle would cut the path of the pole's precession. While this was not possible for constellations lacking an elongated shape such as Cancer (the Crab), the remainder all gave one possible solution corresponding to a date of around the middle of the third millennium BC. Taking the mean of his results, Ovenden obtained a date of 2800 BC ± 300 years. He then asked a colleague, astronomer Archie Roy, to repeat the measurements. Roy obtained a date of 2900 BC ± 500 years.

Having established the era of the constellation makers, Ovenden followed Maunder[14] in noting that 'significant points' on the celestial sphere during this epoch are associated with snakes or dragons. Around 3000 BC, the constellation of Draco (the Dragon) linked the north celestial pole to the pole of the ecliptic, Serpens (the Snake) turned through 90 degrees at the **First point of Libra**, and Hydra (the Water Snake) extended for ninety degrees around the celestial equator. None of these constellations contain any bright stars, so Ovenden concluded that it was their location in the sky that was significant and made them worthy of designating as constellations. To Ovenden, this alone was proof that the constellations were not simply fanciful resemblances to mythical figures, and that they were intentionally established as part of a primitive celestial coordinate system.

In the second part of the paper, Ovenden considered the zone of avoidance, which he found to have been distorted by the 'artistic exuberance' of medieval star maps and was 'irregular sausage-shaped' rather than circular. Accordingly, he referred back to Ptolemy's list of constellations, which was based on a star catalogue that appears in Hipparchus' *Commentary on the Phaenomena of Eudoxus and Aratus.* Written before he discovered precession (see Chapters 4 and 12), *Commentary* is Hipparchus' sole surviving work – but the star catalogue is also not original work. As the title implies, it is based on the work of two more ancient Greeks: astronomer and mathematician Eudoxus of Cnidus (*c.*390-337 BC), and the poet Aratus (*c.*315-240 BC).

Eudoxus wrote two important treatises, the *Phaenomena* ('Appearances') and the *Enoptron* ('Mirror of Nature'), but both these works are now lost. However, we know of them through Aratus' poem, also entitled the *Phaenomena.* About 270 BC he was commissioned by King Antigonus II Gonatus of Macedonia (319-239 BC) to write a poem about Eudoxus' work. The poem supposedly gives us a very good idea of what was in the work of Eudoxus, though without the original we cannot be certain. It has three sections relating to constellations: firstly, a description of 49 constellations; secondly, a list of stars that lie on the celestial tropics and equator; and thirdly, a list of constellations that rise or set at the same time as a zodiacal constellation rises. In total, there are 173 items of astronomical lore[18]. Hipparchus mentions, and apparently had access to, a work entitled *The Sphere of Eudoxus*, in addition to the Aratus poem[10].

If the *Phaenomena* of Eudoxus and Aratus were based on data contemporary with the constellation makers around 2800 BC, then Hipparchus would have been working from a source that in his day was already 2500 years out of date. Thus, Hipparchus could see stars that the putative constellation makers could not, and vice versa because the zone of avoidance had changed since the latter's time. *Commentary* suggests that Hipparchus noticed discrepancies between Aratus' description and what he could see himself observing from the island of Rhodes. Hipparchus discovered precession by comparing the measured positions of stars with those recorded by earlier astronomers, but Ovenden speculated that the discrepancies in Aratus' *Phaenomena* might have put him on the right track. If Hipparchus had considered the possibility that had been a time when Aratus' now old descriptions of the sky was accurate, it could have led him to look for the effects of precession.

Turning to Aratus' *Phaenomena* in more detail, Ovenden considered (1) statements that describe the position of stars with respect to the poles and hence relate to a particular epoch, but are independent of the observer's latitude, and (2) statements that describe rising and setting of stars, which depend on the observer's horizon with respect to the stars, and hence on both the epoch and latitude of the observer. From these, Ovenden arrived at a date of 2600 BC ± 800 years, and a latitude 36° N ± 1.5°. These results were in good agreement both with those from the first part of the paper, and those obtained by Maunder and by Crommelin half a century earlier. They suggested that the statements in the *Phaenomena* refer to a definite latitude, and to a time well before Aratus lived – the time of the constellation makers.

The next question Ovenden attempted to address was why? For what purpose had a group of people sat down and 'designed' the constellations? Based on the numbers of references to ships, storms at sea, and navigation in Aratus' *Phaenomena*, Ovenden suggested that the purpose was to set up a simple navigation system. Though primitive and not very accurate, it would have sufficed for short voyages where ships are mostly in sight of land. It would have been essential for longer voyages along the 36th parallel where land is largely out of sight between Crete and Italy or Malta, or for voyages to Egypt or Libya[1]. The mythology behind the constellations might have predated and inspired the naming of the constellations. It would have served as an *aide-memoire* to sailors, who would have been familiar with the tales, and who could now relate them to patterns in the sky. The *Phaenomena* also contains two agricultural references, suggesting a secondary role for the constellations as an agricultural calendar. They thus encompass two systems: the equatorial constellations (of that era) for navigation and zodiacal constellations agriculture. Ovenden speculates that the zodiacal constellations might be part of an earlier, purely agricultural system.

The longitude of the constellation makers cannot be determined purely from the data, but Ovenden considered the civilisations present in the Mediterranean at the time in question: the Babylonians, Phoenicians, Egyptians, and Minoans. The Babylonians would have voyaged mainly in the Persian Gulf and the Indian Ocean, which are too far south. Although the Egyptians used the Mediterranean, they too would have conducted much of their seafaring in the southern waters of the Red Sea, Gulf of Aden, and Indian Ocean. The Phoenicians and Minoans were both major seafaring civilisations, but Ovenden came down in favour of the latter. He notes that Homer's *Odyssey* contains astronomical references that are consistent with the date and latitude he obtained for the constellation makers. The works of Homer contain vestiges of Mycenaean tradition, and the Minoan civilisation on Crete was eventually absorbed by the mainland Mycenaeans. However, Ovenden admitted that Aratus edited a version of the *Odyssey*, making it a dubious source of evidence. Even more speculative – by Ovenden's own admission – is the suggestion that the Dodecanese island of Astypalia, located at 36°30' N 26°18' E, is where the constellations were devised. The island is locally known as Astropalia, suggesting an astronomical connection. This is usually thought to be a corruption of Astypalia, but Ovenden suggested that this is not the case. He notes that Astypalia is a common name in the Aegean, and that it would be more likely for an uncommon name to be corrupted to a common name rather than vice versa. Overall, if Ovenden's main conclusions are accepted, then the cases for a Minoan or a Phoenician origin appear to be equally strong.

Eighteen years after Ovenden's article in *The Philosophical Journal*, his colleague Archie Roy[19] returned to the subject of the origin of the constellations with an article in the journal

Vistas in Astronomy. He noted the constellations are popularly associated with Greek mythology, but claimed that it can "*easily be shown*" that many of them are far older than the ancient Greeks. He summarised Ovenden's work, noting the good agreement with Maunder and Crommelin. He suggests that the Sphere of Eudoxus was an actual celestial globe (which has been lost) upon which Eudoxus based his *Phaenomena.* Roy believes that it was "*almost certainly a stone or metal globe on which the brighter stars, the ecliptic and equator and the constellation figures were engraved*", and that it dated to the time of the constellation makers.

Roy followed Ovenden in attempting to find a match for the skies as described by Aratus. Although desktop computers were starting to come into use in 1984, they were still fairly primitive by today's standards. However, Roy was able to make use of the planetarium of the Glasgow College of Nautical Studies. He examined the validity of each statement in Aratus' *Phaenomena* against the skies at selected dates between AD 2000 and 5000 BC, seeking a best fit for the orientation of the constellations with respect to the celestial equator. By this means, he obtained a date of 2000 BC ± 200 years, rather more recent than those obtained by Ovenden and himself in 1966. Possibly as a compromise, he subsequently referred to a date of 2300 BC, corresponding to the later stages of the Early Minoan/Prepalatial period. He endorsed Ovenden's conclusion that the constellations were created as a navigation aid to sailors rather than as an agricultural calendar. He likewise noted that the constellations are oriented with respect to the celestial equator rather than to the ecliptic, and that Aratus' *Phaenomena* contains numerous nautical references, but only a few pertaining to agriculture.

However, Roy also pursued a number of additional matters arising. In one passage from the *Phaenomena*, there is a reference to the Earth in equipoise, with a polar axis:

This axis forms on either side a pole;
The one we see not but the opposite
Is high o'er ocean in the north; two Bears
Called Wains move round it, either in her place.

This suggests that Eudoxus, writing in the third century BC, was aware that the Earth was a sphere, suspended in space, rotating on an axis, and with two poles of which only one could be seen. This would be unsurprising, as the spherical Earth was generally accepted by the end of the fifth century BC. But if the *Phaenomena* contains knowledge from a much earlier age, it suggests that the makers of the Sphere of Eudoxus around 2300 BC were already aware of such considerations. As we have seen (Chapter 8), it is unlikely that Bronze Age navigators of the Mediterranean knew that the Earth was spherical, although travellers between distantly separated latitudes might have begun to suspect.

Eudoxus lived just two centuries before Hipparchus, and his Sphere was already over 1,500 years out of date. Many of the stars depicted would not by that time have been visible from Greece. He might have been aware of the mismatch, and perhaps have realised that his Sphere depicted the skies of a much earlier age, but he missed the opportunity to discover precession. Roy asks the question, why did Eudoxus base his description of the heavens on it at all? Could he not have got hold of an up-to-date Sphere? And from where did he obtain the Sphere that he had?

Roy noted that Eudoxus visited Egypt to study astronomy, and he suggested that he had obtained the Sphere from Egyptian priests. The Sphere was "*one of these very ancient navigational star-globes which, after countless voyages, had ended up in the priestly archives of Egypt*". But why was it so out of date? The answer, according to Roy, is that between 2300 BC and Eudoxus' visit to Egypt, something happened to the navigators' civilisation, so there was nobody to make more up-to-date Spheres. Roy then considered four civilisations of the ancient world: the Phoenicians, the Egyptians, the Babylonians, and the Minoans, i.e., the same four civilisations considered by Ovenden. He also reached the same conclusions as Ovenden, though for different reasons. The Phoenicians were at the right latitude, but they were too recent. The Egyptian civilisation, on the other hand, was old enough, and they were mapping the skies during the time in question. Also, Eudoxus supposedly obtained the sphere in Egypt. But there are two problems: firstly, Egypt is too far south, secondly since so much of their material culture has survived, why do we not have more of their Spheres? If they were in widespread use, surely many would have survived in tombs and the like. The Babylonians had a sophisticated astronomical tradition, but the problem is again that they were too far south. By elimination, this leaves the Minoans, who were accomplished navigators, and they were also in the right place at the right time. Crucially, their civilisation collapsed around 1450 BC, putting Sphere-makers out of business. The collapse, once attributed to a massive volcanic eruption on the island of Thera (now Santorini), was probably brought about by internal revolts and economic pressure from the emerging mainland Mycenaean civilisation.

Problems with the void zone argument

Perhaps the most obvious problem with the Sphere of Eudoxus is the idea that Eudoxus had in his possession a Minoan celestial globe that was already two thousand years old, something that reads like the plot from an Indiana Jones movie. The objection Roy makes to the Egyptians being the makers of the Sphere surely applies to the Minoans: if they were making widespread use of celestial globes, why have we not found more of them, either in Crete or the many places visited by Minoan seafarers? In any case, how feasible is it that the Minoans were able to make celestial globes in the first place?

Assuming it existed at all, it seems likely the Sphere was made by Eudoxus himself[11], and that both it and his *Phaenomena* were based either on his own observations or earlier descriptions of the heavens. If the former was the case, then it would imply that his *Phaenomena* was based on inaccurate rather than outdated observations. British Classical scholar David Dicks[20] questions the basic premise that Aratus' *Phaenomena* refers to the skies as they appeared in a much earlier epoch. He rejects the underlying assumption that Eudoxus' data, as reproduced by Aratus, cannot have been simply inaccurate. He suggests that the data was probably the earliest attempt to describe the celestial sphere as it appeared in the fourth century BC, and that it was inevitable that mistakes were made. Hipparchus, two centuries later, was living in an era when astronomical theory and practice were far better understood.

Dicks further notes that to intentionally design constellations to mark the equator, ecliptic, and equinoctial points, or that to position them in a specific relationship to the celestial and ecliptic poles, requires a knowledge of great circles, celestial co-ordinates, and celestial spheres, which did not exist in Egyptian and Babylonian astronomy until after their development by the Greeks in the latter part of the fifth century BC. It would be necessary to assume that Minoan astronomy reached such levels in the third millennium BC, only for the knowledge to be lost. Dicks suggests that assumptions about the sophisticated mathematical concepts required arise almost unconsciously from an inability to distinguish between what is known about early astronomical thinking and more modern concepts.

If we nevertheless accept the argument that Eudoxus was working from earlier and now outdated sources, the question remains, why were they not brought up to date? Even if you accept the rest of his argument, Roy's suggestion that the Minoan collapse was responsible is not convincing. Roy's date of 2300 BC for the constellation makers is 850 years before the Minoan collapse; by that time there would already have been significant changes to the night's sky. Nobody would ever use an outdated celestial coordinate system for navigation at sea, with land out of sight.

One possibility is that the lore and tradition used for practical purposes, such as navigation and rituals entailing actual observations, *was* updated. However, there was also a learned or poetic tradition that remained unchanged, because there was no need to verify that it was still correct. By analogy, present-day astrologers still consider the Sun to reside in Aries at the First point of Ares; in the real world, precession has shifted the First point of Ares into Pisces[1,6].

However, Schaefer[6] noted that there are further issues with Ovenden's and Roy's analyses. The most obvious is the statement by Ovenden that Hipparchus was observing from Rhodes at a latitude 31° N, whereas Rhodes is actually at 36°24' N. Roy repeated the error, claiming that Hipparchus traditionally observed at a latitude 31° N – the latitude of Alexandria. Ovenden and Roy also made systematic errors in ignoring the effects of **refraction** and extinction. Atmospheric refraction amounts to 0.6 degrees (36 arcmin) at the horizon. Thus, if the constellation makers' southern horizon corresponded to the southernmost star at a declination of -50°, then their implied latitude would be 40°36' N rather than 40°00' N. Atmospheric extinction means that a star will not remain visible all the way down to the horizon, and it will fade out when the angle above the horizon drops below a certain angle. This angle, known as the extinction angle, depends on atmospheric conditions and the brightness of the star in question, but typically ranges between two to four degrees. Ignoring extinction thus introduces a systematic northwards bias of two to four degrees into the derived latitudes.

Schaefer then examined the claims made in more detail. He listed 16 constellations for possible orientation on the north celestial pole, but also found that many others have neither parallel nor perpendicular orientations that go even close to the pole. He determined the epochs his 16 constellations would have indicated the north celestial pole and found that there was no epoch when they had all done so. Instead, polar orientations were randomly distributed in time from 10,000 BC to AD 6000.

Next, Schaefer looked for epochs when the group of constellations including Auriga, Perseus, Hercules, and Boötes would have formed a ring along the celestial equator. If Ovenden was correct, the geometric centres of these constellations should have had roughly equal declinations around 2500 BC. In fact, this happened around AD 400 with a spread of 4 degrees. Around 2500 BC, the spread was 33 degrees. Schaefer noted that in any given epoch, there would be several constellations forming an apparent ring; groups of constellations could be chosen to support any epoch.

Finally, Schaefer considered the significant positioning of the constellations of Hydra, Draco, and Serpens. In terms of average declinations, Hydra was coiled around the equator from 6000 BC to AD 1000, with the best fit at around 500 BC. The positioning of Draco and Serpens he dismissed as coincidence: over the last ten millennia, these constellations have indicated or linked solstitial, equinoctial, or polar points on many occasions. Shaeffer's analysis showed that claims of alignments on the north celestial pole, constellation rings, and the significant positioning of Hydra, Draco, and Serpens could all be dismissed.

An outlier

Schaefer nevertheless conceded the validity of the main premise of a void zone around the south celestial pole. However, he made the proviso that even when refraction and extinction were taken into account, it could not be assumed that the constellation makers would populate the sky with constellations right down to the horizon. The zone of avoidance would thus provide only a limiting value to the latitude of the constellation makers.

In order to examine the void argument objectively, Schaefer considered each of the southern constellations touching the void and the key star within in each on a case by case basis. The 'key star' was deemed to be the northernmost star critical to the 'essence' of the constellation, without which it would lose its shape. He then calculated the declinations for these key stars for dates between 3000 BC and the present day, taking refraction an extinction into account, to obtain limiting latitude for observation as a function of year. He applied this approach to Ara (the Altar), Corona Australis (the Southern Crown), Piscis Austrinus (the Southern Fish), Centaurus, and the now-defunct constellations of Argo and The Waters.

The next step was to draw up a list of candidate civilisations that could have been the constellation makers, with dates. Schaefer considered the ancient Greeks (latitude of 38° N at 500 BC), the Babylonians (latitude of 32° +1°/-3° N at 700 BC), the Minoans (latitude 36° N at 1500 BC), the Egyptian Old Kingdom (latitude 30° N at 2700 BC), and the Old Europeans (latitude 43° N from 4000 to 2500 BC). The limiting latitudes were then plotted on a graph; the constellation makers must have lived south of these limits. Schaefer's calculations yielded a date of 690 BC +210/-360 years and a limiting latitude of 33° N, giving a best fit with the Babylonians, who were well south of all of the limiting latitudes throughout the first millennium BC.

As we have seen, the Babylonians are generally accepted to have contributed the zodiacal and para-zodiacal constellations to the ancient Greek system, but of these only Piscis Austrinus features in Schaefer's analysis. Schaefer[18] subsequently attempted to obtain a date and latitude for the origin of the astronomical lore in Aratus' *Phaenomena*; he noted that this was a separate exercise to that of obtaining a date and limiting latitude for the origins of the void-touching constellations themselves. To ensure objectivity, he used all 173 items of lore rather than a selected subset. In addition, 31 items of astronomical lore from Hipparchus, attributed to Eudoxus, were included. From this, he obtained a list of constraints on declinations, right ascensions, rising times, setting time, and lower culmination times (for circumpolar objects). The two major translations of the *Phaenomena* (Mair and Kidd) were compared for possible impact on constraints. Two methods were used to analyse the constraints: a 'method of circle centres' and a more rigorous method based on the standard χ^2 (chi-squared) comparison between models and data. If these two quite different methods yielded similar results, then there was unlikely to be a technical or conceptual problem with the analysis method.

The circle centre method suggests that the observations used to construct the lore of

Eudoxus came from a latitude of 34° ± 2° N and an epoch of 900 BC ± 350 years; the χ^2 comparison method yielded a latitude of 36.0° ± 0.9° N and an epoch of 1130 BC ± 80 years. The first is consistent with the Babylonians, the second with the Assyrian empire to the north. The problem is that both dates are earlier than that of 690 BC +210/-360 years previously obtained by Schaefer for the origins of the void-touching constellations. Clearly, the origin of constellations must precede the lore. Schaffer described this as no more than a constraint on the date of the lore, but for the date obtained by the more rigorous χ^2 comparison method, the ranges only just overlap at 1050 BC.

Thus, having raised very valid objections to the work of earlier researchers and carried out extremely rigorous analyses, Shaefer obtained a series of dates that were of only marginal consistency with each other, and completely at odds with the dates obtained by other researchers. One is left with the feeling that Schaefer's results have demonstrated that the void argument, while theoretically sound, cannot tell us very much about the origin of the non-zodiacal constellations.

A synthesis

Perhaps, though, this is too pessimistic. Roslyn Frank[1] described Schaefer's result as an 'outlier'. She believes that the problem is that the degree of precision claimed by the various researchers is not justified. Along with Spanish physicist Jesús Arregi Bengoa[10,1], she believes that is not possible to narrow down the date and location to anything more precise than a time frame from 4000 to 2500 BC and a latitude band from 36° to 41° N. It is possible that many constellations were created by different Mediterranean seafaring groups at different times over this period, not all of which stood the test of time. The Navigators were not necessarily a homogenous group; they might not even have all been members of one of the high civilisations of the Mediterranean. Nor did they necessarily invent all of the constellations they used: it is entirely possible that they repurposed long-recognised formations. However, it is likely that from these sources are drawn the twenty or so unaccounted constellations from the 48 compiled by Ptolemy. Thus overall, the origin of the constellations can be seen as a synthesis of the gradualist and uniformist models. The 88 constellations that make up the night sky represent an extraordinarily diverse set of traditions stretching from the last Ice Age to the Age of Discovery.

10
Astronomy in ancient Mesopotamia

The land between the rivers

Ancient Mesopotamia, the 'land between the rivers' of the Tigris and Euphrates was home to the world's first urban societies, the distant forerunners of today's nation-states. Increasingly complex societies emerged from around 6000 BC, and around 3200 BC the first city-states appeared in Sumer (present-day southern Iraq). Though these cities frequently came into conflict with one another, multi-city empires and territorial states were a later development. The first of these was the empire of Sargon of Akkad, whose reign is commonly given as 2334-2284 BC, although the exact dates are uncertain. The Akkadian Empire was the first of many to dominate the region: it was followed by the Third Dynasty of Ur (from 2112 BC); the Old Babylonian Empire (from 1880 BC); the Kassites (from 1595 BC); the Assyrians (from 935 BC); the Neo-Babylonian (or Chaldean) Empire (from 626 BC); the Achaemenid (Persian) Empire (from 539 BC); and the Seleucid Empire (from 312 to 63 BC).

Astronomy has a long history in Mesopotamia, and it remained continuous despite the political upheavals[1]. The Babylonians were meticulous observers of the night sky and they made extensive records of astronomical phenomena. Their motivation was not scientific in any modern sense of the word: Babylonian astronomers were primarily astrologers, who sought to interpret their findings in terms of omens and portents. But while we now tend to decry astrology as complete nonsense, it must be recognised that it served as the motivation behind some excellent work that was highly influential on later astronomers. Another important consideration is that in a state-level society, astronomers and mathematicians were usually fulltime specialists, who did not have to concern themselves with such matters as farming or building settlements.

The Mesopotamians also possessed an advanced knowledge of mathematics. Texts dating back to the Sumerian and Old Babylonian periods show that scribe school students copied and learned multiplication tables, were familiar with the use of reciprocals (for division), squares, square roots, and cube roots (though not apparently cubes); and they were taught to solve commodity market rate problems and complete exercises in geometry. Unfortunately, our knowledge of Mesopotamian mathematics is far from complete. Most of the texts we have are table texts and exercises written by scribe school students; a smaller number are advanced exercises probably copied by older students from teachers' original texts. There are very few well organised, original texts produced by the unknown mathematicians who laid the foundations of Mesopotamian mathematics[2]. However, its sophistication is clearly demonstrated by its application to astronomy, as we shall see.

King and temple dominated Mesopotamian life to such an extent that it is often claimed that ordinary individuals were unimportant. The collective nature of this society is often compared unfavourably with the individualism of ancient Greece. Certainly, in contrast to Greece, we know very little about the originators of Babylonian mathematical astronomy. In reality, the notion of a monolithic bureaucracy is probably an oversimplification. There were multiple versions of planetary theories for all of the celestial bodies, suggesting that the scribes tasked with producing lists of upcoming planetary phenomena were testing and fine-tuning the theories rather than mindlessly applying sets of rules. Many tablets from Uruk are signed by scribes who are believed to have belonged to just two families. Many of them were priests, or the descendants of priests. This suggests that in Uruk at least mathematical astronomy was the work of a small number of people, who were often related by family, and whose astronomical work was part of the work of the temples. Two names that have come down to us are the Chaldean astronomers Naburimannu and Kidinnu, with whom later Greek and Roman astronomers were apparently familiar. They might have been responsible for the application of the mathematical systems known as System A and System B (see below) to lunar theory[3]. There are also claims that Kidinnu was aware of **precession**, but these are not widely accepted[4].

Cuneiform texts

The many Mesopotamian texts that have come down to us were originally encoded on soft clay tablets using a stylus in a script known as *cuneiform*, meaning 'wedge-shaped'. The word is due to Thomas Hyde, professor of Hebrew and Arabic at Oxford, who coined it in 1700. However, some years earlier, the German explorer Englebert Kaempfer used the very similar term *cuneatae*. The first attempts at deciphering these ancient texts began early in the nineteenth century, and significant progress was being made by the 1850s[5]. Mesopotamian writing gradually evolved from a means of recording transactions into a fully-fledged logosyllabic system, i.e., a combination of logograms (symbols used to represent words, syllabograms (symbols used to represent syllables), and determinatives (symbols used to add information; for example, the pound (£) sign tells the reader that the quantity represents a sum of money). At the same time, the number of symbols used fell from around 1,200 to just 600, and the symbols themselves became progressively more stylised. Use of a wedge-shaped stylus encouraged a style in which the symbols were made up of combinations of wedges and straight lines, until they bore little resemblance to their original form. So emerged the classic Mesopotamian cuneiform script that remained in use for the next 2,500 years. It was adapted for use in the Akkadian Empire during the reign of Sargon, and later variants are found on Babylonian and Assyrian tablets. Cuneiform only fell into disuse after the conquests of Alexander the Great between 334 BC and his death in 323 BC. The last-known cuneiform text dates to AD 75[6].

The known astronomical tablets are of two main categories: those concerned with mathematical astronomy and those concerned with non-mathematical astronomy. The mathematical texts were published in 1955 by Austrian-American mathematician and science historian Otto Neugebauer in a volume entitled *Astronomical Cuneiform Texts*, usually abbreviated to ACT. The non-mathematical texts are usually referred to as NMAT or Non-Mathematical Astronomical texts. These include the Astronomical Diaries and predictions that were derived from them, and also the Late Babylonian Astronomical Texts (LBAT).

Mesopotamian numbering systems

Instead of our familiar decimal (base-10), the early Sumerians used different numerical systems for different purposes. For example, a mixed-base sexagesimal system where numbers alternate by factors of 10, 6, etc. (i.e., 1, 10, 60, 600, 3,600) was used for counting discrete objects such as animals, people, fish, or implements; a mixed-bases bisexagesimal system where numbers alternate by factors of 2, 10, 6, etc. (i.e., ½, 1, 10, 60, 120, 1,200, 7,200) was used for cereal, bread, and other items distributed as part of a rationing system; and a time system of 1, 10, 30 was used to form a calendar of ten-day weeks, three-week months, and twelve-month/360 day years. A total of thirteen different numerical counting systems have been identified from ancient texts[7].

The Old Babylonians inherited the Sumerian sexagesimal numbering system and used it in their mathematical texts, although it was now a pure sexagesimal (base-60) system[8]. Having 59 distinct numerals would have been impossibly unwieldy, so there were only two: a downward stroke to represent one unit and a left-facing chevron to represent ten units. These were combined to form each digit. This was not a mixed-base system: the decimal element was restricted to aiding the representation of the large numbers of digits required by a base-60 system.

Modern scholars tend to avoid using the downward strokes and chevrons when writing Mesopotamian numbers and instead use the decimal equivalents of each digit, delimited by commas. Thus 1,30,49 in sexagesimal $= 49 + 31 \times 60 + 1 \times 60 \times 60 = 5{,}509$. Fractional numbers were treated in a manner analogous to decimals: $^{1}/_{60}$, 1/3,600, etc. When writing such a number, modern scholars use a semicolon as a radix point (the equivalent of a decimal point); for example, $1{,}15;30 = 60 + 15 + {}^{30}/_{60} = 75.5$. It is common to transliterate Babylonian numbers into a mixed base, with the integral portion in decimal and only the fractional portion in sexagesimal, for example, 29;31,50,08,20 is the standard Babylonian value for the number of days in the **synodic month**. Converting the fractional portion to decimal is liable to produce recurring decimals, so it is often left in sexagesimal in literature. The Mesopotamians lacked the concept of the zero, so when numerals were written in columns a blank space was left as a placeholder. For other situations, a separate symbol to denote an empty place was used.

Like ourselves, the Mesopotamians divided the circle into 360 parts; their word for 1/360 of a circle was the *ush*, which accordingly is usually translated as 'degree'. This system was introduced around 420 BC[1], and it will be appreciated that there is complete equivalence between it and our own system of degrees, arcmins, and arcsecs. What we would write as 3°26'15" would be 3;26,15 in the Mesopotamian system. The day, which began at sunset, was also divided into 360 *ush*; an *ush* of time was thus equal to four of our minutes. There were 30 *ush* to the *bēru*, making the latter equal to two of our hours. These units were used to record the timing and duration of eclipses, although the *bēru* eventually fell out of use and such data was recorded solely in *ush*. Times were recorded as 'after sunset' (sunset to midnight); 'before sunrise' (midnight to sunrise); 'after sunrise' (sunrise to midday); or 'before sunset' (midday to sunset). Although time units of fixed length were used for recording astronomical phenomena and in other scientific and non-scientific contexts, the Mesopotamians also divided the daytime into twelve *simanu* or 'seasonal hours', the length of which depended on the time of the year[9,10,11].

Mesopotamian calendars

The calendar as used in Mesopotamia for civil, religious, and everyday purposes was lunisolar: it tracked the solar year, but the months followed the phases of the Moon. Each month began with the first appearance of the crescent Moon in the evening sky; and the year began in late March or early April, i.e., at the nearest new Moon to the start of spring. The evening crescent Moon appears every 29 or 30 days, meaning that a year of twelve synodic months comprises only 354 or 355 days, and will rapidly fall out of steps with the seasons. To get around this problem and ensure that New Year's Day (1 *Nisannu*) fell close to the spring equinox, an intercalary month was periodically added to the calendar for a year of thirteen months. Either a second *Ululu* (the sixth month) or a second *Adaru* (the twelfth month) would be added[12].

An ideal intercalary scheme should insert seven extra months over nineteen years. This is because the 19-year period represents the lowest common multiple of the mean solar (or **tropical**) year of 365.2421897 days (365 days 5 hrs 48 min 45.19 sec) and synodic month of 29.530589 days (29 days 12 hrs 44 min 2.8896 sec); i.e., 235 synodic months equals 19 tropical years. The 19-year period is known as the Metonic cycle after the fifth century BC Greek astronomer Meton of Athens who made it the basis of the Greek calendar[12]. In fact, it took many centuries for the Babylonians to arrive at such a scheme.

The decision as to when a month was to be intercalated was originally made by royal decree, and it only happened whenever seasonal events such as river levels and crop ripening were getting significantly out of step with the calendar. The problem with this approach is that such factors rely to an extent on the weather and may occur early or late. On average, however, it was necessary for months to be intercalated every three years. By around 1000 BC, the Assyrians were relying on astronomical schemes to determine intercalation. These schemes are recorded in the MUL.APIN astronomical compendium; they were based on the **heliacal risings** of certain constellations and on conjunctions of the Moon with the Pleiades. The best known of these rules is the so-called Pleiaden-Schaltregel rule, which states that: "*If Moon and Pleiades are in conjunction on the first day of Nisannu, this is a normal year; if Moon and Pleiades are in conjunction on the third day of Nisannu, this is a leap year.*" This rule would result in a month being intercalated every third year on average[13,14].

Around 528 BC, soon after the start of the Achaemenid period, a regular cycle of three intercalary months every eight years was introduced. This was based upon the octaeteris, an eight-year cycle in which the phases of the Moon occur on the same day of the year plus one or two days. This was replaced in 499 BC by the more accurate system based on the Metonic cycle, which provided for seven additional months over a period of nineteen years. The extra months were added in years 2, 5, 7, 10, 13, 16, and 18 of each cycle[12].

An oft-overlooked matter is to what extent the Babylonian calendar depended on actual sightings of the new Moon, which could of course be affected by weather conditions. Modern computations show that by and large, Babylonian months did indeed commence with the new Moon. The outgoing month was not, apparently, allowed to continue indefinitely until weather conditions permitted a sighting. However, there was no regular sequence, such as an alternation between 29-day and 30-day months. Letters from astronomers suggest that in Assyrian times, the decision as to when to start the new month was made by the king. Reports were sent from multiple sites, so increasing the probability that weather conditions would permit at least one observation of the new Moon. Where weather had interfered, astronomers would estimate the start of the month from the age of the Moon at the point when it was eventually seen. By the Achaemenid period, new Moon prediction had become sufficiently accurate for it to be regarded as a legitimate means of setting the start of the new month[15].

In addition to the lunisolar calendar, at least two schematic or 'ideal' calendars were in use in Mesopotamia. These comprised 12 months of 30 days each, for a year of 360 days, and were used in administrative contexts to simplify calculation of rations, rates of work, and interest on loans. Although this calendar took longer than the lunar calendar to fall out of step with the seasons, periodic intercalation of a 30-day month for a year of 390 days was still required. However, these calendars eventually were further simplified by assuming that a year always comprised twelve months, with no intercalation. In addition to its administrative use, the 360-day calendar appears in various literary and religious works such as the creation epic *Enuma Elish*. It is possible that the 360-day year reflected a Babylonian belief in an ideal state of the universe[16,17].

The astronomical diaries

It is not clear when the Babylonians began systematically recording astronomical phenomena. The zodiacal constellations are thought to have originated in Mesopotamia, as are the first formal star maps (see Chapter 9). However, the earliest astronomical records we have are Old Babylonian observations of Venus compiled in the time of Ammisaduqa, ninth king of the First Dynasty of Babylon (reigned 1646-1626 BC). The originals, now lost, recorded the dates of the first and last appearances of Venus after sunset and before sunrise throughout the reign of Ammisaduqa. What we have are astrological texts dating to the eighth and seventh centuries BC, which incorporated the already-ancient Venus data. We therefore know nothing about the original observer, why they recorded Venus data, or whether they recorded similar data in other reigns, or for other planets[18].

The British Museum has a collection of astronomical diaries from Babylon, comprising around 1,200 fragments of various sizes and in various states of preservation. The collection was mostly purchased from antiquities dealers in Baghdad in the 1870s and 1880s, at a time when no other institution was acquiring antiquities from Iraq. Unfortunately, none have any archaeological context as they were excavated by workmen whose primary interest was in recovering ancient baked bricks for reuse. The majority of the tablets are broken and re-joining the fragments is time-consuming. While most of the known tablets are held by the British Museum, some fragments are in other museums around the world and in some cases, these have been temporarily reunited with pieces from the same tablet at the British Museum. Not all the fragments have secure dates: those that do are from the Neo-Babylonian, Achaemenid, and Seleucid periods. The earliest dates to 651 BC, the next earliest to 567 BC, and the one after that to 453 BC. The rest date from 385 to 50 BC. However, there are fragmentary cuneiform tablets with records of lunar eclipses going back to around 750 BC; this data was almost certainly extracted from now lost astronomical diaries. Also, Ptolemy's *Almagest* suggests that in the second century AD, Babylonian astronomical records going back to the time of Nabu-nasir (reigned 747-734 BC) were still available[18].

The diaries each cover half of a Babylonian year in periods of six or seven lunar (synodic) months. They contain dates of the various phases of the Moon (with estimates where weather conditions prevented actual observation). Lunar information includes (1) time between sunset and the setting of the first visible evening crescent (*na*); (2) time between moonrise and sunset on the last day that the Moon rises before sunset, i.e., on the last day before full Moon (*me*); (3) time between sunset and moonrise next day, i.e., on the first day after full Moon (*ge*), (4) time between moonset and sunrise on the last day the Moon sets before sunrise, i.e., on the last day before full Moon (*shu*); (5) time between sunrise and moonset next day, i.e., on the first day after full Moon (*na*); and (6) time between moonrise and sunrise of the last visible morning crescent (*kur*) (Babylonian texts use the same symbol for both new Moon and full Moon). These six observations are collectively known as the Lunar Six, and ii-v only as the Lunar Four[18].

Data recorded for outer planets (Mars, Jupiter, and Saturn) included the dates of first visibility in the east before sunrise (*Γ* or *MF*), first stationary point or beginning of retrogression (*Φ* or *BR*), acronychal rising – i.e., rising at sunset, or opposition (*Θ* or *OP*), second stationary point or end of retrogression (*Ψ* or *ER*), and last visibility in the west after sunset (*Ω* or *EL*), along with the position in zodiacal signs (30 degrees) and thirtieths of a zodiacal sign (degrees) in which first and last visibility occurs. The stationary points or stations are where a planet's motion through the zodiac halts at the beginning and end of retrogression, which is a period when the planet appears to move in a retrograde manner against the background stars. Retrogression is due to Earth overtaking or being overtaken by the planet in question, although the Babylonians did not know this. It happens when the Earth, Sun, and planet in question are lined up, and for this reason it is sometimes referred to as the solar anomaly or inequality with respect to the Sun (any departure of a planet from uniform angular motion is referred to as an anomaly or inequality). For an outer planet, retrogression happens when it is at opposition; for an inner planet when it is at inferior conjunction (i.e., between the Earth and the Sun) and hence not visible. For the inner planets Mercury and Venus, the dates and zodiacal signs were recorded for first visibility in the east before sunrise (*Γ* or *MF*), stationary point in the east before sunrise or eastern elongation (*Φ* or *EE*), last visibility in the east before sunrise (*Σ* or *ML*), first visibility in the west after sunset (*Ξ* or *EF*), stationary point in the west after sunset or western elongation (*Ψ* or *WE*), and last visibility in the west after sunset (*Ω* or *EL*). Collectively, these events are known as the synodic phenomena and a complete cycle as the **synodic period**. In a Babylonian context, they are often referred to as the 'Greek-letter' phenomena[18,19].

Detailed records were kept of lunar and solar eclipses, both those observed from Mesopotamia and those reported from elsewhere. For lunar eclipses, the details reported were time and angle of first umbral contact (*U1*) given in *ush* after sunset; time in *ush* from beginning to totality (*U2*) or to greatest magnitude if partial; an estimate of the magnitude if partial; and duration of the various phases in *ush*. If the Moon rose or set with the eclipse in progress, the timing and state of the eclipse at that point was recorded. No record was made of penumbral eclipses, which probably passed unnoticed. For solar eclipses, the same details were recorded, with the timing of first contact given in *ush* after sunrise. A record was also made of planets becoming visible at totality[18,20].

Prior to 562 BC, observed eclipse timings were recorded to a precision of 5 *ush* (20 min), and subsequently to a precision of one *ush* (4 min). However, the accuracy actually achieved was far less; the average error in timing was around half an hour, and it did not significantly improve after 562 BC. Part of the problem, especially for lunar eclipses, was determining the exact moment of contact. Another factor was accuracy of timing. It is not known how Babylonian astronomers obtained their timings, but it is assumed that they used water clocks. These could probably only be read to the nearest two *ush* (8 min)[11,21].

Other phenomena in the astronomical diaries recorded included details of conjunctions of planets with other planets, with the Moon, and with 28 reference stars known as Normal Stars; and the dates on which planets moved from one zodiacal sign into another. In addition, non-astronomical data about weather conditions, river levels, and commodity prices were recorded. Details of current events were sometimes included, which in many cases has facilitated dating the tablets[18,22,19].

The phases of Venus

It is often claimed that the Babylonians were aware of the phases of Venus, the discovery of which is conventionally attributed to Galileo in 1609. In 1915, archaeologist Joseph Offord noted that there are a number of cuneiform texts that describe the planet as having 'horns'; this has been interpreted as evidence that they could discern the tiny crescent with the naked eye. Assyrian and Babylonian texts state that Ishtar (Venus) was the daughter of Sin (the Moon), and Offord suggested that they were depicted as horned deities because they are both able to appear in crescent form[23]. The claim was met with scepticism: American astronomer William Wallace Campbell[24] found it "*difficult to believe that any one has ever seen the crescent form of Venus without telescopic aid*" and suggested that the Mesopotamian references were nothing more than a lucky guess similar to Jonathon Swift's 'prediction' in *Gulliver's Travels* that Mars has two small moons (actually discovered 150 years later). According to Campbell, the logic was that if the Moon periodically presents as a crescent in the evening and early morning skies, then Venus – which is only seen at these times – should also appear as a crescent.

The maximum angular size of Venus at crescent phase is 1.1 arcmin. As noted in Chapter 8, the theoretical maximum resolving power of the human eye is about 0.4 arcmin, although most people cannot resolve better than about 1.0 arcmin. In principle therefore, it should just be possible to see the crescent phase. In practice, though, the glare of Venus is so intense that seeing an actual shape to it is very difficult. Nevertheless, the internet abounds with claims to have seen the crescent with the naked eye. Although no formal studies appear to have been carried out, the late astronomer and broadcaster Sir Patrick Moore[25] cited "*well-authenticated*" cases going back to the mid-nineteenth century of the crescent having been seen with the naked eye. Moore concludes that "*...there seems therefore little doubt that the phase really is visible to people with exceptional eyesight*".

Planetary cycles and goal-year texts

The Babylonians were aware that there are cyclic patterns to the movements of the five naked-eye planets. Babylonian tablets suggest that they were aware of the synodic periods of the planets and the relationships of these to the tropical periods. The tropical period of a planet is the average interval between successive passages through the **First point of Aries** (for an **inferior planet**, it is one tropical year; for a **superior planet**, it is slightly less than the **sidereal period**). Babylonian astronomers obtained the following figures:

Jupiter: 36 tropical periods = 391 synodic periods = 427 years,
Saturn: 9 tropical periods = 256 synodic periods = 265 years,
Mars: 151 tropical periods = 133 synodic periods = 284 years,
Venus: 1,152 tropical periods = 720 synodic periods = 1,152 years,
Mercury: 46 tropical periods = 145 synodic periods = 46 years,
or 388 tropical periods = 1,223 synodic periods = 388 years.

This means, for example, that 71 years corresponds to 65.01 synodic periods of Jupiter or 5.99 tropical periods. Thus, Jupiter's synodic phenomena, as viewed from Earth, will be repeated almost exactly every 71 years[12].

From around 250 BC, these cycles were used in a series of predictive texts known as the Goal-Year texts. These texts contained predictions of the lunar and planetary phenomena for a specific year; the term 'goal year' is a literal translation of the German *Zieljahrtexte.* Predictions included the Lunar Six, synodic phenomena, planetary passages of the Normal Stars, and eclipses. For a particular planet, the predictions were made on the basis of data recorded in an astronomical diary for the year preceding the goal year by the number of years corresponding to the cycle time for that planet. Thus, for the Mesopotamian year x, predictions from Jupiter would be taken from the astronomical diary for years x -71 and x - 83; for Saturn from the diary for year x - 59, and so on[18,12].

In addition to the Goal-Year texts, there were two types of almanac. The Normal Star Almanacs contained the same predictions as the goal-year texts, as well as predictions for the solstices, equinoxes, and Sirius phenomena. The Lunar Six was sometimes replaced by the simplified Lunar Three: the dates of first visibility of the evening crescent, first moonset after sunrise, and last visibility of the morning crescent. The second type, simply referred to as Almanacs, was an expanded version of the Normal Star Almanac, adding the current zodiacal signs of the planets visible at the start of each month, and the dates on which a planet moves into a new zodiacal sign[22].

The predictions contained in the almanacs could not simply be copied from the Goal-Year texts. The latter contained two sources of inaccuracy, which Mesopotamian astronomers needed to correct for. The first is that the Goal-Year texts were calculated in Mesopotamian years, which can contain either twelve or thirteen (synodic) months. Consequently, a planet's goal-year period expressed in Babylonian years might be a month shorter or a month longer than the true period for that planet. Babylonian astronomers were apparently aware of the need to apply a correction of one month where necessary. The second issue was that the periodicities of both synodic phenomena and planetary passages are not quite whole numbers of sidereal years. The solution was to add or subtract a fixed number of days for the planet and event type under consideration; for example, for the synodic phenomena of Venus, four days were subtracted from the date given in the Goal-

Year text. By such means, the Babylonian astronomers achieved predictions that were typically accurate to within one day and rarely out by more than three days. The predictions recorded in the Almanacs and the Normal Star Almanacs are sufficiently consistent to suggest that both originated from the same sources; the few discrepancies were probably due to copying errors[22,26].

Mathematical astronomy

As well as using the cyclical behaviour of the Moon and planets as the basis for making predictions, Babylonian astronomers also developed advanced mathematical methods for calculating their future movements. Our knowledge comes from around 450 Astronomical Cuneiform Tablets from Babylon and Uruk dating from 350 to 50 BC. Two types of tablet are concerned with the calculations. Around 340 of the tablets are ephemerides, listing the calculated data in rows and columns. The remaining 110, known as procedure tablets, explain the calculations themselves. All these use the zodiac as a coordinate system for representing celestial positions. The algorithms for the computations comprise branching chains of arithmetical operations that can be represented as flow charts[27]. Each calculation was intended to predict dates and **celestial latitudes** for successive occurrences of a specific synodic phenomenon for a specific planet; for example, the first appearance before sunrise (*MF*) of Jupiter. Very few tablets give data for planets when between phenomena[19].

The Babylonian use of a lunisolar calendar, with its vagaries of 29 and 30-day months, was a complicating factor when compiling lists of predicted dates. Accordingly, the dates were expressed in *tithis*, or thirtieths of a synodic month, within the month. The *tithi* is thus slightly shorter than one day. By using *tithis* instead of days, Babylonian astronomers avoided the need to compute the varying lengths of future months, which would have been necessary if times were to be expressed as actual dates in the civil calendar[28]. The term *tithi* is Indian, and refers to the use of this interval of time by Indian astronomers. The latter might have borrowed the concept from Mesopotamia, although its exact origin is uncertain (see Chapter 13)[29].

Two systems of calculation were used, known to modern scholars as System A and System B. System A was a zonal system in which the ecliptic was divided into zones. Within each zone, it was assumed that the object in question moved at a constant speed, instantaneously jumping to a new speed whenever it moved between zones. Babylonian astronomers were aware that this 'step system' was only an approximation, but it was one that gave good results in practice. System B was a 'zigzag' system in which motion was assumed to increase at a constant rate from a minimum to a maximum, and back again. System B is easier to use, and it is a better approximation than System A. It is therefore probable that it was a later development[12].

The simplest version of System A used a 'fast zone' and a 'slow zone'; more complex variants employed up to six zones, each with its own rate of motion. In all cases, each zone z was allocated a value σ_z. To calculate the **celestial longitude** of the next occurrence of a given synodic phenomenon of a planet P from the celestial longitude of its most recent occurrence (λ), then for whatever zone in which λ lies (say zone 1), the corresponding value of σ (say σ_1) is added to λ. If this takes P into the next zone (say zone 2), then the portion of the amount added that lies in the second zone is adjusted for that zone's rate by multiplying by σ_2 divided by σ_1. If this now takes P into a third zone, then a similar calculation is performed: the portion lying in the third zone is multiplied by σ_3 divided by σ_2, and so on. There are many zonal systems described in the procedure tablets; others have been deduced from ephemerides for which the actual procedure tablet has not been found. It is also possible to work backwards from an ephemeris to identify the procedure tablet used for the calculations[30,31,32,19].

Although Babylonian mathematicians made frequent use of geometrical methods, it had long been believed that their astronomers relied purely on arithmetic rules. Then, between 2002 and 2008, Dutch science historian Mathieu Ossendrijver transliterated and translated four cuneiform tablets from the collection at the British Museum. Ossendrijver noted that the tablets contained portions of texts describing a trapezoid procedure. One of them also contained references to Jupiter and the others contained additional procedures concerned exclusively with Jupiter. The link between the trapezoid procedures and Jupiter was interesting but tentative. Then, some years later in 2014, Ossendrijver was studying old photographs of an uncatalogued tablet from the British Museum's collection. The previously unpublished photographs were provided by retired Assyriologist Hermann Hunger. The new text contained trapezoid instructions matching those in the other tablets, but it did not mention Jupiter by name. It was probably intended for the use of a private individual familiar with the planet rather than for a temple library - but because references to Jupiter were omitted, the significance of the tablet had eluded scholars for decades[33].

The new text contains a nearly complete set of instructions for computing the motion of Jupiter along the ecliptic in accordance with a scheme designated $X.S_1$. Prior to the discovery of the new text, this scheme was too fragmentarily known to identify its connection with the trapezoid procedures. Scheme $X.S_1$ covers one complete synodic cycle (*MF*, *BR*, *OP*, *ER*, and *EL*). Although it is a purely arithmetic scheme, Ossendrijver[27] showed that $X.S_1$ was mathematically equivalent to the trapezoid procedures described in the four tablets he had examined. Four more texts were subsequently identified that preserved portions of scheme $X.S_1$, though no comparable scheme for any planet other than Jupiter has yet emerged[34].

In modern terminology, the trapezoids represent a graphical depiction of Jupiter's daily angular movement along the ecliptic versus time in days. The tablets describe the motion of Jupiter over the first sixty days of its synodic cycle, during which it traverses 10;45° (10°45'); they then describe how to obtain the time at which Jupiter has traversed half of this distance, i.e., 5;22,30° (5°22'30") by dividing the trapezoid into two smaller regions of equal area; this gives an answer of 28 days.

This procedure, a forerunner of integral calculus, was previously believed to have been first used at Merton College, Oxford in the fourteenth century AD. Although ancient Greek astronomers would later employ geometrical methods, the Babylonian trapezoidal methods differed in that they worked not with configurations in physical space but in a purely abstract space defined by velocity and time.

Although Seleucid-period Babylonian astronomers now possessed mathematical models that enabled them to predict the movements of celestial bodies with great accuracy, they still lacked an underlying kinematical model. There is no evidence for any knowledge of the fundamental concepts of the **celestial sphere**, of the **obliquity of the ecliptic**, or of geographical latitude and longitude[8].

Eclipse prediction

We do not know just when Babylonian astronomers first attempted to predict eclipses. As noted above, the earliest astronomical diaries are thought to be those compiled in the time of Nabu-nasir (reigned 747-734 BC). These records apparently already included eclipse predictions. From half a century later, we have tablets from Nineveh that include reports sent to the Assyrian kings Esarhaddon (reigned 680-669 BC) and Assurbanipal (reigned 668-631 BC) by Babylonian and Assyrian scholars. The reports concern astrological matters, and some describe observations of lunar and solar eclipses. But quite a few refer to predicted eclipses that were not seen. Sometimes, this was simply due to cloudy conditions, but on other occasions the weather was favourable, and the prediction presumably in error. It is not always clear how the predictions were made, and many predictions relied upon ominous events, studies of sheep livers, and other means of divination. However, some texts do make clear that attempts were being made to predict eclipses astronomically[35,36].

The Assyrians were aware that lunar and solar eclipses generally take place within a month of one another; and that eclipses could happen either six or occasionally five synodic months apart. The shorter interval occurs when Moon reaches a node towards the end of an **eclipse season**; when it reaches a node five synodic months later, the next eclipse season will have already begun – but the Assyrians were apparently unsure as to when a five-month interval might occur[35]. However, by the sixth century BC, Neo-Babylonian astronomers had discovered a sequence of periodicities in the Moon's motions known as the saros (see Appendix). The word is Greek, but it was originally applied to eclipse cycles by the seventeenth century English astronomer Edmund Halley (best known for calculating the orbit of the eponymous comet). The term was not used by either the Babylonians or ancient Greeks.

The saros results from a series of coincidences of nature: 223 **synodic months** (6585.3213 days or 18 years 11 days 07 hrs 43 min) is almost exactly the same length of time as 242 **draconic months** (6585.3572 days or 18 years 11 days 8 hrs 34 min), and also coincides with 239 **anomalistic months** (6585.5374 days = 6585 days 12 hrs 54 min). The net effect is that at the conclusion of 223 synodic months from the time of an eclipse, the Moon will again be both at **syzygy** and at a **node**, and it will also be at the same distance from Earth. This will not only give rise to another eclipse, but also for a solar eclipse the **eclipse limit** will be the same. This latter factor is equally important because were the Moon to be at a greater distance from Earth than previously, the eclipse limit would be smaller, and an eclipse might not occur at all. The period of 223 synodic months is referred to as a saros cycle. Eclipses do of course occur more frequently than every 18 years, so there are a number of saros series in operation at any one time. A saros series is a collection of eclipse seasons, each of which is separated by 223 synodic months (i.e., one saros cycle) from the preceding eclipse season.

Saros series are numbered: those involving the Moon's descending node receive even numbers and those involving the ascending node receive odd numbers. Each saros series evolves and has a finite life. For a saros series involving the Moon's descending node, the series begins with an eclipse at the south pole. Each successive eclipse then has a track more northerly than the last, until a final eclipse at the north pole concludes the cycle. For a saros cycle involving the Moon's ascending node, the reverse happens, with the series beginning at the north pole and concluding at the south pole. When one saros series concludes at one pole another one will begin at the other pole. A complete series lasts from between 1,206 to 1,442 years; at any one time there are 43 saros series in operation.

Although the saros can be used to predict both lunar and solar eclipses, it is harder to discern for the latter. The last eight hours of a saros cycle displaces the next eclipse in the sequence a third of the way around the world, and Babylonian astronomers would have had to wait three saros cycles or 54 years 34 days (a period known as the exeligmos) for a solar eclipse to return to the Near East. Even then, there is no guarantee that even a partial eclipse would have been seen as the track would have shifted either to the north or to the south. By contrast, lunar eclipses are visible anywhere in the world where the Moon is above the horizon: successive lunar eclipses in a saros cycle would have a good chance of being observed from the same place, especially in winter when the full Moon is up for longer.

It is not known just how the Babylonians learned about the saros, although records suggest that they were aware of it by no later than 575 BC. With 43 saros series in operation at any one time, identifying a pattern even to lunar eclipses is difficult. It presupposes that there is a pattern in the first place. However, astronomers were aware that the movements of the planets followed predictable cycles, so it would have been logical for them to have assumed that analogous cycles governed lunar and solar eclipses also.

The duration of the saros cycle can be determined by performing a frequency analysis of the periodicities of pairs of partial lunar eclipses with similar magnitude and direction of penetration into the Earth's shadow. Not every pair will belong to the same saros, but peaks will occur at the saros, the double-saros, and the exeligmos. Allowing for eclipses being missed due to adverse weather conditions, it is estimated that around 70 observations of partial lunar eclipses over the course of 300 years would be sufficient to obtain a value for the duration of the saros cycle[21].

The duration of the saros cycle does vary around its average of 6,585.3213 days, and thus the time between an eclipse and the next in its saros series will vary slightly. Over the course of one saros, the excess of time above 6,585 days varies in a roughly linear fashion from a minimum of around 6 hrs 20 min to a maximum of around 9 hours and back. To obtain a more precise estimate of when an eclipse will occur, it is necessary to know the value of the excess on the predicted day of the eclipse. A recently discovered text, tablet BM 45861 from the British Museum's holdings, contains a mathematical 'zigzag' function which increases and decreases linearly between a maximum and a minimum. It has been suggested that this function was intended to represent the excess of the saros over 6,585 days. It yields the required offset to a precision of 5 *ush* (20 min), but eclipses were only timed to this level of precision. In any case, the uncertainty, as noted above, was probably greater. Separate implementations of the function were required for 'odd-numbered' and 'even-numbered' eclipses, as the two cycles are out of phase by half a period. When the saros is at maximum length for 'odd-numbered' eclipses, it is at a minimum for 'even-numbered' eclipses, and vice-versa[37,36].

Babylonian astronomers combined the saros with their knowledge that eclipses are possible every six months and occasionally after five months to obtain a matrix-based saros scheme for predicting eclipse possibilities. During a saros cycle of 223 synodic months, there are 19 **eclipse years** of two eclipse seasons each and hence 38 eclipse possibilities. Of these, x eclipse possibilities are at 6-month intervals and y eclipse possibilities are at 5-month intervals. Thus:

$5x + 6y = 223;$

$x + y = 38;$

Solving these simultaneous equations gives us $x = 33$ and $y = 5$.

Hence within a saros cycle, there are 33 eclipse possibilities at 6-month intervals, and the remaining 5 are at 5-month intervals. Astronomers made the assumption that the 5-month intervals were evenly spaced throughout the cycle, and that each cycle began with a 5-month interval. The 38 eclipse possibilities are divided into five groups, each headed by a 5-month interval. Three groups will have a 5-month interval followed by seven 6-month intervals; the other two will have a 5-month interval followed by only six 6-month intervals. For the whole saros cycle, the 38 eclipse possibilities are grouped 8-7-8-7-8. The sequence is then repeated for the next saros cycle. The choice of where the saros cycle is deemed to begin arbitrary: beginning it at other points would yield groupings of 7-8-7-8-8, 8-7-8-8-7, 7-8-8-7-8, or 8-8-7-8-7[35,36].

The tablet LBAT *1420 contains a record of lunar eclipse observations and predictions for the period 603-575 BC, using the 8-7-8-7-8 grouping. This tablet probably dates to soon after the last entry in 575 BC, thus indicating that the saros was recognised no later than then. The Eclipse Texts LBAT *1414, LBAT 1415 + 1416 + 1417, and LBAT *1419 also use the 8-7-8-7-8 grouping. They date to the later part of the fourth century BC, but they are thought to be part of a large compilation of eclipse records that originally covered the period 746-314 BC. The saros scheme used by the Eclipse Texts (hereafter the Early scheme) is described in LBAT *1418, a single tablet which contains calculations of lunar eclipse possibilities for the period 646-573 BC[35].

However, there is another group of tablets known as the Saros Canon, which comprises LBAT 1428 and two related texts, LBAT *1422 + *1423 + *1424 and LBAT *1425. These tablets date to the later fourth century BC and cover the period 526-256 BC. The Saros Canon describes a saros scheme with different months for the first lunar eclipse possibility in each saros cycle to those of the Early scheme. It has been suggested that the Saros Canon scheme replaced the Early scheme around 526 BC, but it fails to predict several lunar eclipses that were successfully predicted and recorded during the period from 526-256 BC, suggesting that the Early scheme remained in use throughout that time[35].

If so, then what was the purpose of the Saros Canon scheme? One possibility is that it was an attempt to derive the layout of the eclipse possibilities within a saros cycle from first principles[38]. Another possibility is that it was but one of several empirical schemes put forward during the later fourth century BC as a replacement for the Early scheme, although no other variant schemes from this period have so far come to light[35].

The saros was eventually revised at a date between 278 BC (when it was still in use) and 248 BC (by which time it had been superseded), although the Saros Canon was not adopted even then. By this time, the Early scheme was beginning to break down: its prediction for eclipse of 19 April 248 BC would have been a month late, and the eclipse of 11 September 245 BC would not have been predicted at all. Both eclipses were successfully predicted by the new scheme. There were two further revisions of the saros: around 200 BC, and around 110 BC[35].

There are, understandably, far fewer records of solar eclipses than there are of lunar eclipses, although both were treated in exactly the same way. Predictions were based upon a 'Solar Saros', which seems to have undergone revisions at around the same times as the lunar saros. Both used the same 8-7-8-7-8 distribution, with the solar saros always starting four eclipse possibilities (i.e., two eclipse years) earlier than the lunar saros[35].

The eclipse predictions included a number of lunar and solar eclipses that were not expected to be visible from Mesopotamia. Such eclipses would be marked *sa dib*, meaning "*which passed*". Eclipses that were expected to be visible but did not prove so were marked *ki pap nu igi*, meaning "*observed, but not seen*". Predictions for lunar eclipses were successful 55 percent of the time; another 35 percent corresponded to penumbral eclipses, which were not recorded. Thus, only ten percent of the predictions were in error. The mean accuracy of timing for successful predictions was to within 17 *ush* (1 hr 8 min). After 550 BC, this increased slightly to 14 *ush* (56 min)[20].

A total of 61 solar eclipse predictions are known, all of which correspond to an eclipse that was visible somewhere in the world. Of these eclipses, only 28 of these were visible at the latitude of Mesopotamia, and not of all these were visible at its longitude. The mean accuracy of the timing of the predictions was to within 29 *ush* (1 hr 56 min) for these 28 eclipses, and to within 55 *ush* (3 hrs 40 min) for the remaining 31[39]. Predictions were always made for the time when the eclipse was expected to begin (i.e., first contact), although this was always given in relation to sunset or sunrise rather than to a specific time. Consequently, it was necessary to take into account the seasonal change in the length of night and day when making the prediction[35].

By the Seleucid period Babylonian astronomers had developed highly advanced mathematical models of lunar motion (or **lunar theory**) which could be used to predict the times and magnitudes of both lunar and solar eclipses. The methods are dependent upon the elongation of the moon from the ascending node which is assumed to move in retrograde by a constant amount each month. Eclipse possibilities occur at a syzygy at which the celestial latitude of the Moon's centre is closest to zero. There are scattered System A lunar ephemerides that can be extrapolated to cover the whole period from 318 BC until around 50 BC, meaning that a complete set of eclipse predictions can be generated. This shows that the method fails to predict a number of both lunar and solar eclipses for which predictions are recorded in the almanacs. The System B ephemerides cover the period from 205 to 74 BC. In this case, it is not possible to extrapolate over the entire period and only the few where dates are also attested in the almanacs can comparisons be made. Here again, the ephemerides fail to predict all the eclipses recorded in the almanacs. Because these mathematical methods were less effective than the saros, astronomers continued to base their eclipse predictions upon the latter[35].

The periods of lunar motion

The mathematical lunar theories developed by Babylonian astronomers required accurate mean values for the synodic, sidereal, draconic, and anomalistic months. Unfortunately, it is not clear how they obtained these values from their relatively crude observations, but it is straightforward to show how the draconic month and the sidereal month could have been derived in terms of the synodic month. It is probable that these periods had been accurately determined by around 500 BC[40].

To obtain the draconic month:

There are 38 eclipse possibilities in a saros cycle of 223 synodic months, and therefore 19 eclipse years.

In one synodic month, the Moon completes an orbit of the Earth (i.e., a revolution through 360°) and also progresses $(360 \times 19/223)° = 360\ (1 + 19/223)°$ with respect to the nodes.

Movement of Moon in one saros cycle

$= 223 \times 360\ (1 + 19/223)°$

$= 360 \times (223 + 19)°$

= 242 complete rotations with respect to the nodes.

Hence 242 draconic months = 223 synodic months;

1 draconic month = $223/242 \times 1$ synodic month.

To obtain the sidereal month:

From the Metonic cycle, there are 235 synodic months in 19 years. For each synodic month, the Moon has to complete slightly more than one circuit of the Earth in order to catch up with the Sun; this amounts to a complete circuit over the course of a year; hence the number of sidereal months in one year equals the number of synodic months plus 1. The number of sidereal months in 19 years is therefore $235 + 19 = 254$.

Hence 254 sidereal months = 235 synodic months;

1 sidereal month = $235/254 \times 1$ synodic month.

The next step was to determine the value of the synodic month. The standard Babylonian value used in System B lunar calculations, and cited by Ptolemy in the *Almagest*, is 29;31,50,08,20 days or 29 days 191;0,50 *ush* in sexagesimal notation. In decimal, this is 29.530594 days (29 days 12 hrs 44 min 3.3333 sec). This figure differs by less than half a second from the currently accepted mean value of 29.530589 days (29 days, 12 hrs, 44 min 2.8896 sec)[41]. From it, we obtain a value for the draconic month of 27.212076 days (27 days 5 hrs 5 min 23.3664 sec) and for the sidereal month of 27.321613 days (27 days 7 hrs 43 min 7.2632 sec). The accepted values are 27.212220 days (27 days 5 hrs 5 min 35.8080 sec) and 27.321661 days (27 days 7 hrs 43 min 11.5104 sec) respectively.

How did Babylonian astronomers obtain such an accurate value? The synodic month can be measured by timing the interval between successive occurrences of the same phase, but to determine the exact point at which the phase occurs is not as easy as it sounds. Today, an astronomer equipped with a medium-sized amateur telescope might time the interval between successive first-quarter (waxing half) Moons. Dichotomy (i.e., the point at which exactly half the Moon is illuminated) can be timed accurately by using an occulting bar and a Moon filter to determine the moment when the terminator becomes straight. The length of the synodic month can vary by up to $\pm$ 7 hours, so in practice the astronomer would need to time the interval between two first quarter Moons several years apart to obtain a mean value.

The Babylonians, of course, lacked telescopes, occulting bars, and Moon filters, but there is one occasion where the phase of the Moon can be accurately determined without any of these things – a lunar eclipse, which can only happen when the Moon is full. The time interval between any pair of lunar eclipses, divided by the number of elapsed full Moons between them, will yield the average duration of the synodic month over this period. As noted above, the average error in timing was about 30 minutes; for a pair of eclipses, the average error is about $30 \times \sqrt{2} \approx 42$ minutes. However, if the number of months between the pair of eclipses is large, then these errors and variations in the length of the synodic month will become far less significant. For two eclipses 300 years (or 3,711 full Moons) apart, they will amount to an error of 42 minutes divided by 3,711; or less than two-thirds of a second[21].

Although the Babylonians could have extracted a value for the synodic month from their extensive eclipse data, they might also have obtained it by dividing the length of the saros by 223. If so, then based on the *Almagest* value, Babylonian astronomers must have been working from a saros of 29;31,50,08,20 × 223 = 6585;19,20,58,20 days. There is an attested Babylonian value for the saros of 6585;19,20 days, which yields a synodic month of 29;31,50,08,04 days as opposed to the standard value. Since the same tablet also has the standard value of the synodic month, it is possible that 6585;19,20 is simply a truncation of 6585;19,20,58,20. Alternatively, it might have been a mean value obtained from the 'zigzag' functions used to calculate the excess above 6,855 days as described above[41].

To obtain the anomalistic month is a rather more complex matter. It corresponds to the periodicity of the lunar anomaly, that is to say the variation of the Moon's speed along the ecliptic as seen from Earth from a maximum at perigee to a minimum at apogee, and back; but unlike the draconic month it is not obviously derivable from the saros. Yet there is an attested value of 27;33,16,30 days (27.554583 days, or 27 days 13 hrs 18 min 36 sec) used for System A calculations, which is very close to the accepted value of 27.554550 days (27 days 13 hrs 18 min 33.12 sec). There is also a System B relationship of 251 synodic months equals 269 anomalistic months, which gives an anomalistic month of 251/269 × 29;31,50,8,20 = 27;33,16,27 days when using the standard Babylonian value of the synodic month. This was probably rounded to give 27;33,16,30 days[40]. How could this result have been obtained from the kind of observations available to Seleucid period astronomers?

It can be shown[42] that the quantity Σ, defined as the sum of the Lunar Four timings *shu*, *na*, *me*, and *ge*, is related to the angular velocity of the Moon (as viewed from Earth) when it is at opposition (i.e., full). Σ oscillates between a maximum and a minimum and back, corresponding to syzygy at full Moon moving from **perigee** to **apogee** and back. This oscillation is due to the difference between the lengths of synodic and anomalistic months, and its periodicity is about 14 synodic months. In other words, if the Moon is full at perigee, then one synodic month later, it will have already passed perigee; conversely, when it next reaches perigee it will not yet be full. However, after 14 synodic months, the full Moon will be close to (but not exactly at) perigee.

To a first approximation, in 14 synodic months, the Moon will have made 14 + 1 = 15 returns to perigee, hence 14 synodic months ≈ 15 anomalistic months. This value is about 11 minutes too high, a difference that would have become apparent to Babylonian astronomers after three periods, or 42 months. The next step would have been to track Σ over the course of a complete saros of 223 synodic month, and to show that there are just under 239 oscillations in that time. Dividing the saros of 6585;20 days by 239 gives an anomalistic month of 27;33,13,18 days in sexagesimal, or 27.553694 days in decimal (27 days 13 hrs 17 min 19.2 sec). The result is 1 minute 15 seconds less than the accepted value, but as noted above, Mesopotamian astronomers obtained a more accurate figure by using the ratio 251/269 between synodic and anomalistic months. It has been suggested that this ratio was chosen as a mid-point between the 14/15 ratio (too high) and the 223/239 ratio (too low)[40]. It seems likelier that by continuing to track Σ, it was found that the next whole-number ratio beyond 223/239 was 251/269, and that use of this ratio in System B calculations yielded better results.

The legacy of Babylonian astronomy

We do not have to look far to see how the influence of Mesopotamian mathematics persists to this day. Although we do not use the sexagesimal system for calculation, the division of the circle into 360 degrees, and the division of both the degree into minutes and seconds of an arc are a direct reuse of the Babylonian system. The Babylonians did not mark the passage of time in hours, minutes, and seconds, but there is an equivalence of 15 *ush* to the hour, and the division of the hour into minutes and seconds of time is of course a reuse of the sexagesimal system. But the legacy of the Babylonian astronomers and mathematicians goes much further than our measurements of angular distance and time. They bequeathed a legacy of observational data and predictive systems that was of inestimable value to the astronomers of Hellenistic Greece and later. Even today, details of eclipses recorded by the Babylonians have been used to investigate long term variations in Earth's rate of rotation.

However, Babylonian astronomy had its limitations. Its primary objective was to identify periodic relationships for the various celestial phenomena, in order that these might be accurately predicted. These predictions were used for astrological purposes and calendrical purposes. Barring eclipses and stations of planets, Babylonian astronomy was chiefly focused on horizon phenomena, i.e., first and last appearances on the eastern or western horizons, and periods of visibility of the Moon. Unlike the later Greeks (see Chapter 12), there is no evidence that Babylonian astronomers ever attempted to seek underlying explanations for the movements of the Sun, Moon, and planets[43]. There is also little reason to believe that there were any important theoretical advances after the computational schemes had reached the levels seen in cuneiform texts from around 300 BC. Once predictions had become as accurate as observations, there was no real motive for further investigation. Hence the Babylonian astronomy that eventually passed into the hands of Greek astronomers was more a skill than a science: the quality of predictions was excellent, but the practitioners by this time probably knew very little about the origins of the theories and observation upon which their schemes rested[44].

Nevertheless, the development of astronomy in ancient Mesopotamia was one of the most significant events in human history. The desire to predict future events is probably as old as *Homo sapiens*, but hitherto such predictions had relied on divination, interpretation of omens, and other supernatural means. Now this had changed. For the first time, it was possible to use purely scientific methods to predict the occurrence of natural phenomena.

11
Calendars and Clocks of ancient Egypt

A surprising lack of an astronomical tradition

The great civilisation of pharaonic Egypt emerged as a unified state shortly after the rise of the Sumerian city-states of Mesopotamia. Although there is little doubt that the two regions were in contact, the formation of the Egyptian state was a purely African phenomenon, uninfluenced by developments in Mesopotamia. It emerged from Predynastic chiefdoms by around 3000 BC. According to Classical tradition, the unification of Upper and Lower Egypt was achieved by a king named Menes, who founded the 1st Dynasty. The Early Dynastic period (*c.*3000-2686 BC) spans the first two dynasties. It was followed by the Old Kingdom (2686-2160 BC), the Middle Kingdom (2055-1650 BC), and the New Kingdom (1550-1069 BC), which were separated from one another by periods of decline. The Late period (664-332 BC) was characterised by repeated Persian invasions until the arrival of Alexander the Great in 332 BC. The Ptolemaic period lasted until the annexation of Egypt to Rome in 30 BC[1].

Ancient Egypt was a major military power, and it was prominent in such fields as art, literature, architecture, and medicine. It is therefore surprising that it produced very little in the way of an astronomical tradition, certainly not comparable to that of its contemporaries in Mesopotamia. Prior to the Ptolemaic period, when Egypt was open to the influence of the Greek scientific tradition, there is no evidence to suppose that the Egyptians kept detailed records of the movements of the Moon and planets, or of eclipses. There is a complete absence of technical terminology. Egyptian mathematics, while adequate for the needs of everyday life, was far too primitive to enable the development of mathematical treatments of planetary and lunar movements comparable to those of the Babylonians. A 'catalogue of the universe' compiled around 1100 BC by a scribe named Amenope lists only five constellations, including Orion and 'the Foreleg' (Ursa Major). Despite its considerable significance for the Egyptian calendar, Sirius is not mentioned, nor are the planets listed. In the light of this, claims that the Sphere of Eudoxus was an ancient Egyptian star globe (see Chapter 9) look even more dubious. What is of interest, however, is the Egyptian contribution to calendrics and the measurement of time[2,3,4].

Three calendars?

American Egyptologist Richard Parker[5,6,3] believes that the Egyptians used three calendars: two lunar and one civil. From Predynastic times and earlier, Egyptian life and by implication the Egyptian year revolved around the Nile. The annual flooding of the Nile between June and September deposited mineral-rich silt along its banks, upon which Egyptian agriculture depended. The earliest Egyptian calendar was lunar and consisted of three seasons of four months each: *akhet* ('inundation'), *peret* ('emergence'), and *shemu* ('low water or harvest'). Each month was divided into four 'weeks', based on the primary phases, i.e., new, first quarter, full, and third quarter phases. The twelve months were each named for the most important feast days that occurred in them. The first month of the year, the month of the *Thoth* feast, was accordingly named *Thoth*, and the first day of *Thoth*, and of the new year, was *tpy-rnpt*.

Parker[5] has claimed that each month began with the morning after the last appearance of the morning crescent Moon rather than the more common practice of starting from the first appearance of the evening crescent. This, Parker argued, is because the Egyptian day began at dawn rather than sunset. For peoples like the Mesopotamians, whose day began at sunset, a new month was deemed to begin when the evening crescent Moon appeared for the first time after the period of invisibility. But if the day begins at dawn, then the new month is most suitably defined as beginning at the point when the morning crescent Moon fails to appear in the dawn skies. That the Egyptian day did begin at dawn in historical times is well established, and there is no reason to suppose that in earlier times the beginning of the day was ever moved from the evening to the morning. There is nothing in the origin of any of the Egyptian calendrical systems that could have necessitated a shift.

Like all lunisolar calendars, this calendar required the insertion of intercalary months every two or three years to keep it in step with the seasons. Other than the atypical starting point of its month, the original Egyptian lunar calendar would have been undistinguished but for a remarkable coincidence. The annual rising of the Nile occurred soon after the **heliacal rising** of Sirius, associated with the Egyptian goddess Sopdet or Sothis. The reappearance of this bright star in the dawn skies, after a 70-day period of invisibility, thus signalled the coming of the inundation. The first suggestion of a Sirius calendar was made in 1786 by German historian Johann Christoph Gatterer from his studies of Classical texts. Parker[6] believes that the Sirius calendar came into use in Predynastic times. He claimed[5] that an ivory tablet from the 1st Dynasty can be plausibly read as "*Sothis, the opener of the year; the inundation*". If correct, it would confirm that the Sirius calendar was introduced no later than Early Dynastic times. However, the translation of the tablet has been challenged[7].

The heliacal rising typically occurred during *wp-rnpt* ('opener of the year'), the last and normally twelfth month of the year, named for a festival marking the reappearance. If Sirius reappeared during the last eleven days of this month, then an intercalary month known as *Dhwty* was added to the start of the next year, which would therefore be a 'great year' of thirteen lunar months. This ensured that the reappearance of Sirius continued to occur during the last month of the year. However, the first day of *Thy* was still considered to be *tpy-rnpt*. Synchronisation with Sirius rather than the Sun meant that the calendar was luni-stellar rather than lunisolar[5,3].

It should be noted that the heliacal rising can be affected by non-astronomical factors. Under typical conditions, Sirius at heliacal rise will be seen when it is about 6° above the horizon while the Sun is about 5° below the horizon, but variations in atmospheric conditions can affect the **extinction angle**. This, together with observations affected by cloudy skies, can result in errors of up to a few days in the reported date of the heliacal rising in any one year[8]. Even under ideal conditions, the date is affected by latitude. For every degree south of 30° N (the approximate latitude of Heliopolis and Memphis), the helical rising of Sirius occurs one day earlier[5]. However, the effects of **precession** were quite small:

between 3500 BC and 500 BC, as observed from 30° N, the heliacal rising slowly shifted from 16 July to 18 July[8].

The last eleven days rule nevertheless kept the lunar calendar in step with the heliacal rising of Sirius and hence the start of the inundation, but it was not suitable for administrative or fiscal use. Egyptian bureaucrats therefore introduced a civil calendar that was entirely separate from recurring astronomical phenomena: a year comprising three seasons each containing four thirty-day months. Each month comprised three 'decads' or 'weeks' of ten days, referred to as 'first', 'middle', and 'last'. The twelve months were preceded by five epagomenal (intercalary) days or 'days upon the year' for a total of 365 days. The count of 365 days was arrived at either by counting the number of days between successive heliacal risings of Sirius or by averaging the number of days in several lunar years. Consulting data on the length of recent lunar years would have yielded an average of 365 days over the course of 25 lunar years[5].

The problem was that this civil year is about a quarter of a day shorter than the tropical year, and it would soon have fallen out of step with the heliacal rising of Sirius. The difference might not have been noticed at first, but after twelve years the civil calendar would have been off by three days. The solution – an additional epagomenal day every fourth year – was not adopted until Roman times; the civil calendar was allowed to drift, only falling in step with the Sothic calendar once every 1,456 tropical years or 1,461 civil years after taking into account the **proper motion** of Sirius. The two calendars were in step in AD 139, so assuming that they were initially also in step then the civil calendar must have originated in either 1317 BC (New Kingdom), or in 2773 BC (Early Dynastic period), or in 4229 BC (Predynastic period). The earliest literary source mentioning a correlation between the two calendars dates from the reign of Middle Kingdom ruler Senusret III (reigned 1870-1831 BC), so the civil calendar must therefore have been in use by that time[9].

It is unlikely that the social and economic complexity of Egypt in 4229 BC would have necessitated a civil calendar, so by elimination we are left with 2773 BC – or are we? Parker[5] suggests that the discrepancy between the tropical and 365-day years, and the solution of adding an extra intercalary day every fourth year, were both so obvious as to suggest that the two calendars were never intended to be linked in the first place. Accordingly, he argues, the new civil calendar was launched on *tpy-rnpt* of the Sothic calendar year (New Year's Day) rather than *wp-rnpt* (the day of the heliacal rising of Sirius). The former can precede the latter by as few as twelve days (when the need for an intercalary lunar month is just avoided) or by as many as 41 (when an intercalation is just necessitated). Starting from the 2773 BC line-up, and given that *wp-rnpt* moves against the civil calendar by 1 day every four years, then Parker believes that the civil calendar was instigated between 2937 BC (2773 BC - 4 × 41 years) and 2821 BC (2773 BC - 4 × 12 years), probably nearer the latter.

The new civil calendar did not replace the Sothic calendar; the two now ran concurrently, with the latter providing an agricultural and festival year. Given the variability of the Sothic calendar, it would have remained broadly in step with the civil calendar for around fifty years. But the two would have continued to drift apart, until after two centuries the first lunar and first civil months of the year did not overlap by a single day.

According to Parker[5,6], a second lunar calendar was introduced around 2500 BC. As with other lunar calendars, it required intercalary months to keep it in step – not with the tropical year but with the civic year of 365 days. In the fourth century BC, this second lunar calendar was divorced from observation of the Moon and adopted a scheme based on a 25-year cycle. The workings of the calendar are described in a text known as the Papyrus Carlsberg 9. The text provides a scheme for determining the civil calendar start dates of the second and fourth lunar months of each season over the course of 25 years. The basis of operation is that 25 Egyptian civil years (9,125 days) are 309 lunar months (9,124.952001 days) to within just over an hour (1 hr 9 min 7.2 sec). The scheme provided for intercalary months in the 1st, 3rd, 6th, 9th, 12th, 14th, 17th, 20, and 23rd years of each cycle. This was based on the rule that whenever the first day of *Thoth* (lunar) would begin before the first day of *Thoth* (civil), an intercalary month was inserted to ensure that the first day of the lunar year did not precede the first day of the civil year.

The Papyrus Carlsberg 9 dates to AD 144 and lists a number of calculated cycles going back to AD 19, but Parker[5] notes that the dates of lunar months given largely relate to the appearance of the evening crescent Moon rather than disappearance of the morning crescent. Calculations show that the scheme would have yielded the disappearance of the morning crescent Moon with great consistency around 357 BC; but around 856 BC the predicted dates were one day ahead of the disappearance of the morning crescent. It is therefore likely that this second lunar calendar was introduced around that time.

Star clocks

One of the texts in the pyramid of Unas (reigned 2375-2345 BC), the ninth and final ruler of the 5th Dynasty, says "*Unas has regulated (or: made clear) the night, Unas has sent on their ways the hours*". From this, it could be inferred that stars were being used to tell the time at night in the twenty-fourth century BC, but 'star clocks' might have come into use five hundred years earlier, soon after the introduction of the civil calendar[3].

The idea behind the star clock is straightforward enough: at any time of the night, at any time of the year, a particular group of stars will be rising. Assuming you know the date and rising times of constellations for that date, you can determine the hour of the night. The Egyptians used the risings of 36 groups of stars termed 'decans' equally spaced at ten-degree intervals along the celestial sphere, in a band parallel to but south of the ecliptic; they were chosen so as to have the same 70-day period of invisibility as Sirius. The heliacal risings of the decans occur at intervals of ten days, i.e., intervals of one decad. The ten-degree separation of the decans means that one rises every 40 minutes. Thus, Decan x will mark the end of the night until the heliacal rising of Decan y, by which time Decan x will be rising one 'hour' (of 40 minutes) before the end of the night. Ten days later, Decan z will rise heliacally; Decan y will mark the penultimate 'hour' of the night, and Decan x will mark the antepenultimate 'hour'[2,9].

In summer in Egypt, only twelve decans rise in darkness; hence night was divided into twelve hours. Star clocks consist of 36 columns of twelve lines each, giving for each decad and each hour of the night the decan that marks the end of that hour. Each decan moves diagonally upwards from one decad to the next, hence the clock is known as a diagonal calendar. After 120 days, it drops out altogether. To take into account the epagomenal days, another twelve intermediate decans were provided. Such tables have been found on the insides of coffin lids, though they were presumably used by the living as well as the dead. At the latitude of Egypt, seasonal variations in the length of day and night are less pronounced than they are at higher latitudes; the system would have been useless in Europe. There was still the problem that the Egyptian year was ¼ day shorter than the tropical year; every forty years the clock would fall a whole decad behind, so the system had to be revised at regular intervals[2,3].

By the time of the New Kingdom rulers Sety I (reigned 1294-1279 BC) and Rameses IV (reigned 1153-1147 BC), this system had been replaced by one defined by when the decans reached **culmination**. The groupings would have required substantial revision: stars rising an hour apart will make a transit of the meridian at a greater or lesser interval, unless they have the same **declination**. Mathematical and astronomical historian Gerald Toomer[2] suggests that by this time the decans were in any case no longer being used for timekeeping.

In the New Kingdom, a new method was devised for telling the time by the stars. It comprises 24 tables, i.e., two for each month. Each table comprises a depiction of a seated figure and a grid of 9 vertical lines intersected by 14 horizontal lines. The seven inner vertical lines are associated with the figure's left shoulder, left ear, heart, right eye, right ear, and right shoulder. Each of the twelve hours of the night is represented by the position of a star in relation to the figure's features, and also marked on the grid – in other words they are a list of transits. The centre line represents the meridian; the other lines are **right ascensions** to the east and west, as seen by an observer sitting opposite the figure and facing south. We do not know whether the figure was a dummy or an actual man[2].

The system was presumably developed from the second decans scheme, but it was unsatisfactory for several reasons. It would have required the observer to sit in exactly the right place to obtain an accurate reading. The replacement of the ten-day interval with a fifteen-day interval was a major drawback. The twelve-hour night had arisen from the minimum of 12 out of 36 decans being visible at night and the division of the year into 36 decads; this was incompatible with a system of only 24 subdivisions as the decads would move through 1½ places in 15 days (albeit these would correspond to the familiar hours of 60 minutes). It was therefore necessary to either introduce new stars or use additional lines to the left and right of the meridian to ensure at least one line-up every (40-minute) hour. The latter was adopted, which must have made observation even more problematic. The dates and culmination times of Sirius in the tables suggest that this system was introduced around 1500 BC. Examples found in the tombs of the New Kingdom pharaohs Ramesses VI (reigned 1154-1147 BC), Ramesses VII (reigned 1136-1129 BC), and Ramesses IX (reigned 1126-1108) were already out of date[2].

The Egyptians never really mastered theoretical astronomy and their astronomical knowledge does not appear to have progressed much beyond the annual heliacal rising cycles of the stars. On the other hand, they were responsible for the concept of a year of fixed length and the day of 24 hours. The latter arose naturally from the night of twelve hours. The day was divided into ten hours, with an hour on either side for twilight. These hours were at first of unequal length, varying seasonally (seasonal hours), but by no later than 1300 BC the concept of a day of 24 hours of equal length had been established[2].

12
Greek philosophers and astronomers

The first philosophers

Unlike Mesopotamia and Egypt, the early civilisations of the Aegean were not 'primary states' but formed as result of interacting through trade with the civilisations of Mesopotamia and Egypt. The Minoan civilisation of Crete emerged from Bronze Age chiefdoms around 1950 BC and remained a major trading empire until its eclipse by the mainland Mycenaeans around 1450 BC. Mycenaean domination was comparatively short-lived: the mainland kingdoms went into decline during the thirteenth century BC, and Mycenaean power was at an end by around 1000 BC. The collapse was part of a broader malaise that affected the whole of the Eastern Mediterranean during the Late Bronze Age. The causes of the collapse are widely debated, but there were probably a number of factors operating including climate change. In the Aegean, the ensuing Greek Dark Age lasted around 250 years; the period of recovery was rather longer than elsewhere. The Archaic period, which marks the end of the protohistoric period in Greece, is traditionally deemed to have begun with the first Olympic Games in 776 BC.

Barring a possible Minoan origin for the Sphere of Eudoxus, we know nothing about the state of astronomical knowledge in Bronze Age Greece. The earliest insight we have comes from the writings of Homer and Hesiod in the eighth and seventh centuries BC, which were limited to references to but a few stars and constellations. In Hesiod's case, his agricultural almanac *Work and Days* provided recommended times for agricultural activities in terms of **heliacal risings** and **settings**, **achronycal risings**, and **cosmical settings** of stars and constellations (see Chapter 9). However, this was about to change. The impact of the ancient Greeks upon the Western intellectual tradition is entirely beyond the scope of this work, but their achievements in the field of astronomy must, in the words of David Dicks[1] "*...rank among the most enduring which even that gifted race has bequeathed to posterity*".

The Pre-Socratics is the general term for the Greek thinkers who were active from the sixth century BC up until and during the time of Socrates in the second half of the fifth century BC. They include the group centred in Miletus, Ionia in the sixth century BC later known as the Ionian School, which included Thales, Anaximander, Heraclitus, and Anaxagoras. Other Pre-Socratics include Parmenides, Xenophanes, and Empedocles. Unfortunately, our knowledge of their work comes from sources that are secondary or tertiary at best, and often considerably further removed from the original. Consequently, little of what is attributed to them is known to be true with any degree of certainty[1,2].

What we do know is that the early Greek philosophers were living in an intellectual climate unlike any that had existed before. They were, for the first time in human history, unfettered by mythological tradition and religious dogma. Instead, their thinking was informed purely by logical reasoning. But their early theories were little more than speculations. They were not scientific insofar as there was no attempt that we are aware of to interpret observational data in terms of their models[1].

In any case, early Greek astronomers were at a disadvantage in comparison to their Babylonian counterparts. The centralised government and stable bureaucracies of ancient Mesopotamia were conducive to the keeping of extensive records of astronomical phenomena. By the time Babylonian astronomers began developing predictive planetary theories, they already had an extensive observational database to test their theories against. The absence of such records and a propensity to philosophising over observation handicapped Greek astronomy[3]. The Babylonians also had a head start in numerical methods which, though purely predictive rather than explanatory, could have been used to test Greek geometrical models. In the event, it was not until around 150 BC that Hipparchus became the first Greek to make extensive use of the observational data and mathematical methods of the Babylonians[4].

Early theories of the Pre-Socratics

One of the Pre-Socratics main beliefs was in the existence of a fundamental element or first principle (*ἀρχή*) of which everything was ultimately composed. Thales of Miletus (*c.*624-545 BC) believed that the fundamental element was water and that the Earth floated upon it. Heraclitus of Ephesus (*c.*535-475 BC) believed that the fundamental element was fire. To Diogenes of Apollonia (active *c.*430 BC), it was air. Anaximander (*c.*610-546 BC) also believed in a fundamental substance, but that it was not any known element – otherwise it would 'conquer' the others. Empedocles (*c.*494-434 BC) argued that there were four fundamental elements: earth, air, fire, and water. Anaxagoras of Athens (*c.*510-428 BC) suggested that the number of fundamental elements is infinite. Finally, Democritus (*c.*460-370 BC) proposed the essentially modern paradigm that the universe is made up of atoms.

The modern view of a spherical Earth likely began with Pythagoras (*c.*570-495 BC) or at any rate the Pythagorean school, although this is uncertain. When Socrates (470-399 BC) was in his youth, the matter was still apparently being debated, although by the end of the fifth century BC no Greek writer of repute believed in a flat Earth (see Chapter 8). Philolaus (*c.*470-385 BC), a member of the Pythagorean school, should be credited for the first non-**geocentric** model of the Solar System: he proposed a central fire around which the (presumably spherical) Earth moves, screened from it by an unobserved body known as *Antichthon* or counter-Earth. Outside the Earth's orbit are, in order, the Moon, the Sun, the visible planets, and finally the fixed stars. The Earth orbits the central fire in a day (night falls when the Earth and the Sun are on opposite sides of the central fire); the Moon orbits in 29½ days; and the Sun in a year. With so many large celestial bodies moving around the central fire at great speed, it is inevitable that their motions will produce noise. Each body produces a different tone, according to its distance from the central fire. We do not hear the resulting 'harmony of the spheres' because we are all exposed to it from birth, and hence subconsciously filter it out[5].

The problem with the scheme was, it proposed that the orbits of the Earth and other bodies around the central fire were coplanar; whereas to account for the observed phenomena it would be necessary for them to be inclined with respect to one another and to the Earth's equator. As such, Dicks[1] regards it as a prime example of the issues that bedevilled Pre-Socratic astronomical thinking in that theories were devised with little consideration for whether or not they explained observational evidence.

Nevertheless, in addition to the spherical Earth, many other modern concepts were beginning to emerge. Among the more notable of these, Heraclitus recognised that the stars remain in the skies by day despite being hidden by the glare of the Sun; Parmenides of Elea (512-400 BC) recognised that Hesperus and Eosphoros are evening and morning apparitions of the same planet (Venus), and that Moon shines by reflected light; Empedocles suggested that the Moon circles the Earth and is responsible for solar eclipses, and that the Sun is larger and further away than the Moon. Anaxagoras was the first to give the true explanation of the Moon's phases; he also recognised that the Earth is responsible for lunar eclipses. Curiously, though, he supposedly maintained that the Earth is flat. He also believed that the Sun was "*larger than the Peloponnese*" (which is correct, but is something of an understatement). By the latter part of the fifth century BC, at least some astronomical thought was shifting from speculative theories to a more scientific and observation-driven approach[1,3].

Did Thales predict a solar eclipse?

The best-known achievement attributed to the Pre-Socratics is the eclipse prediction of Thales. According to Herodotus (*c.*484-425 BC), he predicted a solar eclipse which occurred in the middle of a battle between the warring Medes and Lydians, resulting in both sides laying down their weapons and agreeing to a truce. But Thales only predicted the year and not the exact date of the eclipse, and the actual eclipse he predicted is uncertain. That of 28 May 585 BC is usually cited – but is the story even true, and if so, then how did Thales make his prediction? The story has been dismissed as a myth, or at best a fortuitous use of the saros[6]. The problem is that in order to invent a story about a solar eclipse prediction, it is necessary to accept that eclipses are something that can be predicted in the first place rather than non-deterministic acts of the gods. Moreover, Herodotus is not the only historian to have recorded Thales' prediction. Diogenes Laertes (AD 180-240), in his biography *Lives of ancient philosophers* notes that Xenophanes (*c.*570-475 BC) was also aware of it. He also notes testimonies by Heraclitus and Democritus. Diogenes Laertes' information comes from *History of Astronomy*, by Eudemus of Rhodes (*c.*370-300 BC). Eudemus was an early science historian and a pupil of Aristotle (384-322 BC). Given that Aristotle was cautious when it came to the received information on Thales, it is unlikely that Eudemus would have credited Thales with predicting a solar eclipse without being very confident of his sources. That he cited Xenophanes and Heraclitus suggests that he sought independent confirmation of Herodotus' account[7].

All things considered, the story is very likely to be true, leading on to the question of how was the prediction made? The obvious possibility is that Thales was in contact with Mesopotamia, where astronomers were using the saros to predict lunar eclipses no later than 575 BC, and probably rather earlier (see Chapter 10). But the saros is less useful for predicting solar eclipses, and it is likely that Thales used another cycle: either the exeligmos (54 years) or the octon (47 lunar months). The problem is that the eclipse of 28 May 585 BC could not have been predicted by either cycle. Instead, it seems that the eclipse predicted by Thales was either that of 21 September 582 BC or that of 16 March 581 BC. As to why Thales did not predict an exact date, it is possible he felt this was safer given the vagaries of intercalary months in the records of past eclipses. Another possibility is that with two eclipses predictions just six months apart, he decided to bet on at least one of them being correct. Military campaigns were more likely to be fought at the end of the summer than late winter, so the 21 September eclipse is the likelier of the two. However, it is quite possible that the story of the eclipse occurring during an actual battle was a later embellishment[7].

Thales is commemorated by a 32 km (20 mile) diameter lunar crater, and by the asteroid 6001 Thales.

Meton and Euctemon

By around 450 BC, the Greeks were compiling astronomical and meteorological almanacs known as *parapegmata*, and they continued to do so for around 300 years. They were originally engraved stone or wooden tablets, listing astronomical and meteorological phenomena for each day of the month. Every day had a hole alongside the predictions for that hole; a movable peg was used to mark the current day. 'Parapegmata' means 'to fix alongside'. The Athenian astronomers Meton and Euctemon are frequently cited in early *parapegmata*. The two were active in 432 BC (though we do not know their dates of birth and death). They made observations in Thrace, Macedonia, and the Cyclades, as well as Athens[1,2].

Euctemon is known to have devised a star calendar, which listed the appearances and disappearances of prominent stars over the course of the year. In this, he was likely attempting to replace or at very least improve upon the Greek civil calendars. Every city-state had its own calendar, complete with its own month names. The start of the year varied from city to city, and no two cities followed the same practices for periodically intercalating a thirteenth month. Unlike the Mesopotamians, the Greeks never did instigate a regular scheme of intercalation. Hence it was impossible to communicate a time of year between cities by giving a day and month. By contrast, a stellar phenomenon such as the achronycal rising of Arcturus was unambiguous[3].

Meton is known for the eponymous Metonic cycle, although the same cycle was used in Mesopotamia from 499 BC (see Chapter 10). We do not know if Meton discovered the cycle independently, or simply learned of it from the Mesopotamians. As implemented by Meton, the cycle contained 235 lunar months, of which seven were intercalary. Of these, 110 were 'hollow' months of 29 days and 125 were 'full' months of 30 days, for a total of 6,940 days. If the relationship is assumed to be exact, the mean **synodic month** is 6,940/235 = $29\ ^{25}/_{47}$ or 29.531915 days (just under two minutes too long) and the **tropical year** is 6,940/19 = $365\ ^{5}/_{19}$ or 365.263158 days (just over half-an-hour too long). Just over a century later, Callippus of Cyzicus (*c.*370-300 BC) proposed a 76 year/940 lunar month cycle consisting of four Metonic cycles less one day, i.e., 27,759 days. This corresponds to a year of exactly 365¼ days (about eleven minutes too long) and a lunar month of 29.530851 days (22 seconds too long)[8]. However, it is possible that the Callippic cycle was devised with the primary intention of setting the tropical year of the Metonic cycle to exactly 365¼ days. If so, then the improved accuracy of the synodic month was purely fortuitous[9]. Regardless, astronomers subsequently used the Callippic cycle for calendrical purposes. The first year of the Callippic cycle began at the summer solstice of 330 BC.

Notwithstanding claims for observations in the Neolithic (see Chapters 4-6) or even Upper Palaeolithic (see Chapter 3), Meton and Euctemon are the earliest names associated with observations of the equinoxes. Euctemon determined the lengths of the season, starting from the spring equinox, to be 93 days (spring), 90 days (summer), 90 days (autumn), and 92 days (winter). The present-day figures are 94.1 days, 92.2 days, 88.6 days, and 90.4 days, so Euctemon's results were not particularly accurate. However, they do demonstrate that he was aware that the seasons are of unequal length, and of the non-uniformity of the Sun's annual motion along the **ecliptic**[8,1].

The concept of the equinoxes is far more sophisticated than a simple knowledge (as opposed to a scientific understanding) of the solstices. Awareness of the latter, as we have seen, predates Neolithic times. Present-day hunter-gatherer groups are widely aware of the northerly and southerly limits of the Sun's rising and setting points; but these groups show little awareness of the equinoxes (see Chapter 1). Unlike with the solstices, the equinoxes are not marked by a reversal of solar movements, nor by maxima or minima of shadow lengths at noon. Dicks[8] argues that a scientific understanding of the equinoxes implies an understanding of the spherical Earth; the celestial sphere with equator and **tropics**; the ecliptic as a great circle traversed by the Sun and inclined to the equator; that the plane of the equator is inclined to the horizon; and that the **meridian** is a plane passing through the observer's **zenith** and the **celestial pole**, intersecting the horizon at right angles. Barring claims by Alexander Thom (see Chapter 5) and Michael Ovenden (see Chapter 9), there is no evidence that such concepts were understood prior to the latter part of the fifth century BC.

Dicks[8] goes on to claim that unlike the solstices, the equinoxes cannot be determined purely by observation, or at least not with anything available to the ancient Greeks. This is not entirely true; both may be determined without recourse to calculation, but only to the nearest day. It would simply be a matter of determining the days on which the Sun rises and sets midway between its northerly and southerly limits, i.e., when it rises due east and sets due west. The only requirements would be clear horizons. Due east and west are simply the midpoints between north as defined by the north celestial pole and south as defined by the direction of the Sun at noon. Nevertheless, Dicks is correct to assert that a full understanding of these phenomena corresponded to a mathematically advanced stage of astronomical thought of the type that emerged in the last decades of the fifth century BC.

Meton is commemorated by a lunar plain comprising several merged craters. To the east of this feature, a crater has been named for Euctemon. Callippus also has a lunar crater named for him.

Determining the lengths of the seasons

As noted, Euctemon's lengths of the seasons were not very accurate. Around 330 BC, Callippus obtained values of 94 days (spring), 92 days (summer), 89 days (autumn), and 90 days (winter); these are correct to the nearest day. By Hipparchus's time, it was well known that the length of the year was very close to 365¼ days, and equinoxes and solstices were being recorded to the nearest quarter of a day: morning, midday, evening, or midnight. Thus, if the winter solstice occurred at midday one year, in the next year it would occur in the evening[2]. How might such precision have been obtained in an era before it was possible to determine with any great accuracy the moments when the Sun was at maximum and minimum **declinations**, or when it was on the **celestial equator** – even assuming that it was even above the horizon at these points?

The *Almagest* describes a means to determine the moment that the Sun crosses the local meridian, and its **altitude.** A stone block is set up with one face lying in a north-south direction, with a quadrant affixed to it. An angular scale is engraved on the quadrant, and a small peg is fixed at the point where the two radii meet. Prior to local noon, the shadow of the peg falls across the scale; but it disappears completely as the Sun crosses the meridian. The scale reading at that instant corresponds to the altitude of the Sun at local noon.

Such an instrument is used to find the declination of the Sun at noon by subtracting the distance from the zenith (90° - altitude) from the observer's latitude. The maximum (summer) or minimum (winter) noon declination will correspond to the days of the summer and winter solstices. To obtain quarter-day precision, it is necessary to find two points, either side of the solstice, when the Sun is at the same declination. The midpoint between the two is the moment of the solstice. If possible, two days are noted – one before (*D1*) and one after (*D2*) the solstice – when the declinations at noon are equal. The solstice will be midway between *D1* and *D2*, either at midday (if there are an even number of days between *D1* and *D2*) or midnight (if this number is odd). If this is not possible, then the noon declination at an arbitrary day (*D1*) some days prior to the solstice is noted. The noon declination will continue to increase (in summer) or decrease (in winter) until the solstice, then it will reverse direction. The last day (*D2*) at which the noon declination is still greater (in summer) or is still less (in winter) than that on *D1* is then noted. On the next day (*D3*), the noon declination will be less (in summer) or greater (in winter) than that on *D1*. Approximate midnight between *D2* and *D3* as the point where the Sun's declination would be the same as that at noon on *D1*. The midpoint from here to noon on *D1* (which will be either morning or evening) is taken to be the point at which the solstice occurs. This method of halving intervals gives a precision of a quarter of a day or six hours. The choice of intervals is important – too close to the solstices and the day by day change is too small to record accurately; too great and the non-uniformity of the Sun's motion will induce errors. The optimal interval is 90 days, i.e., 45 days on either side of the solstice[10,11].

A similar method is used for the equinoxes. The altitude of the Sun at noon is recorded for the days around the equinox as it crosses the celestial equator. If the Sun is on the equator at noon, then the equinox occurs at noon on that day. If there are two readings equidistant from the equator at noon, then the equinox occurs at the midpoint, either noon or midnight. If not, an arbitrary day (*D1*) before the equinox is chosen. Let the distance of the Sun from the equator at noon be x. The last day (*D2*) following the equinox on which the distance of the Sun from the equator at noon is still less than x is noted. At noon on the next day (*D3*), the distance will be greater than x. The point at which the distance of the Sun from the equator was the same as at noon on *D1* is approximated to be at midnight between *D2* and *D3*. The equinox is taken to be at the midpoint between here and noon on *D1* (either morning or evening), again to quarter-day precision. The equinoxes are easier to determine than the solstices because the declination of the Sun is changing more rapidly[10,11].

Eudoxus

We have already encountered the astronomer and mathematician Eudoxus of Cnidus (*c.*390-337 BC). As noted in Chapter 9, his two important treatises, the *Phaenomena* and the *Enoptron*, are now lost. We know of his work only through later authors, including Aristotle (384-322 BC) and Simplicius of Cilicia (*c.*AD 490-560). Eudoxus devised a method of describing the motions of celestial bodies in terms of a series of concentric rotating spheres, with Earth at the centre. The Sun, Moon, and the planets were each fixed to a sphere, while the fixed stars all shared the same, outermost, sphere. The model assumed that all bodies remain at a constant distance from the Earth.

The Moon's motion is described by three spheres. The outermost sphere is pivoted at the celestial poles and rotates from east to west once every twenty-four hours, representing the Moon's **diurnal motion**. The second sphere is pivoted within the first sphere at the poles of the ecliptic and rotates from west to east (i.e., in the opposite direction to the first sphere), representing the longitudinal motion of the Moon through the zodiac once every **sidereal month**. The third sphere is pivoted within the second sphere, with its axis inclined to it. It rotates in the opposite direction to the second sphere, and at a much slower speed. This third sphere represents the cyclical variation of the Moon's **celestial latitude**, and the westward drift of the points of maximum celestial latitude.

Eudoxus was undoubtedly aware of the value of the **obliquity of the ecliptic** (and hence the inclination of the second sphere to the first) was around 24°; and that the maximum celestial latitude of the Moon (and hence the axial inclination of the third sphere) is around 5°. The period of rotation of the third sphere was presumably the time taken for the points of maximum deviation to make a complete circuit of the ecliptic (i.e., the **nodal cycle**). The problem with this model, as described by Simplicius, is that the swings in the Moon's celestial latitude occur over the course of a sidereal month (27.32 days), not the nodal cycle (18.61 years). For this reason, it has been suggested that Simplicius confused the period of the second sphere with that of the third. If the second sphere rotates once every nodal cycle and the third once every **draconic month** (27.21 days), then the motion of the Moon in celestial latitude will be correct[2]. However, Dicks[1] is critical of such interpretations. He doubts whether Eudoxus knew anything about the nodal cycle or distinguished the draconic month from the synodic month. It does seem unlikely, though, that such a gross error in Eudoxus's system could have gone unnoticed[12]. In any case, there are still problems with the model, in that it assumes that the Moon's velocity around the ecliptic and its distance from Earth are both constant. The Babylonians were aware that the Moon's velocity is variable; the apparent diameter of the Moon at **perigee** is 14 percent greater than when it is at **apogee**, and this difference is detectable[1].

The Sun's motion is also described by three spheres: the first two of which fulfilled the same rules as their lunar counterparts, except that the rotational period of the second sphere was a year rather than a month. The third sphere represents a supposed deviation of the Sun from the ecliptic. Given that the Sun (by definition) does not leave the ecliptic, the third sphere is redundant. The difference between the apparent diameter of the Sun at perihelion and at aphelion is very small, but the assumption that it moved at a constant speed was known to be false. As we have seen, Euctemon had by this time determined that the seasons are of unequal length.

For the five visible planets, Eudoxus had to additionally model the periods of retrograde motion (i.e., **solar anomaly**). Accordingly, he assigned each planet four spheres, to make a total of 26 spheres for the model as a whole (excluding the sphere of the fixed stars). As before, the first sphere represents the diurnal revolution of the heavens and the second sphere represents the longitudinal motion of the planet along the ecliptic. The second sphere is again pivoted within the first, with its axis inclined at an angle of 24°. For an outer planet, this sphere rotates once every **tropical period** of the planet, in a west to east direction. For Mercury and Venus, the periodicity is one year.

The third and fourth spheres (let us call them S_3 and S_4) are intended to cater for **retrogression.** S_3 is pivoted within the second sphere on an axis perpendicular to the axis of the second sphere (i.e., on the ecliptic). Inside it, S_4 is pivoted to turn on axis inclined to that of S_3; the angle of inclination varies with each planet. S_3 and S_4 both rotate once every **synodic period** of the planet, but they rotate in opposite directions. If the two spheres were coaxial, then their rotations would cancel out; however, if they are not quite coaxial the rotations will not quite cancel out and the planet will describe a figure of eight or lemniscate (Eudoxus termed it a *hippopede*, meaning 'horse fetter') as it travels around the ecliptic, centred longitudinally on it. The up and down motion reflects the planet's orbital inclination, and when the planet moves back along the *hippopede* it is in retrogression.

The length and breadth of the *hippopede* is determined by the angle between the rotational axes of the third and fourth spheres; for example, for Saturn an inclination of 6° gives a *hippopede* length of 12° and a breadth of 18' and for Jupiter an inclination of 13° gives a *hippopede* length of 26° and a breadth of 1°28'. These yield reasonably accurate values of 6° for the retrogression arc and 9' for deviation from the ecliptic for Saturn; and 8° for the retrogression arc and 44' for deviation from the ecliptic for Jupiter. Unfortunately, the model is less successful for the other three planets: Mars and Venus can never become retrograde unless S_3 and S_4 rotate in the same direction; for Mercury, the retrogression arc and deviation from the ecliptic are about half the correct values[1]. As with the Moon, the model fails to explain the clearly visible variations in brightness of the planets (up to 25 times in the case of Mars) due to their constantly varying distance from Earth[2].

Callippus, a student of Eudoxus, later tried to address the some of the issues with the model, in particular the variable speed of the Moon (i.e., **lunar anomaly**) and the inequality of the length of seasons due to the variable speed of the Sun (i.e., **zodiacal anomaly**). Callippus added two extra spheres each for the Sun and the Moon, and one extra sphere each for Mercury, Venus, and Mars, to make a total of 33 spheres for the model as a whole[8].

Eudoxus of Cnidus has craters on Mars and the Moon named for him, in addition to the asteroid 11709 Eudoxus and the African butterfly *Charaxes eudoxus*. Callippus of Cyzicus has had to make do with a small lunar crater.

Aristotle

Aristotle (384-322 BC) is one of the most pivotal figures in the history of Western science and philosophy. Bertrand Russell[13] notes that Aristotle's *Phusike* ('Physics') and *De Caelo* ('On the Heavens') "*dominated science until the time of Galileo*". However, the term 'science' as we now know it did not come into use until the nineteenth century. From Aristotle's time onwards, the study of nature was referred to as natural philosophy, and its practitioners were referred to as natural philosophers rather than scientists. Natural philosophy, as its name implies, was considered a subdiscipline of philosophy. It is considered to be the forerunner of natural science, a discipline that encompasses subjects including physics, chemistry, geology, astronomy, and biology. These subjects were rarely studied in isolation, either from each other or from mathematics.

Like his mentor Plato, Aristotle was not primarily an astronomer and his view on the subject have come from largely incidental references in his works. The *De Caelo* is not an astronomical text as such, despite the title. Instead, it is a philosophical discussion of cosmological principles, intended to show how astronomical thought can be accommodated within Aristotle's world view. In the *De Caelo*, Aristotle accepts the standard view that all terrestrial matter is made up of the four elements of earth, air, fire, and water, but he postulates a fifth element, aether or the 'first body'. Aether is an invisible and frictionless substance, and it is the basic constituent of the celestial regions. In his treatise *Physics*, Aristotle states that the natural motion of aether is circular and eternal, as evidenced by the circular movement of the heavenly bodies. The natural motion of the terrestrial elements is either to or from the centre of the cosmos depending on their heaviness; and heavier objects will fall faster than lighter ones. Objects made of earth move naturally towards the centre, as may be seen by dropping any piece of earth. It follows that Earth (the planet) must be at the centre of the cosmos, because the particles of which it is composed all strive to reach the centre. Fire, on the other hand, is lighter than earth and hence moves upwards, away from the centre. Aristotle also distinguishes between such natural motions and 'violent' or forced motions. The forced motion of a body is linear, but it will gradually come to a halt if left alone. Aristotle believed that the Earth was stationary and that the heavens rotate around it. Any movement of the Earth would have resulted from a forced motion, and its effects would therefore be temporary[1,3].

The concept of regular circular motion was a paradigm from which the ancient Greeks never managed to escape. The origin of this view is unclear, with later workers attributing it to the Pythagorean school or Plato. Nor does the surviving literature does not offer a definition of 'regular'; though in the theories of Eudoxus and Hipparchus, it always involved motion around a circle at a constant speed[2].

As far as we know, Eudoxus and Callippus regarded their system as a purely mathematical description, and they did not believe that the spheres actually existed. However, Aristotle believed that the spheres were real bodies, composed of aether. This introduced a complication into the Eudoxus/Callippus scheme in that the spheres for bodies inwards from Saturn could not directly pivot within the sphere of fixed stars due to the presence of the spheres for other celestial bodies intervening. In *Metaphysics*, Aristotle gets around this problem by introducing 'unrolling' spheres to counteract the motion of all but the first sphere in each set. Thus, for each set, the number of unrolling spheres is one fewer than the spheres which combine to produce the motions; this makes a total of 55 spheres of which 22 are unrolling spheres (the Moon, being the nearest to the Earth, does not require any counteraction)[1,2].

Aristotle is commemorated by the lunar crater Aristoteles, located to the immediate north of Eudoxus. The asteroid 6123 Aristoteles, discovered in 1987, is also named for him. Both take the Classical Greek form of his name.

A heliocentric view

With Aristotle's update to the sphere model, the geocentric view of the Earth at the centre of the universe was well-entrenched, and indeed it held sway until the sixteenth century AD. It did not, however, remain unchallenged. We have already seen that Philolaus suggested that the Earth might not lie at the centre of the universe, albeit revolving around a hidden 'central fire' rather than around the Sun. Heraclides Ponticus (*c.*390-310 BC) might have attended some of Aristotle's lectures, but seems not to have accepted his conclusions as he was one of the first to suggest that the stars are fixed and that the Earth rotates on a central axis in a west to east direction[1]. Unfortunately, none of Heraclides' scientific writings have survived, but his claims are attested by two much later writers, Aëtius (1st-2nd century AD) and Simplicius of Cilicia[3]. Some decades later, Aristarchus of Samos (*c.*310-230 BC) proposed the first **heliocentric** view of the Solar System with the Sun at the centre and the Earth and other planets revolving around it in circular orbits. On this view, only the Moon continued to circle the Earth.

A major problem for the heliocentric theory was the apparent lack of any stellar **parallax.** Over the course of a year, the positions of the closer stars should move against the background of the more distant stars as the Earth orbits the Sun and the viewing angle changes. The Earth could not be moving around the Sun unless even the nearest stars were immeasurably further away than the Sun. This is of course the correct explanation. The effects of stellar parallax are way beyond naked-eye visibility; not until 1838 did the German astronomer and mathematician Fredrich Bessel use the method to determine the distance to the star 61 Cygni with reasonable accuracy.

Another problem, at least from the Aristotelian point of view, was the assumption that any motion must be directly perceptible to the senses. If the Earth was rotating and moving, why was nobody aware of it? The idea of an Earth that rotates on its axis was eventually revived in the fifth century AD by the Indian astronomer Āryabhata (see Chapter 13). Aristarchus's heliocentric model was forgotten until the sixteenth century AD. In 1651, the Italian astronomer Giovanni Battista Riccioli named a 40 km (25 miles) diameter lunar crater in the Oceanus Procellarum for Aristarchus. He is also commemorated by the asteroid 3999 Aristarchus.

Lunar and solar distances

Aristarchus examined the relationship between the Earth, Moon, and Sun in a work entitled *On the Sizes and Distances of the Sun and Moon*, which is his only surviving work and indeed the earliest surviving astronomical treatise from ancient Greece. Aristarchus believed that astronomy should follow geometry in being based upon simple axioms that could be taken for granted. He listed six axioms, as follows:

1. The Moon is illuminated by the Sun;
2. The Moon circles the Earth;
3. At half Moon, the great circle dividing the illuminated and dark portions points straight towards Earth;
4. When its visible disk is halved (i.e., at dichotomy), the Moon's angular distance from

the Sun is a thirtieth of a right angle less than a right angle (i.e., $^{29}/_{30} \times 90° = 87°$);

5. The width of the Earth's shadow (as seen during a lunar eclipse) is twice the width of the Moon;
6. The apparent size of the Moon is a fifteenth of a sign of the zodiac (i.e., $^{1}/_{15} \times 360/12° = 2°$).

Aristarchus deduced that:

1. From axiom 4, the Sun is 18 to 20 times further from the Earth than is the Moon;
2. From the unstated assumption that as seen from Earth, the Sun and the Moon appear to be the same diameter, the Sun's diameter is greater than that of the Moon by the same factor;
3. From axioms 5 and 6, the diameter of the Sun is 6⅓ to 7⅙ greater than that of the Earth.

Aristarchus was working before the invention of trigonometry, which would have considerably simplified his calculations. At dichotomy, from axiom 3, the Earth (E), Moon (M), and Sun (S) form a right-angled triangle. From axiom 4, the angle $\angle SEM = 87°$ and the angle $\angle ESM$ must be 3°.

The distance $EM = ES \sin(\angle ESM) = ES \sin(3°)$.

Hence, $ES/EM = 1/\sin(3°) = 19.1$.

i.e., the Sun is 19.1 times further away than the Moon; and hence its diameter is 19.1 lunar diameters.

The absolute sizes can be calculated from axioms 5 and 6. The angle of **parallax** (P_M) between the Moon (M) as seen on the horizon by observer (A) and by a hypothetical observer (O) at the centre of the earth is given by $\sin(P_M) = r/OM$, where r is the radius of the Earth. For the Sun (S), the relationship is $\sin(P_S) = r/OS$. Since OS is 19.1 times greater than OM, P_M will be 19.1 times greater than P_S.

The angular radius of Earth's shadow is τ. From axiom 5, the diameter of the shadow is two lunar diameters; hence τ is one lunar diameter, which from axiom 6 is 2°.

The angular radius of the Sun is σ; given that the angular diameters of the Sun and the moon are the same, $\sigma = 1°$.

It can be shown that $\sigma + \tau = P_S + P_M$ (See Evans[3] for proof).

Hence $\sigma + \tau = (19.1 + 1)\, P_S = 20.1\, P_S$;

$P_S = (1 + 2)°/20.1 = 0.15°$ and $P_M = 19.1 \times 0.15° = 2.85°$.

But $\sin(P_S) = r/OS$ and $\sin(P_M) = r/OM$;

$OS = r/\sin(P_S)$ and $OM = r/\sin(P_M)$.

Hence $OS = r/\sin(0.15) = 382$ Earth radii and $OM = r/\sin(2.85) = 20.1$ Earth radii.

Diameter of Sun $= 2\,OS \sin(\sigma) = 2 \times 382 \sin(1) = 13.3\, r$ or 6.65 Earth diameters.

Diameter of Moon $= 2\,OM \sin(\sigma) = 2 \times 20.1 \sin(1) = 0.70\, r$ or 0.35 Earth diameters.

Aristarchus's reasoning and working were sound, but his starting figures were hopelessly inaccurate, as consequently were his results. The angular separation of the half Moon from the Sun ($\angle SEM$) is actually less than a right angle by just nine arcmin, way beyond the precision of angular measurements possible at the time; it would also have been difficult to determine the exact moment of lunar dichotomy (see Chapter 10); and measuring the angular distance between the centres of extended objects such as the Sun and Moon is also problematic (and in the case of the Sun, hazardous). More puzzling is the value Aristarchus used for the apparent diameter of the Moon, which is four times the correct value of about half a degree. It is not particularly difficult to obtain a good value; Aristarchus might simply have guessed a figure for the purpose of demonstration. During the third century BC, Greek astronomers placed more emphasis on the method than on the actual result[3].

The Sun is actually 400 times further from the Earth than is the Moon, and hence 400 times its diameter; the mean lunar distance is about 60 Earth radii, and the mean solar distance is 24,500 Earth radii. The Sun's diameter is 110 times greater than that of the Earth, and the Moon's diameter is about 0.273 Earth diameters.

Subsequent attempts to improve on these estimates ran into the same problems, but later astronomers, notably Hipparchus and Ptolemy, abandoned methods relying on lunar dichotomy. Hipparchus produced a work entitled *On sizes and distances*, but unfortunately it has been lost and our knowledge of it is confined to Ptolemy's *Almagest* and a commentary on the *Almagest* by the Greek mathematician Pappus of Alexandria (*c.*AD 290-370). Pappus notes that in Book 1 of *On sizes and distances* Hipparchus attempted to make a direct measurement of the lunar distance by the parallax method. He used a solar eclipse (thought to be that of 14 March 189 BC) that had been total near the Hellespont but only four-fifths of the Sun's diameter was eclipsed at Alexandria. He therefore took one-fifth of the Moon's angular diameter as its parallax between the Hellespont and Alexandria; the Moon's angular diameter he took to be 1/650 of a circle (i.e., 33'14"). Assuming that the Sun's parallax between the two was effectively zero and could be neglected, he obtained a lunar distance of between 71 to 83 Earth radii, with a mean of 77 Earth radii[14,15].

In Book 2, however, Hipparchus made a different calculation. The lunar and solar distances (D_M and D_S), and the angular radii of the Moon (θ) and Earth's shadow (φ) are related by the following equation:

$D_M = D_S / ((\varphi/\theta + 1) D_S \sin(\theta) - 1)$

Using a lunar angular diameter of 360/650 and an Earth shadow of 2½ Moon angular diameters, Hipparchus obtained a lunar distance of between 62 and 72⅔ Earth radii, with a mean of 67⅓ Earth radii. The solar distance was given 490 Earth radii, but it has been suggested that he simply assumed a solar parallax of 7 arcmin as the least perceptible parallax and used this to derive a minimum distance for the Sun (a distance of 490 Earth radii corresponds to a solar parallax of 7 arcmin). If this solar distance is assumed, then the lunar distance comes out as 67⅓ Earth radii. However, Hipparchus's value for the lunar distance was a considerable improvement over that of Aristarchus[15,14].

Ptolemy attempted to measure the Moon's parallax by using his **lunar theory** to calculate its position relative to the centre of the Earth and then comparing the result with direct observations made from Alexandria. From this, together with estimates for the angular diameters of the Sun, Moon, and the Earth's shadow, he obtained values for the mean lunar distance = 59 Earth radii; the mean solar distance = 1,210 Earth radii; the diameter of the Moon = 0.292 Earth diameters; and the diameter of the Sun = 5.5 Earth diameters. He obtained a good result for the lunar distance and a reasonable one for the lunar diameter, but his values for the solar distance and diameter were still hugely underestimated[3].

Wheels within wheels

As with many of the early Greek mathematicians, we know very little about the personal life of Apollonius of Perga (*c.*262-190 BC). Known as 'the Great Geometer', his work was very influential in the history of mathematics. His work *Conics* introduced familiar concepts such as the parabola, ellipse, and hyperbola. He was also responsible for introducing one of the most influential concepts of Greek astronomy: the deferent-and-epicycle model of planetary motion. Intended only to be broadly descriptive, it was gradually refined over the next three centuries into a highly successful quantitative theory[3].

In the basic version of the theory, the motions of a planet (*P*) are represented by a *deferent circle* and an *epicycle*. The deferent circle represents the steady eastwards motion of the planet around the ecliptic, and the epicycle represents its periodic retrogressions. The deferent circle is centred on Earth (*E*), with a point (*K*) moving along it in a prograde direction (from west to east) at a constant speed. The epicycle is a second circle, centred on *K*. The **celestial longitude** of *K* increases at a steady rate and is known as the mean longitude ($\bar{\lambda}$). The planet *P* travels around the epicycle in a prograde direction at a constant speed, completing one revolution every synodic period. For the inner planets, the periodicity of *K* around *E* is one year; for the outer planets, the periodicity of *K* around *E* is the sidereal period of the planet. The difference between $\bar{\lambda}$ and the actual celestial longitude λ of *P* is known as the equation of centre (q). To account for variations in latitude (i.e., departures from the plane of the ecliptic) for the inner planets, the deferent circle is coplanar with the ecliptic, but the epicycle is inclined to it; for the outer planets, the deferent circle is inclined to the plane of the ecliptic, but the epicycle is parallel to it[16,3,17].

The deferent-and-epicycle model was firmly rooted in the concept of uniform circular motion. Successive versions of the deferent-and-epicycle model moved centres of deferents away from the Earth, and then moved centres of uniform motion to points known as equants. All of these refinements were necessary in order to make the models match the observed movements of the celestial bodies.

A pivotal figure

Hipparchus of Nicaea (*c.*190-120 BC) was a Greek astronomer, mathematician, and geographer, who has been described as the greatest astronomer of Classical antiquity. He was instrumental in the development of astronomy as a mathematical science; and he was the first Greek to make use of trigonometry, sexagesimal arithmetic, and to measure angles in degrees.

As we have seen, he worked on the problem of lunar and solar distances and diameters, considerably improving on Aristarchus's results for the Moon. Hipparchus is best known for his discovery of the **precession of the equinoxes**. He obtained accurate values for the tropical and sidereal years, noting that they differed. He speculated that the proper motions of stars might be detectable over a very long period of time. Other work included a star catalogue and theories to explain the motions of the Sun and the Moon.

As a geographer, Hipparchus introduced the concept of a coordinate system, used star observations to determine latitude, and suggested that longitude could be determined by simultaneous observation of solar eclipses (an early attempt to solve the so-called Problem of Longitude). Hipparchus is also thought to have been familiar with a branch of mathematics known as combinatorics.

Despite his pivotal contributions to both astronomy and mathematics, only his *Commentary on the Phaenomena of Eudoxus and Aratus* (see Chapter 9) has survived, and we know of his work largely through the writings of others, notably Ptolemy. Hipparchus lived in Rhodes for much of his life, but very little else is known about his personal life. We do not know if he married, had children, or anything about his physical appearance. Images of him appeared on Nicaean coinage between AD 138 and 253, some 250 years after his death. As with the Chandos portrait of Shakespeare, we have no idea whether or not they were a true likeness.

Hipparchus is commemorated by a 150 km (95 mile) diameter crater on the Moon, a 93 km (58 mile) diameter crater on Mars, and by the asteroid 4000 Hipparchus. In 1989, an astrometric satellite launched by the European Space Agency was given the rather contrived acronym HIPPARCOS (HIgh Precision PARallax COllecting Satellite). During its four-year mission, it obtained high-precision positional data on over 100,000 stars.

Table of Chords

In geometry, a chord of a circle is a straight line connecting two points on its circumference (the diameter is the special case of a chord which passes through the centre of the circle). If the angle subtended by a chord AB from the centre C is θ, then the length l of the chord is given by the equation:

$l = r \operatorname{crd}(\theta)$

where r is the radius of the circle and crd is the chord function.

The angle $\theta/2$ is the angle between the radius AC and the perpendicular bisector of AB; and AC, the half-chord AD, and the perpendicular bisector CD form a right-angled triangle;

$AD = AC \sin(\theta/2)$;

Hence $l/2 = r \sin(\theta/2)$ or $l = 2r \sin(\theta/2)$.

In terms of the more familiar sine function, $\operatorname{crd}(\theta) = 2 \sin(\theta/2)$.

For this reason, the sine function is sometimes referred to as the half-chord function. Chords were extensively used in the early days of trigonometry, predating the three trigonometric functions (sine, cosine, and tangent) and their reciprocals (secant, cosecant, and cotangent) commonly used today. Hipparchus compiled a chord table to assist him in his calculations of planetary orbits. His table divided the circumference of a circle into 21,600 arcminutes (i.e., $360° \times 60$), with entries at 7½° intervals. He is the earliest mathematician known to have used trigonometric tables[18].

The Hipparchan cycle and the length of the tropical year

As we have seen, in Hipparchus's day, the length of the year was known to be very nearly 365¼ days and solstices and equinoxes were recorded to the nearest quarter of a day. As noted, the Callippic year of 365¼ days is about eleven minutes too long. In the *Almagest*, Ptolemy quotes from two of Hipparchus's lost works: *On the Length of the Year* and *On Intercalary Months and Days*. In the first of these works, a value for the length of the year of 365¼ minus 1/300 days is cited. This is an improvement on the Callippic year, albeit still about 6½ minutes too long.

If the length of the year is exactly 365¼ days, then the timings of the solstices and equinoxes will shift forward by a quarter of a day each year; for example, from noon to evening. But if the relationship is not exact, then over a sufficient period of time the discrepancy will show up as the solstices and equinoxes occur either earlier or later than the predicted times. Ptolemy notes that Hipparchus compared his timing of the summer solstice with a timing obtained by Aristarchus 145 years earlier. The interval between them fell half a day short of what it would have been were the year to be exactly 365¼ days long. Half a day every 145 years is a shortfall of one day every 290 years, and hence a year length of 365¼ minus 1/290 days. From this, it is commonly assumed that Hipparchus rounded to 365¼ minus 1/300 days[2].

But is this view correct? Hipparchus was apparently dubious about the supposed quarter-day accuracy of solstice timings. In the worst possible case, simple observational errors could account for the half-day shortfall over 145 years. It is possible that he simply performed the calculation to verify a figure he had obtained by other means not involving observation at all. He was aware of the standard sexagesimal value for the synodic month of 29;31,50,08,20 days as used by Mesopotamian astronomers in System B lunar calculations (see Chapter 10), and had he used this value in the 76 year/940 lunar month Callippic cycle he would have obtained 940 × 29;31,50,08,20 = 27758;45,30,33,20 days ≈ 27,758¾ days, i.e., a quarter of a day shorter than the standard Callippic cycle of 76 years of 365¼ days each or 27,759 days. The correction amounts to one day every 4 × 76 = 304 years or 1/304 days per year. This fraction could have been rounded up to give a tropical year of 365¼ minus 1/300 days[9].

However, there might be a chicken-or-egg situation. In *On Intercalary Months and Days*, Hipparchus set out a lunisolar calendar containing 112 intercalary years and 192 standard years in a 304-year cycle, in which the tropical year of 365¼ minus 1/300 days was either derived or confirmed. Did Hipparchus derive the calendrical cycle from the length of the year, or was it the other way around?

If Hipparchus had considered the Metonic relationship of 19 years = 235 synodic months, then he would have obtained a year of 235 × 29;31,50,08,20 / 19 = 365;14,48,33,35,47 days ≈ 365;14,48 days from the Mesopotamian value for the synodic month. Equating this approximate value for the year to the Callippic cycle, we obtain 76 × 365;14,48 = 27758;44,48 days ≈ 27,758¾ days, i.e., one Callippic cycle less a quarter of a day. If we multiply by four to obtain a whole number of days, we obtain the 304-year Hipparchan cycle of 4 × 940 synodic months = 3,760 synodic months, or 4 × 27,758¾ = 111,035 days as reported in *On Intercalary Months and Days*; from this it follows that the tropical year is 111,035/304 = 365¼ minus 1/304 days, which Hipparchus would then have approximated to 365¼ minus 1/300 days[9].

Hipparchus also considered the possibility that the year might vary in length; his reasoning was that if the length of the month varies, the same might be true for the length of the year. Indeed, the length of the tropical year can vary by up to half an hour; for example, the interval between the spring equinox in 2015 and 2016 was 365 days, 5 hrs, 44 min, 56 sec; the corresponding 2016-17 interval was 365 day, 5 hrs, 58 min, 36 sec; and the 2017-18 interval was 365 days, 5 hrs, 46 min, 41 sec. These differences are due to the gravitational effects of other planets on the Earth's orbit. However, they are undetectable given that the precision of timings available to Hipparchus was only to the nearest quarter of a day. After recording the timing of spring and autumn equinoxes over a 33-year period from 161 BC, he concluded that any variation in the length of the tropical year could not be greater than a quarter of a day[2,3].

Hipparchus and the Sun

Any **solar theory** has to account for the zodiacal anomaly, or irregular movement of the Sun along the ecliptic. It speeds up and slows down gradually; reaches maximum speed at the same time of the year; and the time from minimum to mean speed is less than the time from mean to maximum speed. Hipparchus chose to account for this using only uniform circular motion. He proposed a circular orbit for the Sun (S) where the Earth (E) does not lie at the centre (C). Such an orbit is eccentric in the sense the term is used by engineers to describe a wheel with its rotating axle offset from its centre. It differs from the present-day astronomical sense of the word, which is used to express the degree of ellipticity of an orbit. To calculate the position of the Sun at any time, four basic parameters are required: a 'start position' for the Sun, i.e., its celestial longitude at one particular moment; the length of the tropical year; the solar apogee (A); and the solar eccentricity. The solar apogee is the point of the orbit furthest from Earth; the solar eccentricity (e) is the distance from Earth to the centre of the Sun's orbit (EC) divided by the radius of the Sun's orbit (CS). If a parallelogram is drawn with sides EC, CS, $S\bar{S}$, and $E\bar{S}$, where $CS = E\bar{S}$ and $EC = S\bar{S}$, then the point $\bar{S}$ will revolve around E at the same average speed as the Sun. For this reason, $\bar{S}$ is referred to as the mean Sun[2].

As with a planet, the celestial longitude of the mean Sun is known as the mean longitude ($\bar{\lambda}$), and it increases as a steady rate. The difference between the actual celestial longitude of the Sun λ and $\bar{\lambda}$ is known as the equation of centre or solar equation (q). The value of q depends on the mean Sun's angular distance from apogee, a quantity known as the mean anomaly ($\bar{\alpha}$). If the longitude of the apogee is A, then $\bar{\alpha} = \bar{\lambda} - A$. Similarly, the Sun's actual angular distance from apogee is known as the true anomaly (α), given by $\alpha = \lambda - A$. Thus, $q = \alpha - \bar{\alpha}$.

q is zero when the mean and actual Sun are at apogee or perigee; it is positive from perigee to apogee (real Sun leads mean Sun) and negative from apogee back to perigee (real Sun trails mean Sun). The largest value of q, q_{max}, is known as the maximum solar equation. It occurs when the Sun is 90° away from apogee, i.e., when $\alpha = 90°$ or $270°$:

Hence $\sin (q_{max}) = EC/CS = e$;

$q_{max} = \arcsin (e)$.

For any value of $\bar{\alpha}$, $q = \arcsin (e \sin (\bar{\alpha}) / \sqrt{(1 + 2e \cos (\bar{\alpha}) + e^2)})$.

Provided A and e are known, we can calculate q and hence λ for any value of $\bar{\lambda}$. Using the modern value of 0.0334 for e, we obtain $q_{max} = \arcsin (0.0334) = 1°55'$, meaning that the mean Sun and actual Sun are never as much as two degrees apart. Note that the equation of centre and the maximum solar equation are not equations in the same sense as an algebraic equation, though the etymology is the same: *aequatio* (*Lat.* 'to make equal')[3].

Ptolemy reports that Hipparchus used two intervals to obtain a value for the solar eccentricity: from the spring equinox to the summer solstice (spring), 94½ days; and from the summer solstice to the autumn equinox (summer), 92½ days. From this, assuming that the equinoxes and solstices are spaced at 90° intervals on the ecliptic, it may be determined that the solar eccentricity e is 143/3,438 (= 0.0416), hence the maximum solar equation q_{max}

is 2°23'. The solar apogee is located 24°30' west of the summer solstice, or at celestial longitude 65°30' (the correct value in Hipparchus's time was 66°). The lengths of the other two seasons are autumn 88⅛ days and winter 90⅛ days (see, for example, Thurston[2] for the full details of this calculation) to give a total of 365¼ days.

Of course, the Sun does not move around the Earth in an eccentric circle, but Hipparchus's model is nevertheless extremely effective and yields celestial longitudes that differ from those calculated by modern methods by less than half a degree. Furthermore, the differences are due to inaccuracies in the values of the basic parameters Hipparchus used. If more accurate values are used, the differences drop to no more than one arcmin[2].

Hipparchus and the Moon

For the Moon's movements, Hipparchus employed a similar model to that used for the Sun, but there was one difference. Instead of being fixed, the centre of the eccentric orbit (*C*) moves in a circular path around the Earth (*E*) with a periodicity of one sidereal month. The interval between successive returns of the Moon (*M*), *C*, and *E* to a straight line (*MCE*) is one **anomalistic month**. This may be recast as an epicyclic system similar to that proposed by Apollonius. In this case, the Moon moves in a retrograde (east to west) direction around an epicycle, the centre of which (*C*) in turn circles the Earth (*E*) in a non-eccentric, prograde (west to east) path along a deferent. *C* is sometimes referred to as the 'mean Moon'. The deferent makes one rotation every sidereal month; the epicycle makes one rotation every anomalistic month. The angle $\angle CEM$ is known as the *prosthaphaeresis*: it is the angle that must be added to or subtracted from the longitude of *C* to give longitude of the Moon (or added to or subtracted from the longitude of the mean Moon to give that of the real Moon). The plane containing the epicycle and deferent is inclined at an angle of 5° to the plane of the ecliptic, and the intersection (the lunar nodes) make a slow retrograde motion about the ecliptic in line with the Moon's 18.61-year **nodal cycle**[19,2].

The next step was to obtain the lunar eccentricity for the first model, and the ratio of the radius of the deferent to that of the epicycle for the second model. Given that the two models are equivalent, both values should be the same. A problem is that accurate observations of the Moon are affected by parallax. In Hipparchus's solar and lunar theories, *E* is defined to be the centre of the Earth, not a point on the surface. Hipparchus needed data from three eclipses for his calculations; these can only occur when the Moon is on the ecliptic. He used one trio of eclipses for the eccentric model and another trio for the epicycle model. His results differed significantly. For the eccentric model, the result was 327⅔/3,144 (= 0.1042); for the epicycle model he obtained 247½/3,122½ (= 0.0793). Unfortunately, there is no complete record of the data used by Hipparchus to arrive at these results. Ptolemy suggests that there were small errors in his calculations of the time intervals between the eclipses and the corresponding changes in the Sun's celestial longitude. That the differences were calculated is not in doubt, as the eclipses in question occurred long before Hipparchus was born. Eclipse reports gave only times, not celestial longitudes. However, Ptolemy obtained far more consistent values. Taking the radius of the deferent to be 60, his results were 5;13/60 (= 0.0869) for the eccentric model and 5;14/60 (= 0.0872) for the deferent model. He took 5¼/60 to be the correct value[19,2].

Hipparchus and the Babylonians

The traditional view is that Babylonian astronomy was arithmetical and Greek astronomy was geometrical. As we have seen (Chapter 10), from 500 BC onwards the Babylonians devised elaborate arithmetical sequences such as System A and System B in order to predict the apparent motions of the Sun, Moon, and planets; these models were descriptive rather than explanatory. By contrast, the Greeks sought to explain the cosmos in terms of kinematic models entailing complex circular motions. Prior to the time of Hipparchus, this view was likely to have been correct. Ptolemy's approach relied heavily on numerical methods to set the parameters for his epicycles and deferents, and the *Almagest* makes frequent reference to Hipparchus having also used such methods. That nobody else is similarly credited suggests that it was Hipparchus who instigated this methodology in Greece; there is certainly no evidence that anybody else previously made use of trigonometry or sexagesimal arithmetic, or used degrees, minutes, and seconds for angular measurement. The last two of these were of Mesopotamian origin[4].

Ptolemy lists relationships for the synodic, anomalistic, draconic, and sidereal months that were *"known to the ancients"* . They were (with the exception of the sidereal month) derived from the exeligmos or triple saros, and they feature in Babylonian System A calculations. Ptolemy then goes on to quote 'improved' relationships supposedly obtained by Hipparchus – but the 'improvements' were actually made by the Babylonians. The figure for the synodic month is the same as the Babylonian System B value; the relationship for the draconic month is also taken from System B, and the relationship for the anomalistic month is the same as that for System B multiplied by 17. Analysis of the *Almagest* suggests that Hipparchus also used Babylonian arithmetical schemes to calculate lunar and solar celestial longitudes. Unlike later Greeks, Hipparchus had no earlier, reasonably accurate, geometrical model he could use as a first approximation for determining the parameters for his schemes. Use of the Babylonian schemes greatly facilitated the task[20,4,2].

The discovery of precession

The discovery for which Hipparchus is best known was again reported by Ptolemy. Hipparchus noted that the first magnitude star Spica (α Virginis) lay about 6° west of the **First point of Libra**, whereas observations by Timocharis of Alexandria (*c.*320-260 BC) 150 years earlier put the star at 8° west. Spica lies very close to the ecliptic, and today is to be found about 22° east of the First point of Libra.

To determine the celestial longitude of Spica was not trivial: there is a need for a fixed reference point. Hipparchus probably employed the following method[3]: At any given time, the celestial longitude of Spica is equal to the celestial longitude of the Sun plus the longitudinal distance from the Sun to Spica. The celestial longitude of the Sun may be computed from Hipparchus's solar theory, but to measure an arc from the Sun to Spica is problematic as the latter cannot be seen during the day. Hipparchus's method was to wait for a lunar eclipse, then measure the arc from Spica to the Moon at mid-eclipse. At this point, the Sun and the Moon will be 180° apart, so after making a small correction for the Moon's parallax, the celestial longitude of the Sun, and hence that of Spica can be determined.

To obtain the celestial longitude of Spica in an earlier epoch, Hipparchus would have used historical lunar eclipse data noting the date and time of the eclipse, and the location of Spica relative to the Moon at mid-eclipse. This is the type of data that Timocharis and others recorded, as it was useful for working out a lunar theory. But Hipparchus was able to use it for an entirely different purpose: from the date and time recorded, he calculated the position of the Sun by solar theory, and hence the celestial longitude of Spica as before.

Hipparchus claimed that the solstitial and equinoctial points move westwards with respect to the stars by "*not less than 1/100°*" per year. This is a slight underestimate as the true figure is about one degree every 72 years; but Hipparchus seems to suggest that this figure is merely a minimum value[3]. A problem for him was that even using the method outlined above, the positions obtained for Spica are not very accurate. His two measurements, taken in 145 BC and 134 BC, differ by 15 arcmin; and his values calculated from Timocharis's data for 293 BC and 282 BC differ by 10 arcmin.

It has been noted[21] that some ancient Greek sources ascribed to Hipparchus a value for the length of the year that differs from 365¼ minus 1/300 days. The physician Galen of Pergamon (AD 129-200) claimed the figure is 365¼ plus 1/288 days; Vettius Valens (AD 120-175), a contemporary of Ptolemy, gives a figure of 365¼ plus 1/144 days. This corresponds quite closely to the length of the sidereal year implied by Hipparchus's rate of precession of 1/100° per year. The excess over the length of Hipparchus's tropical year is 1/300 + 1/144 = 111/10,800 days, or 14 min 48 sec. As derived from this rate, the figure is 14 min 36.6 sec. The origin of the year length cited by Vettius Valens is not known. He attributed it to the Babylonians, but it is not supported by any surviving cuneiform tablets. However, if the Babylonians did have a value for the length of the sidereal year, then it is possible that Hipparchus used it to derive his estimate for the rate of precession[9].

Ptolemy determined stellar celestial longitudes by measuring the longitudinal distance between the Sun and the Moon, with both above the horizon, shortly before sunset. As soon as the star became visible, he measured the longitudinal distance between it and the Moon. The celestial longitude of the Sun was again calculated by solar theory; the celestial longitude of the star was given by this amount plus the longitudinal distance from the Sun to the Moon, plus the longitudinal distance from the Moon to the star, plus a correction for the Moon's movement between the two observation, and a correction for the change in the Moon's parallax between the two observations. The method has the advantage of avoiding the necessity for an eclipse. Ptolemy used it to obtain the celestial longitude of Regulus (α Leonis), which he compared with a measurement made by Hipparchus 265 years earlier. He found an eastward shift of 2°40', corresponding to a rate of precession of very nearly 1/100° per year; and he obtained similar shifts for Spica and other prominent stars close to the ecliptic. He also confirmed that the motion was only with respect to the solstitial and equinoctial points; the distances of these stars with respect to one another had not changed since Hipparchus's time[3].

It was therefore clear that the fixed stars were indeed fixed, but that the whole was slowly rotating from west to east. What was less clear was the axis of rotation. Did it pass through the **poles of the ecliptic** or the poles of the equator? By analogy with the easterly motion of the Sun and planets, Hipparchus suggested that the axis was through the poles of the ecliptic. If so, the celestial latitudes of stars would remain fixed, but their declinations would change. Stars near the **First point of Aries** would move northwards (i.e., increase in declination);

those near the First point of Libra would move southwards (i.e., decrease in declination). The greatest changes in declination would be shown by the stars nearest these equinoctial points, whereas those near the solstitial points would show little change. Ptolemy reported that Aldebaran (α Tauri) and the Pleiades, both of which lie close to the First point of Aries, shifted northwards by more than 1°; Spica (α Virginis), Regulus (α Leonis), Arcturus (α Boötis), and Antares (α Scorpii) near the First point of Libra all moves southwards by a similar amount. By contrast, Altair (α Aquilae), which is close to the winter solstitial point, showed a northwards motion of just 0°02'. These results confirmed that the axis of precession is through the poles of the ecliptic[3].

Ptolemy and the *Almagest*

Claudius Ptolemy (*c.*AD 100-170) was an astronomer, mathematician, geographer, and author of the pivotal *Almagest.* He lived in Alexandria, in what by that time was the Roman province of Egypt. He was unrelated to the Ptolemys of Egypt's Ptolemaic dynasty, which had ruled Egypt prior to their defeat at the Battle Actium in 31 BC. Ptolemy's thirteen-volume compendium of the West's astronomical knowledge up to that point runs to 500 pages in its modern translation. The work was originally titled *Mathēmatikē syntaxis* ('Mathematical compendium'). It became known as *Megiste syntaxus* ('Greater compendium'), which in Arabic is *Al majasti.* It is through the *Almagest* that much of our knowledge of the now lost works of Hipparchus, Eudoxus, and others is derived.

In the first part of this work, Ptolemy states that the Earth is spherical and located at the centre of the spherical heavens. It had by now long been accepted that the Earth was spherical, but Ptolemy cited three facts confirming sphericity: the local time recorded for a solar eclipse differs between observers, becoming later as one moves from east to west, implying that the Sun rises later for observers further west; secondly, more northerly observers cannot see stars that can be seen from more southerly latitudes; and thirdly, the disappearance of a ship over the horizon (by Ptolemy's day, merchant ships were large enough for the phenomenon to be clearly observable under good conditions). The Earth must also be very small in comparison to the heavens because of the lack of stellar parallax.

In contrast to these well-reasoned arguments, Ptolemy justified his geocentric view by noting that if the Earth was not at the centre of the heavens, the latter would not be perfectly hemispherical; and that the natural movement of a heavy body is towards the centre, so the Earth would move to the centre if it was not already there. The Earth must be stationary and non-rotating, or there would be considerable disturbances to anything on the surface.

Subsequently, the *Almagest* covers spherical trigonometry, solar and lunar theory, the distances and sizes of the Sun and Moon, precession of the equinoxes, the motions of the planets, and the original 48 of the 88 present-day constellations. In a follow-up work entitled *Planetary Hypotheses*, Ptolemy put forward his cosmological speculations (see below). His *Handy Tables* was an ephemeris - a collection of astronomical tables largely adapted from tables in the *Almagest*, and it was the forerunner of the *zījes* (astronomical handbooks) associated with Arabic astronomy.

Ptolemy's other major work is the *Geographia*, a gazetteer and atlas of the known world of his day. He used a grid system to plot the latitudes and longitudes of around 8,000 cities, mountains, rivers, and other landmarks onto maps. His *Optics* survives only as a poor Arabic translation in which he wrote about the properties of light, including reflection, refraction, and colour. In this work, he also put forward a theory of vision in which the eyes emit rays that are reflected back to enable objects to be perceived. Ptolemy also wrote a four-volume astrological treatise, the *Tetrabiblos* ('four books'), that was as influential in astrology as the *Almagest* was in astronomy; indeed, it remained an important text long after the *Almagest* was superseded by heliocentric models of the Solar System. Ptolemy is commemorated by a 154 km (95 mile) diameter lunar crater, the similarly sized Martian crater Ptolemaeus, and the 5 km (3 mile) diameter asteroid 4001 Ptolemaeus. Sir Patrick Moore owned a black cat named Ptolemy.

Meridian quadrant and equatorial ring

In the *Almagest*, Ptolemy described an instrument known as a meridian quadrant, which is used to determine the moment that the Sun crosses the local meridian, and its altitude at that point. A stone or wood block is set up with one face lying in a north-south direction, with a quadrant affixed to it. An angular scale is engraved on the quadrant, and a small peg is fixed to the corner. The instrument is levelled with the aid of a plumb line suspended from the peg. Prior to local noon, the shadow of the peg falls across the scale; but it disappears completely as the Sun crosses the meridian. The scale reading at that instant corresponds to the altitude of the Sun at local noon. The greatest noontime altitude of the Sun will occur on the day of the summer solstice and the minimum on the day of the winter solstice. The meridian quadrant can also be used to determine the equinoxes, which will occur when the Sun's noontime altitude falls exactly midway between the solstitial extremes.

Readings are taken from the centre of the shadow rather than the tip. This reduces uncertainty due to the indistinct edges of the shadow, although it cannot address the intrinsic difficulty of timing the solstices noted above. The rate of change of the Sun's declination is greatest around the equinoxes, and by using extrapolation, the method could determine the equinoxes to a precision of a quarter of a day. We do not know the size of this instrument but based on later descriptions of similar instruments, the radius of the quadrant was probably 450-900 mm (1 ft 6 in - 3 ft).

Another method of determining the equinoxes is with a specialist instrument known as an equatorial ring. It comprises a metal ring about 450-900 mm (1 ft 6 in - 3 ft) in diameter. The ring is placed in the plane of the equator. During autumn and winter, the Sun will illuminate the ring's lower face but never the upper face; during spring and autumn, it will be the other way around. When the declination of the Sun is zero, i.e., at the equinox, the shadow of the upper part of the ring will fall centrally on the lower part. The advantage of the equatorial ring is that it is not restricted to noon: provided the equinox occurs by day its timing can be determined to much greater precision than by use of the meridian quadrant. The drawback is that the alignment of the ring on the equator must be precisely determined and maintained. Another problem is that refraction will affect the accuracy of readings taken when the Sun is close to the horizon. The meridian quadrant, which only takes readings at noon, does not suffer from this drawback because refraction is then at its minimum.

The Armillary sphere

Ptolemy also described an instrument he termed an *astrolabon*, though it is now usually referred to as an armillary sphere and is unrelated to the medieval astrolabe. The Latin word *armilla* means 'bracelet'. An armillary sphere is a model of the sky consisting of a series of concentric rings, which are pivoted so as to be able to rotate freely within one another. The outermost ring (representing the meridian) is either fixed or (if the sphere is intended to be portable) adjustable so that its pivots are aligned on the celestial poles. Within, other rings represent the horizon, the celestial equator, the ecliptic, and the tropics, though many spheres implement only some of these details.

Generally speaking, spheres intended for demonstration purposes have more rings than those intended to make actual measurements of the positions of astronomical objects. The latter type will be precisely graduated, with either an equatorial ring (for measurements of **right ascension** and **declination**) or an ecliptic ring (for measurements of **celestial longitude** and **latitude**). Measurements are taken by lining up a sighting tube on the object of interest.

Chinese astronomers made extensive use of armillary spheres (see Chapter 15), but it is not known just how early they came into use in Greece. Ptolemy described a sphere with seven rings; a fixed meridional ring (7) carrying a ring adjustable for geographical latitude (6), within which a further five rings are nested. These may be rotated as a unit to simulate diurnal motion. The innermost ring (1) turns within ring 2 and carries a sighting tube used for measuring the celestial latitudes of stars. Ring 3 represents the ecliptic. Ring 4 represents the **solstitial colure** and is fixed rigidly to ring 3. Rings 2 and 5 may be rotated independently of one another within ring 4 around the pole of the ecliptic. Rings 5, 3, and 2 are used for measuring the celestial longitudes of stars.

To measure the longitude of a star, the sphere is adjusted for latitude and meridional alignment. The ecliptic ring 3 is set using the Sun; it is correctly aligned when the shadow of its upper part falls on its inner surface. Ring 5 is set to the longitude of the Sun along ecliptic ring 3 either by direct observation or by **solar theory**. To make an observation, ring 5 is turned until the shadow of its upper part falls on its inner surface. Next, ring 2 is sighted on the Moon (if the Moon is not up during the day, then the position of the sun must be calculated) and its celestial longitude read off ring 3. After sunset, rings 2 and 5 are used to measure the longitudinal distance between the Moon and the target star.

Having thus determined the absolute celestial longitude of this star, it can be used as a reference star to obtain the longitudes of other stars. First, ring 5 is turned on the ecliptic axis until it is set to the celestial longitude of the reference star along the ecliptic ring 3. The entire inner nest is then rotated until ring 5 lines up on the reference star. The sphere is now correctly set. Ring 2 is now lined up on the target star. The absolute celestial longitude of the target star is read off from ring 3. The sighting tube of ring 1 is then used to measure the celestial latitude of the target star.

The Astrolabe

The astrolabe is an analogue computer intended to provide a working model of the heavens.

The celestial sphere is stereographically projected onto a flat surface; the astrolabe is thus the two-dimensional equivalent of the armillary sphere. The word *astrolabe* means 'star taker'. In its most basic form, it consists of a *mater* ('mother board'), one or more *latitude plates*, and a *rete* (pronounced 'reety'), all of which were typically made from brass. The rim of the mater is engraved with scales graduated in hours (0-24) and degrees (0-360).

Affixed to the mater is a latitude plate, which is specific for the latitude of operation. A complete astrolabe was typically supplied with anything up to a dozen plates, which could be stored in the mater with the plate currently in use at the top. At the centre of the latitude plate is a hole corresponding to the north celestial pole. The plate is engraved with three concentric circles centred on the pole and representing the Tropic of Capricorn, the celestial equator, and the Tropic of Cancer. The zenith is indicated by a point above the pole separated by $(90° - l)$ where l is the latitude for which the plate is intended. The horizon is represented by an arc centred on the zenith, and the meridian by a straight line running down the centre of the plate, passing through the pole and the zenith. The plate is marked with a grid of almucantars (circles marking altitude) and azimuths.

The rete is placed on top of the latitude plate, and is free to rotate around the point marking the north celestial pole. It represents the celestial sphere, and is engraved with an off-centre circle representing the ecliptic and indicators to the brightest stars. The ecliptic ring is marked in degrees and divided into the twelve signs of the zodiac. Around the perimeter of the rete are engraved a scale of right ascension marked in hours and a calendrical scale. The appearance of the sky at any desired date and time may be shown by turning the rete so that the desired date on the calendrical scale lines up with the desired time on the mater's hour scale. Atop the rete and also free to rotate is a rule with a scale marked with declinations. If the rule is turned so that it lines up with the indicator for a particular star, the declination and right ascension may be read off from the scales.

Additional functionality was often provided on the reverse side of the mater; for example, calendrical and zodiac scales for determining the celestial longitude of the Sun on any day of the year, a quadrant and alidade (sighting device) for observing the altitude of a celestial body, a shadow box to determine the altitude of the Sun from gnomon readings, and a series of arcs for determining the **seasonal hour** from the altitude of the Sun[3].

The astrolabe was invented by the ancient Greeks and there is some evidence that stereographic projection was invented by Hipparchus. However, earliest surviving astrolabes are Arabic and date to the ninth and tenth centuries AD. They possessed only the basic functionality described above, without the additional features provided on the reverse side of the mater. Over the next thousand years, supplementary scales and other functions were added to simplify a variety of tasks, although a ninth century astronomer would have had little difficulty in understanding an astrolabe manufactured in the nineteenth century[3].

The Equation of Time

Ptolemy made use of the concept of the mean solar day or the average length of the solar day over the course of a year (i.e., 24 hours). An apparent solar day is defined as the time between successive crossings of the meridian by the Sun, but it is not constant in length. Ptolemy distinguished between the mean and apparent solar days and identified the two causes necessitating a distinction. Firstly, the Earth's orbital speed varies slightly over the course of a year; secondly, the Sun moves along the ecliptic and equal sections of the ecliptic do not cross the meridian in each 24 hours. Only over the course of a year does the solar day average out to the familiar 24 hours.

Consider an imaginary Sun going around the celestial equator at a constant velocity, completing a circuit every tropical year and meeting the real Sun once a year at the spring equinox. This imaginary Sun will be on the meridian at local noon every 24 hours, whereas the real Sun will not. The imaginary Sun is known as the equatorial mean Sun. This should not be confused with the (ecliptic) mean Sun above. The difference between apparent solar time and local mean solar time is known as the equation of time and is defined as the difference between the right ascension of the equatorial mean Sun and that of the actual Sun. It may be expressed as an angular distance or converted to units of time. As with the equation of centre, the equation of time is not an equation in the same sense as an algebraic equation.

Epicycles and cranks

Ptolemy revisited Hipparchus's work on solar and lunar theory. He obtained the same or similar values as Hipparchus for the length of the year (365¼ - 1/300 days), the longitude of solar apogee (65°30'), and solar eccentricity (2½ /60 = 0.0417)[2]. Ptolemy's year length was derived from comparisons of his timing of an autumn equinox with an observation by Hipparchus and his timings of the spring equinox and summer solstice with old observations by Meton and Euctemon. Rather puzzlingly, his timings for the equinoxes are out by more than a day, and the solstice by a day and a half. It has been suggested that Ptolemy did not time the solstices and equinoxes at all but calculated when they would occur on the basis of the 365¼ - 1/300 day-year as obtained by Hipparchus[11].

This would explain why Ptolemy obtained the same result as Hipparchus for the longitude of solar apogee, which by his time was 71°. If he were using Hipparchus's data, he would get Hipparchus's results. We have absolutely no idea why Ptolemy would not have made the observations himself, but it could also explain why his supposed results for the obliquity of the ecliptic agreed with those of Eratosthenes, despite being out by more than a third of a degree (see Chapter 8). The section of the *Almagest* dealing with the Sun concludes with a set of tables with which to find the celestial longitude of the Sun. But his inaccurate value for the length of the year means that the tables are slow by 6½ minutes each year; if you calculate the celestial longitude for a date in Ptolemy's lifetime using his tables, the result will be 1¼° too low, corresponding to an error of 30 hours[2].

Having left Hipparchus's solar theory essentially alone, Ptolemy moved on to consider the lunar theory. He twice recalculated the radius of the lunar epicycle: once using data from ancient eclipses and once using data from three eclipses that he had personally observed. As noted above, he obtained more consistent results than Hipparchus and settled on a value of 5¼ where the radius of the deferent circle is 60. His observations showed that while Hipparchus's model works well for when the Moon is full, it is less successful for the half Moon. The observed *prosthaphaeresis* (see above) at half Moon is about 50 percent greater than that predicted by the theory. To compensate, Ptolemy moved the centre of the deferent away from the centre of the Earth (*E*) to a point *D* that revolves around the Earth at the same speed as the mean Moon (*C*), but in the opposite direction. When the (real) Moon is in conjunction with the mean Sun, *D* is on the line joining them to *E*. *D* revolves around the Earth at such a rate that at the next opposition, it is back on the line from *E* to *C*. At half Moon, *D* is on the opposite side of the Earth to the mean Moon, so *C* is nearer to Earth than it is at full Moon. This 'crank' solution makes the epicycle look larger, and so increases the *prosthaphaeresis* at the half Moon without affecting the value at the full Moon[2]. Unfortunately, it also causes the Moon's distance from Earth to vary from a maximum of 64 $^{10}/_{60}$ Earth radii when new or full to minimum of 33 $^{33}/_{60}$ Earth radii when half; the model thus predicts that the Moon's angular diameter will vary by a factor of almost two, which is clearly not the case[22].

Ptolemy's planetary theory

Ptolemy extended the deferent-and-epicycle theory to describing the movements of the planets. Like the solar theory, the planetary model had to cater for zodiacal anomaly as the planets also move along the ecliptic at a non-uniform speed. In addition, the model had to cater for retrograde motion (i.e., solar anomaly), which does not affect the Sun. Apollonius's simple model as outlined above works well as an approximation, but it fails to account for the variable lengths of the retrogressions of some planets; for example, retrogressions of Mars range between fewer than 60 to more than 80 days in duration. It also fails to account for zodiacal anomaly.

To address these issues, Ptolemy introduced two refinements. The first was an eccentric deferent, moving its centre away from the centre of the Earth (*E*) to a point *D*, similar to the lunar model, although *D* remains fixed rather than revolving around the Earth (Mercury was an exception; here Ptolemy found it necessary to retain the contra-rotating 'crank' mechanism). The second was a new concept known as an equant. Instead of moving at a uniform angular speed as seen from *D*, the centre of the epicycle (*C*) moves at a uniform angular speed as seen from a third centre, *Q*, known as the equant point. *Q* lies at the same distance as *E* from *D*, but on the opposite side[2,3]. The model describes the motions of the planets so accurately that it was used by planetarium projectors such as the Zeiss ZKP2 until the introduction of digital systems in the 1990s[23]. However, it was controversial as the introduction of the equant contravened the ancient Greeks' long-standing view that all celestial motion should be composed of regular circular motion[2].

In addition to these basic mechanics, as described in the *Almagest*, Ptolemy set down his cosmological speculations in another, much shorter, work entitled *Hypotheseis ton planomenon* ('Planetary Hypotheses'). Although deferents and epicycles are all that is required for calculation, he followed Aristotle in his belief that the spheres carrying the heavenly bodies are actual physical entities. The Ptolemaic Solar System is made up of a series of nested spheres, composed of aether. Rather than two-dimensional circles, the deferents and epicycles are the equators of these spheres. Ptolemy also assumed that the cosmos contains no empty space. For each planet, the deferent circle and epicycle just fill a spherical shell; the thickness of which is determined by the eccentricity of the planet's deferent and the radius of its epicycle. The epicycle itself is a solid sphere, rotating inside a recess within the spherical shell. The Ptolemaic system thus combined the deferent and epicycle model with elements of the Eudoxus solid-sphere model[3].

In *Planetary Hypotheses*, Ptolemy described a model with 41 spheres, of which eight are 'movers' that are responsible for the diurnal motion of the Sun, Moon, five planets, and fixed stars. Each 'mover' contains a system of subsidiary spheres: the fixed stars, 1; Saturn, Jupiter, Mars, and Venus, 5 each; the Sun, 1; Mercury, 7; and the Moon, 4. Ptolemy stated two principles of celestial motion: firstly, a sphere will move another sphere lying within it only if they do not share a common axis (aether being frictionless means that only if the inner sphere is off-centre can the rotations of the outer sphere affect it and cause it to be pulled around in circles); secondly, a sphere may be moved by another, larger sphere if it is entirely contained within the latter – for example, an epicycle is entirely enclosed in its deferent shell and a planet in its epicycle. He rejected Aristotle's 'unrolling' spheres. In accordance with the first principle, the 'movers' independently rotate from east to west about the poles of the celestial equator, completing one revolution per day. In accordance with the second principle, they carry their associated deferents, epicycles, and planets along in the direction of the whole. Within the eighth and outermost 'mover', the sphere of the fixed stars rotates very slowly in a west-to-east direction to account for precession[24].

Ptolemy also described a model in which most of the spheres were truncated into thin equatorial sections or 'sawn-off pieces', which he believed would reduce the number required to 29. Other models dispensed with the 'movers' except for the outermost, reducing the counts to 34 and 22, respectively. Since the axis of the outermost 'mover' is shared only with the other 'movers', the latter are redundant. All the subsidiary spheres must by the first principle share the diurnal motion of the outermost 'mover' regardless of the other 'movers'. These reduced models required a sphere of 'loose aether' to fill in the gaps. This sphere extended from the inside of the sphere of fixed stars to the upper atmosphere of Earth. Although Ptolemy perceived these models as actual descriptions of the Solar System, he also intended them to be manufactured as planetariums by instrument makers[24].

Together, the *Almagest* and *Planetary Hypotheses* set out what was the Theory of Everything of the ancient world. A major weakness was that Ptolemy believed that celestial physics was completely separate from terrestrial physics, and thus could not be observed and tested. Ptolemaic astronomy would nevertheless remain largely unchallenged until the Islamic Golden Age seven hundred years later[25].

13
India

The historical context

The first urban society to emerge in the Indian subcontinent was the Harappan civilisation, which flourished between 2600 and 1900 BC in the floodplains of the Indus and the now-dry Ghaggar-Hakra rivers. Centred on the two great urban sites of Harappa and Mohenjo-daro, it extended over a wide area of Pakistan, southern Afghanistan, and northwest India. Though it is less well known to the general public than contemporary civilisations in Mesopotamia (Akkadian empire and the Third Dynasty of Ur) and Egypt (Old Kingdom), it dwarfed them in terms of land area and population. Mesopotamian records suggest trading contacts, although the earliest astronomical records from there postdate the collapse of the Harappan civilisation, and we do not know what if any transmission of astronomical knowledge there was between the two.

Despite efforts over many decades, the writing systems of the Harappans have not been deciphered. Some scholars dispute whether the Harappan script genuinely was a writing system[1], although this is a minority view[2,3]. Even the language spoken remains uncertain, although it is thought to be a Dravidian, Indo-Aryan, or Munda language[4]. Conceivably, many languages were spoken throughout the Harappan world[1]. The Dravidian languages are mainly spoken in southern India and Sri Lanka, though there are speakers in Pakistan, Afghanistan, and Bangladesh. Indo-Aryan is a branch of the Indo-European language family, spoken in Pakistan, northern India, Bangladesh, and Nepal. The Munda languages are a subgroup of the Austroasiatic family and are mainly spoken in eastern India and Bangladesh.

Consequently, the Harappan civilisation remains enigmatic to this day. The collapse of Harappan society around 1900 BC was probably due to a variety of economic and environmental factors, principally the drying-up of the Ghaggar-Hakra, but until we can understand their written records our understanding will remain incomplete.

The period that followed the collapse is known as the Vedic period. The name comes from the Vedas, which are liturgical texts that nevertheless give an indication of what life was like at the time. Neither the Vedas nor their associated texts are astronomical texts, but it is claimed that they encode astronomical knowledge from the Vedic period, or even from the Harappan period. They are written in an ancient form of Sanskrit known as Vedic Sanskrit. Like Latin, Sanskrit is no longer in day-to-day use, but it remains one of the 22 official languages of India. Vedic Sanskrit is descended from proto-Indo-Aryan; the Indo-Aryan languages were brought to the Indian subcontinent from Central Asia by Bronze Age pastoralists whose ultimate origin has been traced to the Pontic-Caspian steppe[5]. For a long time, it was believed that the post-Harappan period was characterised by violent Indo-Aryan invasions, but this hypothesis was abandoned in the 1960s as theories postulating invasions fell out of favour[6].

Early Vedic period society was semi-nomadic; the economy was pastoral with limited agriculture. The later Vedic period from about 1200 BC saw more settled societies, the introduction of iron tools, and from around 600 BC, larger urban societies reappeared[6]. In 520 BC, the Persian king Darius I extended the Achaemenid empire into the northwest of the subcontinent, resulting in the transmission of Babylonian techniques and knowledge to Indian astronomers. In 326 BC, following his defeat of the Achaemenid empire, Alexander the Great brought the region under Macedonian control. In subsequent years, under the Seleucid empire, Greek astronomical knowledge was added to the mix. From the fifth century BC, a number of Indian empires rose and fell, of which the most significant were the Mauryan Empire (325-185 BC) and the Gupta Empire (AD 320-550); the notable astronomers Āryabhata (AD 476-550) and Varāhamihira (AD 505-587) lived during the Gupta period. Brahmagupta (AD 598-668), another important figure, was slightly later. Eventually, the Arab conquests of the eighth century AD brought Arabic astronomy into contact with this fusion of indigenous Indian, Babylonian, and Greek astronomy. The chronological development of Indian astronomy after Harappan times is considered to fall within three periods: Vedic (c.1200-400 BC), post-Vedic (400 BC-AD 400), and Classical and Medieval (AD 400-onwards)[7,8,9].

Harappan stamp seals

Pictorial stamp seals have been found at many Harappan sites. They have also been found as far afield as Mesopotamia, at Akkadian sites dating to around 2300-2000 BC. Usually square or rectangular, measuring 20-30 mm on one side, and made from steatite, the seals depict animals, deities, and geometrical motifs. The animals include single-horned bulls (often described as 'unicorns'), bull-antelope chimeras, and goat-antelope chimeras which presumably formed a part of the Harappan mythological tradition. Real animals are also depicted including short-horned bulls, zebus, rhinoceroses, tigers, buffaloes, hares, elephants, and goats.

The stamp seals are conventionally believed to have served as tokens of clearance through customs or taxation formalities for goods exported or imported through the various trade routes of the Indus Valley. More controversially, they have been cited as evidence that astronomical concerns played a part in Harappan society. It has been suggested that the stamp seals indicate the seasons or even specific months and dates on which perishable agricultural produce was meant to be released for shipment. The animals, both real and mythological, and deities depicted on the seals might have corresponded to the zodiacal groups known as *nakshatras*[8].

One of the most elaborate Harappan seals is No. 2340, found during the course of excavations at Mohenjo-daro by British archaeologist Ernest Mackay between 1926 and 1931. The square seal depicts a large ram with long wavy horns and a human face; a deity with a horned headdress and bangles on both arms, standing within the boughs of a pipal tree and looking down at a kneeling priest. Opposite the priest on the other side of the pipal tree is a small square marked into one half and two quarters, while beneath one of his knees is a sacrificial offering on an altar. Along the left-hand upper margin of the seal is a row of five pictographs, of which the last one (when read from right to left) clearly represents a man. On the back of the ram is an upright fish sign. At the bottom of the scene are seven figures in procession, each wearing a plumed headdress, bangles on both arms, and long skirts.

Although as noted the Harappan script remains undeciphered, the upright fish is thought to represent a star or a planet. In most Dravidian languages, the word for 'fish' is 'meen' or 'min', but there is also a homophone (a word having the same pronunciation but different meaning) meaning 'star' or 'planet' (literally 'to shine'). The fish is implied to be a *rebus* (*Lat.* 'by means of things'). Where two or more words are homophones, if one can be easily represented pictorially then it is used to represent another which cannot. For example, the first person singular ('I') can be represented by a picture of an eye. The rebus is a principle employed in many logosyllabic scripts.

Asterisms are represented by a fish symbol combined with vertical strokes numerically equal to the number of stars involved. Many asterisms are so named; for example, *aru-min* ('six-star') is the Old Tamil name for the Pleiades. Planets are represented by combining strokes with the fish symbol itself. For example, a diagonal stroke across a fish has been interpreted as dividing it in two, or **pacu* in proto-Dravidian. In Old Tamil, *paccai* ('green') is also used to refer to the planet Mercury (the planet is supposedly greenish-yellow in appearance). The fish in seal No. 2340 contains a single downward stroke; this is interpreted as *por-kol* ('golden star') or Jupiter, and tentatively identifies the deity as the proto-Brahma (Brahma is the Hindu creator god). The Dravidian word for pipal is *aracu*, and its derivative *aracan* ('king') also refers to Jupiter. Sanskrit sources also associate Jupiter with the pipal tree, and they refer to Jupiter as Brahma, Prajapati, Vacasapati, Brhaspati, or the composite name of Brahmanaspati[8].

While this interpretation of seal No. 2340 incorporates astronomical symbolism, there are more radical astronomical interpretations. The annual monsoon would have been crucial to the Harappan economy, which was predominantly agrarian. Around 2500 BC, the appearance of Arcturus (α Boötis) on the eastern horizon after sunset would have preceded the monsoon season; after the start of the monsoon a few weeks later, Vega (α Lyrae) would have become prominent. In later years, precession would have shifted the evening appearance of Vega to a point where it presaged the monsoon. It has accordingly been suggested that the Harappans had a similar relationship to Arcturus as the Egyptians did to Sirius; but in later times, the focus shifted to Vega. It has been claimed that seal No. 2340 depicts the northern evening sky as it would have appeared at the onset of the monsoon season. In this interpretation, the seal's various elements are identified with the up to eight first-magnitude stars that would have been visible in the northern sky at that time of the year[10].

A Harappan calendrical observatory?

On the basis of the above, there is good reason to suppose that the Harappan people possessed a degree of astronomical competence, and indeed it would have been surprising if a mature state-level society had not. But is there any more direct evidence of a concern for astronomical matters? A fuller understanding must await decipherment of the Harappan script, but there are possibilities in the archaeological record.

Dholavira is a large Harappan site in the province of Gujarat where, it is claimed, a solar observatory operated[11]. The site is located at a latitude of 23°53'18.98" N, less than half a degree north of the Tropic of Cancer. At noon on the day of the summer solstice, the Sun is almost at the zenith; this might have been a factor in siting the observatory. In common with other large Harappan cities, Dholavira is divided into several functional sectors including a walled citadel and, to its immediate west, a structure referred to as a 'bailey'. The term comes from a type of castle known as 'motte and bailey', which was introduced into England and Wales by the Normans. They featured a heavily fortified keep ('motte') and a more lightly defended enclosure ('bailey'). The Harappan citadels are not thought to have served a military purpose, and the terms 'citadel' and 'bailey' are purely illustrative. Some prefer the more neutral term 'acropolis'[12]. The Dholavira bailey measures 120 × 120 m (395 × 395 ft).

The supposed observatory is located in the bailey. It consists of a plinth on which sits the foundations of what was probably a 13-room structure, which included two circular rooms among its otherwise rectangular rooms. It is unusual in as much as Harappan architects favoured rectangular designs, and most other residential and workshop buildings in Dholavira are strictly rectangular. The circular rooms themselves are distinct from other rooms in the complex: they lack bathing areas and other amenities that are present in adjacent rectangular rooms, and they are too small to be residential areas. The bailey area slopes, rising from the south to the north at a steep angle of nearly 23°30'. The angle of slope corresponds to the latitude of the site. Thus, for an observer at the bottom of the slope, the north celestial pole would be at the top, and only circumpolar stars would be visible.

The structure was built on top of an earlier Harappan structure, suggesting that the entire bailey area was filled in and reconstructed, thereby acquiring its present shape. The motte obscures the view to the west, but flat, featureless horizons are visible without obstruction in other directions. These views would probably also have been clear when the city was occupied, as other buildings on the site were probably only single-storeyed.

The two circular rooms are located in the north and the west of the complex. The northern room is spiral-like in plan: the outer surface of its wall lines up with the inner surface at the room's northernmost point. At this point, a straight wall 0.5 m (1 ft 8 in) in thickness extends 4.0 m (13 ft 1 in) into the room in a north-south direction. A wedge-shaped segment 1.5 m (4 ft 10 in) on two sides, and bounded by the curvature of the room's wall, is situated in the southwestern quadrant of the room. The western room is more or less exactly circular, with an interior diameter of 3.4 m (11 ft 1 in) and a wall thickness of 0.75 m (2 ft 6 in). A straight wall of width 1.3 m (4 ft 3 in) joins from the west; beyond this are two east-west oriented walls, 2.18 m (7 ft 2 in) in length and 4.79 m (15 ft 9 in) apart. While the city of Dholavira is aligned 6° ± 30' from true north, features associated with the two circular rooms point to within ± 30' of the west and the north.

Researchers Mayank Vahia and Srikumar Menon[11] investigated the possible astronomical function of the site. They assumed that the walls of the structure were originally 2.5 m (8 ft 2 in) high. The entrance to the northern circular room was assumed to be via a break in the circular wall where it is penetrated from the north by the north-south facing straight wall. The latter was interpreted as a walkway 0.6 m (2 ft) high. The room was assumed to be covered by a flat roof, with a circular opening 0.50 m (1 ft 8 in) in diameter located directly above the southern end of the walkway.

Vahia and Menon simulated the movement of the Sun on the day of the summer solstice. As expected for the overhead noontime Sun, the roof aperture casts a circle of light that moves down the circular wall in the west and across the floor; at local noon it falls on the southern end of the walkway at local noon; it then continues across the floor and up the eastern portion of the circular wall. Less obviously, when the simulation was run for the winter solstice, the circle of light moves down the northwest part of the circular wall and when it reaches the walkway, its northern edge just touches the bottom edge of the circular wall. In addition, the position of the Sun on the walkway at noon would vary systematically throughout the year, and the room could thus function as a calendrical observatory.

Carrying out the same exercise for the western room, Vahia and Menon assumed that the entry was via a break in the circular wall where the straight wall joins from the west. This wall was again assumed to be a walkway 0.60 m (2 ft) high. A flat roof was again assumed for the structure, with a circular opening 0.50 m (1 ft 8 in) in diameter at the southern extreme of the room.

As expected for the summer solstice simulation, the circle of light cast by the roof aperture moves down the southwest of the circular wall and reaches the walkway at local noon, when the southern edge grazes the bottom of the wall before continuing up its southeast portion. For the winter solstice, the circle of light travels down the northwest part of the circular wall and when it reaches the walkway, its northern edge passes close to the bottom of the circular wall. In addition, the two sections of east-west oriented walls to the west of the west circular room frame the setting Sun at the summer and winter solstices, as seen from the entrance to the room; this means that the shadow of the northern walls touches the northern extremity of the entrance at sunset on the day of the summer solstice and that the shadow of the southern wall touches the southern extremity of the entrance on the day of the winter solstice. Finally, if a marker is laid in a north-south direction it would show the changing position of the Sun at noon during the year, so this room too could serve as a calendrical observatory.

If one accepts the assumptions made by Vahia and Menon regarding the heights of ceilings and the positionings of the openings, then there is a good case for supposing that the Dholavira bailey structure was constructed specifically in response to the solar geometry at the site, and that the two circular rooms in the structure were intended for solar calendrical observations. While no clear justification is put forward for assuming a ceiling height of 2.5 m (8 ft 2 in), the overall astronomical case for the site is strengthened by solstitial markers outside the western room, and the 23°30' incline of the bailey land surface. If we accept this conclusion, then the Dholavira bailey structure is the first Harappan structure to have been identified as being intended for astronomical purposes. Vahia and Menon note that Dholavira was an important commercial centre, where the need to keep track of the passage of time would have been essential. If their interpretation is correct, it is likely that similar observatories await discovery at other major Harappan cities.

Astronomy and the ancient Sanskrit texts

The four Vedas ('wisdom') are the most authoritative of the Hindu sacred Sanskrit texts and considered to be *shruti* ('what is heard'), i.e., the product of divine revelation. The oldest of these is the *Rigveda* ('the wisdom of verses'), containing ten books known as *mandala* ('circles') with 1,028 hymns (*sūkta*) totalling 10,552 verses (*mantra*). Most of the hymns are dedicated to specific deities. The date when the *Rigveda* was compiled is uncertain, but it is thought to postdate the Harappan collapse around 1900 BC and it predates the introduction of iron working around 1200 BC[13]. Many Vedic scholars favour a date around 1500-1300 BC[14]. However, an earlier origin has been suggested. The Sarasvati River, referenced in the *Rigveda*, has not been definitely identified, but one possibility is the dried-up Ghaggar-Hakra. If this is correct, it would push the origin of the *Rigveda* back into Harappan times[15]. Despite its great antiquity, the *Rigveda* is learnt and recited with exactly the same content and sequence throughout India.

The other three Vedas are the *Yajurveda* ('the wisdom of sacrificial texts'), a collection of sacrificial rites in two sections *Krishna* ('dark') and *Shukla* ('bright'); the *Samaveda* ('the wisdom of chants'), comprising the melodies and chants required for sacrificial rites and based almost entirely on the same texts as the *Rigveda*; and the *Atharvaveda* ('the wisdom of the fire priest'), comprising occult formulae and spells for such purposes as obtaining children, prolonging life, and warding off evil. The *Yajurveda* and the *Samaveda* are more recent than the *Rigveda*, but they still predate the introduction of iron working around 1200 BC; the *Atharvaveda* is the most recent, with references to iron and hence no earlier than 1200 BC[13].

Attached to each of the Vedas are the *Brahmanas,* the *Samhitas*, the *Aranyakas*, and the *Upanishads*. The *Brahmanas* are commentaries on rituals, ceremonies, and sacrifices and include the *Aitareya Brahmana* and the *Satapatha Brahmana*; the *Samhitas* are mantras and benedictions; the *Aranyakas* are texts on rituals, ceremonies, sacrifices, and symbolic sacrifices; and the *Upanishads* are texts discussing meditation, philosophy, and spiritual knowledge. The Vedas are also accompanied by the six *Vedangas*, which cover the auxiliary disciplines require for their study. All these ancient texts are written in an ancient form of Sanskrit known as Vedic Sanskrit.

Although many astronomical references are claimed for the Vedic texts, the earliest known dedicated astronomical and calendrical text is one of the six *Vedangas*: the *Vedanga Jyotisha*. It is a brief work that has come down to us in two recensions (revisions): a 36-verse recension associated with the *Rigveda* (the Rk-recension) and a 43-verse recension associated with the *Yajurveda* (the Yajur-recension). The *Rigveda* recension is believed to be the older of the two and is traditionally attributed to an author named Lagadha; 30 of its verses are incorporated into the Yajur-recension, though not in the same order. The Yajur-recension is the version typically used by modern scholars[16,17]. It is believed that what we have now is a relatively late work intended to be taught orally and learned by rote. Accordingly, it was written in verse to aid students in memorising its contents[18].

The date of its true origins is uncertain: various quotations within suggest a period between 1200 and 600 BC[18]. The Rk-recension states that the winter solstice occurs when the Sun is at the beginning of *Dhanishta* (the 23rd *nakshatra*, corresponding to the region from α to δ Delphini), which would have been the case a little before this period, at around 1400 BC[19]. Various parameters and formulae stated in the Rk-recension suggest that the *Vedanga Jyotisha* is based upon astronomical observations in northern India[20], but it has also been suggested that the mathematical astronomy outlined in the Rk-recension is of Mesopotamian origin, and that it was transmitted to India during the Achaemenid occupation of northwest India in the period after 520 BC[16].

By about the end of the fifth century AD there were at least five great texts (*siddhāntas*) and about thirteen minor texts (*siddunthikas*) on astronomy. The astronomer Varāhamihira, who lived about this time, lists five *siddhāntas*: the *Paulisa*, *Romaka*, *Vasishta*, *Surya*, and the *Pitamaha*. He summarised them into his *Pancha siddanthika* ('five texts') and acknowledged that their common source lay in the *Pitamaha Siddanrha* ('system of the grandfathers'). Quotations from *Pitamaha Siddanrha* are very similar to the *Vedanga Jyotisha* and an analysis of the language indicates that the core of this text might date back to much earlier times[18]. The *Surya siddhānta* supposedly described the theory of solar eclipses in great detail, including the need to correct solar and lunar **celestial longitudes** for **parallax**[21]; this implies a considerable sophistication for Indian astronomy at an early date.

The *Puranas* ('ancient texts') are a large corpus of Hindu literature considered to be *smrti* ('that which is remembered'), and thus are less authoritative than the Vedas. They date from around AD 400 and contain encyclopaedic collections of myth, legend, cosmology, and genealogy; and, it is claimed, references to astronomical matters. There are traditionally eighteen major *Puranas*, eighteen lesser *Upa Puranas*, and a large number of *Sthala Puranas* or local *Puranas*; however, there is no definitive list.

Origin of the Hindu calendars

Various lunisolar calendars, collectively known as Hindu calendars, have been in use on the Indian subcontinent since ancient times. Understanding their origins provides an insight into the early history of Indian astronomy. The present-day Hindu lunar calendar is divided into 12 *māsa* (**synodic months**) each of 30 *tithi* (lunar days); the *tithi* is defined as the time for the Moon to move through 12° of celestial longitude with respect to the Sun, i.e., one-thirtieth of a synodic month. The concept of the *tithi* (but not the term) was used by Babylonian astronomers (see Chapter 10), and might have originated in Mesopotamia[7].

Variations in both lunar and solar velocity over the course of a *māsa* mean that the length of a *tithi* is not fixed and varies between around 21 to 26 hours. The *divasa* (solar or civil day), on the other hand, runs from sunrise to sunrise. If a *tithi* starts and ends during the course of one *divasa*, a day is omitted from the calendar; for example, the date may jump from the 5th to the 7th of the month. On the other hand, if a *tithi* encompasses two sunrises, the day number is repeated. In that case, two consecutive days are assigned the same number[22]. There are two traditions regarding the start of the month: the Amanta tradition, where the *māsa* begins with the new Moon; and the Purnimanta tradition, where the *māsa* begins at the full Moon. The Purnimanta tradition is widespread in northern India, whereas Amanta predominates in peninsular and northeast India. Both traditions divide the *māsa* into two *pakshas* of 15 days each: the *Shukla paksha* ('bright') when the Moon is waxing and the *Krishna paksha* ('dark') when it is waning.

There is also a solar calendar of twelve months, each of which is named for and corresponds to the movement of the Sun through a *rāśi* (sign of the zodiac). The solar calendar, as defined, exactly follows the **sidereal year**[23]. The *Vedanga Jyotisha* states that the solar year was taken to be 371 *tithis*, or 12 $^{11}/_{30}$ synodic months; this corresponds to 365.188 days, or 1 hr 20 min too short[24].

The lunar calendar does of course require the periodic insertion of intercalary months to keep it in step with the solar calendar. A *sankrānti* ('transition') occurs when the Sun moves from one *rāśi* (sign of the zodiac) to the next. If there are no *sankrāntis* during the course of a *māsa* (i.e., if the whole of a synodic month falls within the same solar month; more likely at times of low solar velocity), then that month is deemed to be intercalary, and it is named *Adhik māsa* or *Purushottam māsa* ('extra month' or 'Vishnu month'). The *māsa* that would have occurred in the normal course of events is then 'bumped' to the next synodic month. Such intercalations occur every 2.7 years on average. Conversely, if there are two *sankrāntis* within a *māsa* (i.e., if the whole of a solar month falls within the same synodic month, which is possible at times of high solar velocity), then that *māsa* is removed from the calendar and is known as a *Kshaya māsa* ('omitted month'). This is rare, but usually intercalary months occur both shortly before and shortly after a *Kshaya māsa*. Thus, a year with a *Kshaya māsa* will not contain fewer than 12 *māsa*[23,25]. The system of the omitted month is unique to the Hindu calendar, but it was not a feature of early Indian calendars. The earliest reference to *Kshaya māsa* is in the work of the twelfth century AD mathematician Bhāskara[25].

The *Rigveda*, *Brahmanas*, and *Samhitas* all describe a twelve-month year with 360 days, or 720 pairs of day and night. This is very similar to the Mesopotamian 'ideal' calendar (see Chapter 10). One possibility is that this calendar came to India from Mesopotamia; but it could just as well have been an independent Harappan invention. A 360-day year is the natural consequence of a twelve-month year, if a month is taken to be 30 days[26].

The *Vedanga Jyotisha* describes a lunisolar calendar with a five-year cycle of intercalation, in which a *yuga* of five years was taken to be 62 synodic months; or 1,830 days; or 1,860 *tithis*. This corresponds to a synodic month of 29.5161 days, which is 20 minutes too short. The same calendar is also described in later works going up to the first century AD. One month inserted in the middle and one at the end of the cycle. This system is thought to be post-Vedic, but it is very crude and will be out by 3¾ days each cycle. Again, a Mesopotamian origin seems unlikely, because by that time more accurate systems of intercalation were in use there[7,8].

However, any form of lunisolar calendar with a system of intercalation would have been of little use to the semi-nomadic societies of Early Vedic times. Accordingly, it has been argued that they derive from the Harappan period: a functional calendar would have been crucial for administrative purposes in the Harappan cities[10].

The Nakshatras

Another significant feature of Indian astronomy is the division of the zodiac into 27 or 28 asterisms known as *nakshatras*. They mark the daily progress of the Moon from west to east over the course of a **sidereal month** and correspond to the 'lunar mansions' of the Western astrological tradition. The sidereal month is 27.32 days, so the Moon will 'reside' in a different *nakshatra* each night. The 27-*nakshatra* system results in exactly 13°20' of celestial longitude for each asterism, although most present-day astrological systems use the 28-*nakshatra* system. There are also systems in which the *nakshatras* are of unequal size.

Based on their names, it is thought that originally there were only 24 *nakshatras*, corresponding to half-months of the year and implying a solar rather than lunar connection. This is consistent with an emphasis in Vedic texts of splitting cycles into bright and dark halves: the *nychthemeron* consisting of day and night; the *māsa* consisting of the *Shukla paksha* and the *Krishna paksha* (waxing and waning of the Moon as above); and the year consisting of the Sun's auspicious *uttarāyana* ('northern course') from winter solstice to summer solstice, and its ominous *daksināyana* ('southern course') from summer solstice back to winter solstice. Only later were three of the 24 nakshatras split in two, to provide the Moon with a 'lodge' for each night of its sidereal cycle[26]. As known from the *Atharvaveda*, the *nakshatra* solar calendar was based on the Moon at opposition (i.e., full); at this time, the Sun will be in the *nakshatra* 180° away from one that the Moon is in conjunction with. As a seasonal calendar, the *nakshatras* were similar in function to the Babylonian *Three Stars Each* calendar (see Chapter 9) and the Egyptian 'diagonal calendars' (see Chapter 11)[8].

The *nakshatra* system is first referenced, albeit obliquely, in the *Rigveda*. Incomplete listings of the *nakshatras* occur in the *Yajurveda* and *Atharvaveda*, but the first complete listing appears in the *Taittiriya Samhita*, which dates to between 1000 and 700 BC. The *nakshatras* remained a persistent feature of Indian astronomy until well into the Classical period[8]. The Vedic period *nakshatra* listings began with *Krttikā* (the Pleiades), which was close to the **First point of Aries** around 2240 BC. Their heliacal rising at the spring equinox marked the new year. Around AD 80, due to the effects of precession, the listings were revised so they began with *Aśvinī* (β Arietis), which was closest to the First point of Aries between 655 BC and AD 300. *Krttikā* is now the third *nakshatra*[26].

The Babylonian astronomical compendium MUL.APIN contains a list of stars lying in "*the path of the Moon*" which, like the oldest *nakshatra*, starts with the Pleiades (see Chapter 9), and it has long been claimed that the Indians received the *nakshatras* from Babylon[27,28]. However, all that this actually suggests is that the MUL.APIN and the *nakshatras* are both derived from traditions originating at a time when the Pleiades marked the First point of Aries. Furthermore, the zodiac had not yet been fully consolidated in the MUL.APIN, which lists 18 zodiacal groupings; whereas the *nakshatras* were 24 in number at this stage and as noted above, were zodiacal. In addition, there is no evidence for any cuneiform equivalent of the *nakshatras*; nor indeed is there even any evidence for contact between Mesopotamia and Vedic period India[26].

Although the fragmentary early references to the *nakshatras* suggest that the *nakshatra* calendar evolved over time, there is reason to believe that it was already fully formulated at the start of the Vedic period, and that its origin is Harappan. The *Brahmanas* and *Samhitas* of the *Yajurveda* list the nakshatras in the context of building a fire altar. Fire altars exist in many variants in the Vedic tradition. A common type was built from 10,800 bricks in five layers and represents the Vedic deity Prajapati ('lord of creation'), and hence the cosmos and the year of 10,800 *muhūrta*. A *muhūrta* is 48 minutes; there are 30 *muhūrtas* in a day and 10,800 *muhūrtas* in a 360-day year. The *nakshatras* and the new and the full Moon are connected with specific *śarkarā* ('pebbles') laid down in two rows around the centre of the uppermost layer. The building of fire altars is described in the Vedic *Śulvasūtras*. These are mathematical texts that describe in detail the required shape and area of the altars, which in some cases involve combinations of rectangles, triangles, and trapezia; and they imply the existence of an advanced tradition of geometry. There is no reference to such a tradition or to the brick-built fire altars in the *Rigveda*; nor could it have arisen in the relatively short time separating the *Rigveda* from the *Samhitas* of the *Yajurveda*. It is more likely that the geometrical tradition arose with the Harappans, who made extensive use of mudbricks of standard sizes in the construction of their cities[26].

The *nakshatra* tradition could even predate the Harappan civilisation. Aldebaran (α Tauri) and Antares (α Scorpii) are 180° degrees apart, and in the *Taittiriya Samhita* both are referred to as *Rohini*[28]. Aldebaran was closest to the First point of Aries around 3055 BC. The two post-Vedic Sanskrit epics the *Mahābhārata* and the *Ramayana* both refer to a time when *Rohini* was 'the first of the stars', suggesting that Aldebaran and Antares marked the equinoctial points and that the heliacal rising of Aldebaran marked the new year in India[26]. 3055 BC lies towards the end of the Era of Regionalisation, a period which saw the appearance of increasingly complex proto-urban settlements, presaging the full-blown urban settlements of the Harappan civilisation, and characterised by formally planned street layouts enclosed by massive mudbrick walls.

More trouble with the Taurid Complex

The four Vedas are not, in themselves, dedicated astronomical texts and attempts to imply Vedic period astronomical knowledge from them are, in many cases, speculative. Firmly in the realms of speculation is the claim that certain passages in the *Rigveda* may be interpreted as references to comets, meteorites, and destructive meteor storms. Thus, Rigvedic descriptions of *maruts* (storm deities) killing people on Earth are interpreted as meteor storms; the birth of *agni* ('fire', i.e., the Vedic fire god) and the Horse in the sky as comets; and *vritra* (an enveloping serpent) as cometary dust veils blocking out the Sun, or an impact winter. All of these effects could have been caused by bombardment from the Taurid Complex (see Chapter 4)[19].

The problem is that as Emmanuel Velikovsky[29] showed, it is very easy to interpret ancient texts through the prism of supposed cosmic disturbances (see Chapter 4). If you are looking for evidence for coherent catastrophism in such texts, it can readily be found. Take, for example, the opening of the sixth seal as described in the *Book of Revelation*:

Revelation 6:12-17

1. *And I beheld when he had opened the sixth seal, and, lo, there was a great earthquake; and the sun became black as sackcloth of hair, and the moon became as blood;*
2. *And the stars of the heavens fell unto the earth, even as a fig tree casteth her untimely figs, when she is shaken of a mighty wind.*
3. *And the heavens departed as a scroll when it is rolled together; and every mountain and island were moved out of their places.*
4. *And the kings of the earth, and the great men, and the rich men, and the chief captains, and the mighty men, and every bondman, and every free man, hid themselves in the dens and in the rocks of the mountains;*
5. *And said to the mountains and rocks, Fall on us, and hide us from the face of him that sitteth on the throne, and from the wrath of the Lamb:*
6. *For the great day of his wrath is come; and who shall be able to stand?*

This famous passage has conventionally been interpreted as relating to the Siege of Jerusalem by the Romans in AD 70; or to the collapse of the Roman Empire and invasion by the Goths and Vandals between AD 375 and 418; or as an eschatological text about the Second Coming of Jesus – but earthquakes, the Sun turning black, the Moon turning blood red, and the stars of heaven falling to earth could easily be interpreted as a description of a meteor storm and its aftermath, with the atmosphere laden with dust. If the neo-catastrophist/Taurid Complex view should ever be confirmed, then ancient Indian and Biblical texts could be seen in a new light. But for the present I remain sceptical of such interpretations.

Precession

Islamic scholar al-Bīrūnī (AD 973-1050) noted that Hindus claim that the pole star *Dhruva* is in a constellation called Śiśumāra, which resembled a four-footed aquatic animal. He added that the name is similar to the Persian constellation Susumar (the Great Lizard), which is the modern constellation Draco. The word *dhruva* means 'fixed', 'stationary', or 'unchanging', and it occurs many times in the *Rigveda* and other Vedic texts.

The first millennium AD saw mathematical astronomy in India reach its peak – yet none of the *siddhāntas* of that time mention the constellation Śiśumāra. There is a good reason for this: there was no bright star at the north celestial pole during this period and for much of the time, the pole lay roughly equidistant between Polaris and Kochab (β Ursae Minoris). Indeed, there had not been a conspicuous pole star since the period 3200-2400 BC, when the third-magnitude star Thuban (α Draconis) was close to the north celestial pole. Thuban came to with 10 arcmin of the north celestial pole around 2800 BC; it is likely that it was the star referred to as *Dhruva*.

Al-Bīrūnī lived over a millennium after Hipparchus' discovery of precession, but is there anything in the texts he studied indicate that Indian astronomers had an earlier awareness of the phenomenon? By 1900 BC, Thuban had moved 5° away from the north celestial pole, and by 1500 BC it was 8° distant. It would have been obvious by then that *Dhruva* was no longer living up to its name. In the *Maitra Aniya Upanishad*, King Brihadratha Maurya (ruled 187-180 BC) asked a sage named Śākāyanya "*why Dhruva drifts, why the air strings holding the celestial bodies dip*", which may be interpreted as a realisation that the pole star had changed its' position, to the bewilderment of the king[19].

More speculatively, in the *Mahābhārata*, a dialogue is quoted between Indra and Kartikeya (the god of war) which refers to the star *Abhijit* (Vega), which has "*slipped down from the sky*". This has been interpreted to refer to the time when Vega became the pole star, which would have been the case around 12,000 BC. The 'slipping down' refers to the **altitude** of the north celestial pole, which from the latitude of India is only about thirty degrees above the horizon. The implication is that ancient Indian astronomers had already been observing the night skies for thousands of years by this time. But Indra also refers to a time when *Dhanishta* had been the first *nakshatra* (i.e., when it had been at the First point of Aries) and to an even more remote period when *Rohini* (Aldebaran) was at the summer solstitial point. *Dhanishta* was at the First point of Aries around 20,000 BC, and Aldebaran was at the summer solstitial point around 23,000 BC[30].

While not impossible, it seems highly unlikely that an oral tradition could have survived from the later stages of the Upper Palaeolithic. It would also be necessary for these early Indian astronomers to have been aware that the Sun will be 'in' a particular star grouping at any given time of the year, even though the stars themselves cannot be seen. As noted in

Chapter 4, this might not have become common knowledge until after early Mesopotamian times.

Vedic texts and the saros

Could Vedic period texts encode a knowledge of eclipse cycles? There is documentary evidence that the Babylonians were aware of the saros no later than 575 BC (see Chapter 10); even if they were aware of it earlier, this date is much later than the likely composition of the Vedas. If Vedic or Harappan period astronomers were aware of the saros, it was probably an independent discovery. The number 3,339 occurs twice in the *Rigveda*, and it also appears in other texts. It is traditionally interpreted as the number of gods honouring *agni* ('fire', Vedic fire god), but such a large and at the same time specific number suggests a physical significance.

There is a passage in the *Brahmanda Purana* in which 3,339 has been interpreted as a period of time: a number of *tithis* – but counting only during the *Krishna paksha* when the Moon is waning. If the *Shukla paksha* when the Moon is waxing were included, the number would be 6,678 *tithis*. As noted above, the *Vedanga Jyotisha* states that the solar year was taken to be 371 *tithis*, or 12 $^{11}/_{30}$ synodic months; 6,678 *tithis* is exactly 18 solar years, or 222.6 synodic months.

This interval is marginally smaller than the saros of 223 synodic months (or 6,690 *tithis*). It has therefore been suggested that the number 3,339 relates to an eclipse cycle and that the ancient Indians were aware of the saros. It is also claimed that a correction was later made to add the additional *tithis*. The *Brahmanda Purana* states that six or seven *pitrs* (spirits of departed ancestors) consume *soma* (a ritual drink described in the *Rigveda*) on the fifteenth day of certain months governed by a set of rules that suggest a connection to the five-year lunisolar intercalary cycle. Six *pitrs* plus 3,339 fire-honouring gods gives 3,345 *tithis* during the *Krishna paksha* or 6,690 overall; this corresponds to the saros of 223 synodic months. The convenient round number of eighteen years was retained for religious purposes[24]. It would be natural for astronomers to initially assume that eclipses would be governed by an exact number of solar years; hence the number 3,339. Later, when it became clear that the relationship was not exact, the *pitrs* correction was applied.

Āryabhata's *Aryabhatiya*

Āryabhata of Kusumapura was one the foremost astronomers of India's Classical age. He was revered as a teacher, although little else is known about his personal life. The 22 km (14 mile) diameter lunar crater Āryabhata is named for him. He was the author of at least two major works on mathematics and astronomy, but only the *Aryabhatiya* has survived. An earlier work, the *Āryabhata siddhānta* is known only through the writings of others. The *Aryabhatiya* is an improved, more mature work, but Āryabhata was still only 23 when he wrote it. Āryabhata had two slightly different systems. The first, put forward in the *Aryabhatiya*, is known as the Āryapaksa system and it continued to be studied in southern India into the nineteenth century. The other system, the Ārdharātrika, was popular in northern India[31,32].

The *Aryabhatiya* is a concise work of 121 *sutras* (stanzas) of Sanskrit verse containing an introduction (*Dasagitaka*) followed by chapters on mathematics (*Ganita*), the reckoning of time (*Kalacriya*), and the celestial sphere (*Gola*). The work begins with an invocation to Brahma[33].

The *Dasagitaka* outlines a system for expressing numbers by letters of the alphabet. The first 25 consonants of the Sanskrit alphabet represent the numbers from 1 to 25; the next eight represent the decades from 30 to 100; and the nine vowels represent powers of 100. Three large units of time are described: the *kalpa*, the *manvantara*, and the *yuga*. The *yuga* in this context is 4,320,000 years, a period of time also referred to as a *mahāyuga* (great *yuga*, see below; note *mahā* is cognate with the Greek *mega*, now an English loanword). The *Dasagitaka* lists the number of revolutions of the Sun, Moon, and planets in a *yuga*; and the diameters of the Earth, Sun, Moon, and planets in *yojana*. The *yojana* is defined as 8,000 times the height of a man; the diameter of the Earth is 1,050 *yojanas*, the Sun 4,410 *yojanas*, and the Moon 315 *yojanas*. Taking the height of a man in India in the fifth century AD to be 5 ft (1.5 m), the *yojana* is approximately 12 km (7.5 miles). This gives the diameter of the Earth at 12,600 km (7,875 miles), the Sun 52,920 km (33,075 miles), and the Moon 3,780 km (2,362 miles). The figures for the Earth and Moon are reasonably accurate, but the Sun's diameter is more than 26 times larger than Āryabhata's figure[34,33,31].

The *Ganita* covers extraction of square roots and cube roots, fractions, arithmetic progression, sums of squares and cubes, quadratic equations, area and volume finding, geometry, and trigonometry[33]. There are *sutras* describing the construction and use of 'half-chords' (sines). Sines were adopted by nearly all later Indian mathematicians in preference to chords. However, the first use of sines in India occurs in Varāhamihira's *Pancha siddanthika* in what is believed to be a summarisation of the *Paulisa siddhāntas*[35]. It is notable that early Indian sine tables used 225 arcmin (3¾°) intervals. This suggests that they were derived from the Hipparchan chord table, which used 7½° intervals, with the relationship crd $(\theta) = 2 \sin (\theta/2)$[36] (see Chapter 12).

The *Kalakriyapada* includes *sutras* on the **tropical year**, synodic month, **solar** and **sidereal days**, and intercalary months. Astronomical topics include the order in which the planets are arranged, solar, lunar, and planetary theory. Āryabhata's solar and lunar theories were similar to those of Hipparchus and Ptolemy. His planetary theories replaced the Ptolemaic eccentric deferent and equant (see Chapter 12) with two epicycles: the *manda* (slow) and *sighra* (fast). These catered for the zodiacal and solar anomalies respectively[31,37].

The *Gola* considers aspects of the celestial sphere, the ecliptic, and the Earth as a sphere suspended in space, rotating on its axis. There are descriptions of the behaviour of the Sun and Moon at the poles, and the differing appearance of the night sky as one moves from the pole to the equator. The section also goes into the causes of solar and lunar eclipses in some detail, with descriptions of what is to be seen at each stage of an eclipse; procedures are also given for calculating the magnitude and duration of eclipses. Āryabhata was aware that motion is relative; and he believed that the Earth rotates on its axis. The *Aryabhatiya* contains a remarkable passage: "*Just as a man in a boat moving forward sees the trees on the bank move in the opposite direction, just so are the stationary stars seen by the people on earth as moving precisely towards the west*"[34,33,31]. His estimate for the length of the sidereal day was 23 hrs 56 min 4.1 sec, which is very close to the accepted value of 23 hrs 56 min 4.091 sec[32].

Born in AD 598, Brahmagupta lived a century after Āryabhata. He is known for two works: the *Brahmasphuta siddhānta* ('Correctly Established Doctrine of Brahma') and the oddly named *Khandakhadyaka* ('a preparation made with candied sugar'). The *Brahmasphuta siddhānta* is a mathematical treatise written entirely in verse around AD 628. In it, Brahmagupta is critical of Āryabhata's *Aryabhatiya*. However, the *Khandakhadyaka*, written in AD 665, was a reworking of the *Āryabhata siddhānta*. It followed the Ārdharātrika system, which it simplifies and explains. Brahmagupta claimed that it gives results useful in everyday life, birth, marriage, etc. quickly and simply, and is written for the benefit of students. Perhaps the unusual title was meant to suggest that the contents were concise and easy to digest. Unfortunately, Āryabhata's original work was subsequently lost[31,32,38].

Notions of great time depth

Hindu cosmology contains references to very long periods of time. For example, the *Puranas* are said to describe an oscillating universe with a period of 8,640 million years[39]. The figure of 8,640 million years is not far from the current estimate of 13,772 million years for the age of the universe, but it is only a fraction of the life of Brahma which is said to constitute 72,000 *kalpas* or 311,040 billion years[7].

Why such vast, yet precise numbers? The *kalpa* of 4,320 million years contains 1,000 *mahāyugas* of 4,320,000 years each, which in turn comprise four smaller *yugas*. The smallest is the *kaliyuga* of 432,000 years; then the *dwaparayuga* of 864,000 years (2 *kaliyugas*), the *tretayuga* of 1,296,000 years (3 *kaliyugas*), and finally the *satyayunga* of 1,728,000 years (4 *kaliyugas*). One *mahāyuga* is thus ten *kaliyugas* and a *kalpa* is 10,000 *kaliyugas* or 4,320 million years[7]. There is also a *manvantara* consisting of 71 *mahāyugas*, or 306,720,000 years[15].

The base unit of this vast period of time, the *kaliyuga* of 432,000 years, appears to be of Babylonian origin, and this is the key to the precision of these amounts. Written in sexagesimal, 432,000 is a nice, round number: 2,0,0,0. This number is given in the Babylonian histories of Berossos and Abydenus as the length of time that the Babylonian kingdom had existed before the Flood[7]. While we can treat with a degree of scepticism any notion that the *Puranas* contain an actual knowledge of the age of the universe[40], or of the early geological history of the Earth[41], it is nevertheless interesting to contrast these notions of great time depth with seventeenth century cleric James Ussher's estimate that the universe came into being as recently as 4004 BC.

Indian planetary theory

In the late fourth and early fifth centuries AD, Indian astronomers combined the *kalpa* and other large units of time with the Greek theory of epicycles. The motions of the Sun, Moon, and planets may all be described as integer numbers of revolutions within a given period of time provided that period is long enough: at the end of that period, all will return to the same positions they occupied at the start of the period. The *kalpa* and *mahāyuga* are periods of time sufficiently long for this purpose, though a shorter period of $^1/_{24}$ *mahāyuga* or 180,000 years was also employed[7].

If the number of revolutions of a planet within period p is n, then at time t after the start of p, the planet will have completed nt/p revolutions; if the fractional part of this number is *f*, then the planet's **mean longitude** $\bar{\lambda}$ will be 360 *f*. The fixed starting point for this system (and the beginning of the *kaliyuga* or smallest and most recent of the four *yugas* within the *mahāyuga*) was taken to be the date of a grouping of the Sun, Moon, Mercury, Venus, Mars, Jupiter, and Saturn close to the First point of Aries either at midnight on 17-18 February 3102 BC, or at sunrise on 18 February. This conjunction could not have been actually observed because modern calculations show that it never happened: only Saturn would have been visible in the predawn sky, and it was more than 40° away from Jupiter. The starting point must have been obtained by calculating the date of the mean conjunction, i.e., the point at which mean longitudes of all these bodies were exactly or nearly zero. In any case, there were no detailed astronomical records being kept around 3102 BC; the earliest records we have are the Old Babylonian observations of Venus compiled in the time of Ammisaduqa around 1650 BC (see Chapter 10)[42].

The problem was that the period of 180,000 years was too short to permit the use of very accurate parameters. Even the use of the full *mahāyuga* posed difficulties. The mean conjunction of 3102 BC was taken to mark the beginning of the *kaliyuga* of only 432,000 years. But the *mahāyuga* also had to begin with such a conjunction; and it was therefore necessary to fit the parameters into a tenth of a *mahāyuga* so that the conjunction at the start of the *kaliyuga* could be accommodated. Effectively, astronomers were working with a period of $^1/_{10}$ *mahāyuga* rather than $^1/_{24}$ *mahāyuga*, so the gain was only by a factor of 2.4 rather than twenty-four-fold. Āryabhata tried to solve the problem by subdividing the *mahāyuga* into four equal *yugas* of 1,080,000 years, but this redefinition of the *mahāyuga* was unpopular with traditionalists in India[7]. Despite the difficulties of working with such enormous numbers, Brahmagupta and his followers later chose to adopt the *kalpa* of 1,000 *mahāyugas* as the fundamental period[42].

Regardless of the fundamental period chosen, the use of a fixed starting point had the advantage that tables of planetary movements could easily be corrected if a more accurate figure for a planet's orbital period became available; for example, Āryabhata revised the number of revolutions of Jupiter in a *mahāyuga* from 364,220 to 364,224: in Āryabhata's epoch 3,600 years after the start of the *kaliyuga*, this amounted to a correction of 1°12' to be added. The drawback of the method was that the mean longitudes obtained were not very accurate, and it was necessary to add corrective terms known as *binja* every one or two centuries[42]. It should in any case be appreciated that the mean longitude of a planet at any given time does not correspond to the actual position of that planet; it was no more than a starting point for calculations involving epicycles.

Hindu numerals and the invention of the zero

The Hindu numeral system was the basis of our familiar 'Arabic' numerals, which are therefore more correctly referred to as Hindu-Arabic. The numerals are descended graphically from the Brahmi numerals, which are attested from around the middle of the third century BC. However, these predate the invention of place-value notation in India and were not used with a zero. Instead, there were separate symbols for each of the decades 10 to 90, and for 100 and 1,000; these were conjoined with the appropriate numeral to denote 200, 300, 2,000, etc.[43].

The Brahmi numerals gradually evolved through time into a multitude of numeric systems, but during the Gupta period a common set of numerals were used throughout the Gupta Empire. They are known (unsurprisingly) as the Gupta numerals; but from around the seventh century AD these in turn began to evolve into another set of numerals, known as the Nagari or Devanagari numerals. The Nagari numerals were probably transmitted to the Arabs during the reign of Caliph Mansūr (reigned AD 754-775)[44] (see Chapter 14), although the sexagesimal system of the Babylonians remained in widespread use for centuries after. Uptake was also slow in the West, even after Latin translations of Arabic scholarly works reached Europe. The Hindu-Arabic system was eventually popularised by Leonardo Bonacci of Pisa ('Fibonacci') (*c.* AD 1170-1240), who travelled extensively around the Mediterranean world in order to study under the leading Arabic mathematicians of his day[45].

We do not know exactly when a place-value system came into use in India, or when the concept of zero as a numerical operand came into being. The latter is traditionally attributed to Brahmagupta. In the *Brahmasphuta siddhānta*, he introduces the concepts of positive numbers ('fortunes') and negative numbers ('debts'). The zero he defined as the result of subtracting any number from itself. He then set out arithmetic rules for handling zeros, positive, and negative numbers:

A debt minus zero is a debt.
A fortune minus zero is a fortune.
Zero minus zero is zero.
Zero minus a debt is a fortune.
Zero minus a fortune is a debt.
Zero multiplied by a debt or a fortune is zero.
Zero multiplied by zero is zero.
The product or quotient of two fortunes is a fortune.
The product or quotient of two debts is a fortune.
The product or quotient of a debt and a fortune is a debt.
The product or quotient of a fortune and a debt is a debt.

The use of a circle '0' to denote zero was thought to be a later development, first attested on an Indian inscription dating to AD 867[46].

In fact, it is likely that both the place-value system and the numerical zero came into use well before the time of Brahmagupta and go back as least as far as the early centuries AD. Texts of that time are known that describe a decimal place-value system with locations for units, tens, hundreds, and thousands; they suggest that it was clearly understood that a digit could take different values depending on position. Rather earlier, around the second or third century BC, the mathematician Pingala used a symbol known as *śūnya* ('empty' or 'void') in his mathematical treatise *Chanda sastra*. The *śūnya* was later used as one of the standard terms for zero. The earliest definite evidence for the use of zero as a numerical operand has now been dated to around 75 years prior to the *Brahmasphuta siddhānta*. An astronomical work, dating to around AD 550, defines a particular constant as having a value of 'sixty minus zero', i.e., sixty[47].

In 2017, it was claimed that a text known as the Bakhshali manuscript provided evidence that a zero symbol was in use as early as the third or fourth century AD. The Bakhshali manuscript is a mathematical text written on birch bark that was found in 1881 at Bakhshali, near Mardan in Pakistan. The text uses numerals with a place-value system, with a dot as a place holder for zero, but its date is uncertain. The document is kept at the Bodleian Library, Oxford. Three samples were radiocarbon dated 'inhouse' at the Oxford Radiocarbon Accelerator Unit. Confusingly, the three samples yielded three completely different dates: AD 224-383, AD 680-779, and AD 885-993 – but the internal consistency of the document and the continuity of its contents makes it unlikely that it is anything other than a single unified work. It should be noted that no tests were performed on any inked portion of the manuscript, and the Bodleian Library was also criticised for releasing their results via the *Guardian* newspaper and a YouTube video rather than the usual academic channels. It is therefore difficult to draw any meaningful conclusions at this stage[47].

Although the use of zero as a number in its own right is an Indian invention, the place-value system was employed by Old Babylonian scribes around 1800 BC, with a blank space as a placeholder (see Chapter 10). However, Indian mathematicians were the first to combine place-value notation with a decimal system. Also, unlike the zero, the blank space used by the Mesopotamians as a placeholder could not be used alone or at the end of a number.

Following the Arab conquests of the eighth century AD, many Indian texts were translated into Arabic; these texts were crucial in the development of Arabic astronomy. Today, India has one of the largest economies in the world and a successful space program. The Indian space agency has had an indigenous satellite launch capability since 1980 and has sent robotic space probes to the Moon and to Mars. India looks set to continue its lengthy astronomical tradition into the twenty-first century.

14
The Islamic Golden Age

A lasting influence

The fall of the Western Roman Empire is usually dated to AD 476, when the barbarian ruler Flavius Odoacer deposed the last Roman emperor, Romulus Augustulus, and proclaimed himself to be King of Italy. While medieval Europe was not the intellectual desert of popular imagination, there were no important advances in astronomy until the sixteenth century. But during this period, a new astronomical tradition arose following the Islamic conquests of the period AD 622-750.

Under the leadership of the Prophet Muhammad (AD 570-632), the Arabian peninsula was unified into an Islamic polity. Subsequently, the Rashidun (AD 632-661) caliphs drove the Byzantines out of Syria and Anatolia, before defeating Sasanian Persian Empire. Control of the caliphate then passed to the Umayyad dynasty (AD 661-750), whose first ruler, Mu'awiyah, moved the capital from Medina to Damascus. By AD 732, the expansion had reached its maximum extent and the empire encompassed the Iberian peninsula south of the Pyrenees, North Africa, the Arabian peninsula, Mesopotamia, Persia, portions of the Caucasus and Central Asia, reaching the borders of China and the Indian subcontinent. In area, it was larger than either the Roman Empire at its height or the empire of Alexander the Great. Despite this, the Umayyads were deposed after just 90 years, and the Abbasid rulers (AD 750-1258) took power.

The Islamic Golden Age was a period of intellectual flourishing that began with the founding of a new capital at Baghdad in AD 762 by the second Abbasid caliph, Abū Ja'far Abdallah ibn Muhammad al-Mansūr (reigned AD 754-775). During this period, the major centre of learning in the Islamic world was a great library in Baghdad known as the *Bayt al-Hikmah* ('House of Wisdom'). Science historians usually refer to the scientific tradition of this era as 'Arabic', although the Islamic caliphates and emirates included not only Arabs but also Persians, Turks, Moors, Kurds, and others[1].

The Golden Age peaked around AD 1000, and then began a long slow decline over the next five hundred years. Traditionally, the decline has been attributed to factors such as the rise of a more conservative interpretation of Islam, as exemplified by the work of theologian and philosopher Abu Hamid al-Ghazāli (AD 1058-1111); or to the sacking of Baghdad and destruction of the House of Wisdom by the armies of Hūlāgū Khan in 1258. Such explanations are probably over-simplistic, and one significant factor might have been the reluctance of the Muslim world to adopt the printing press. The cursive nature of Arabic script did not lend itself to typesetting, but there was also an aversion to reducing calligraphy to a purely mechanical process. In the Islamic world, calligraphy was and still is of immense cultural importance[2].

Arabic astronomy was kickstarted by the translation of Indian and Classical Greek texts into Arabic[2]. As a result, Ptolemy's *Almagest* acquired the name by which it is now generally known (from *al-Majasti*). The translation of Ptolemy's *Planetary Hypotheses* proved to be crucial, as much of this work has survived only as the Arabic translation. The greater part of it would otherwise have been lost[3]. Another important development was the adoption of the Hindu numeral system, which was probably transmitted to the Islamic world during the reign of al-Mansūr[4].

The English language contains a large number of loanwords from Arabic, including *alcohol, chemistry, sofa, safari,* and *mattress.* Mathematical and astronomical terms include *algebra, algorithm, azimuth, nadir, zenith,* and *zero.* The proper names of many stars, too, are of Arabic origin including Algol ('head of the ghoul'), Rigel ('leg' or 'foot'), Betelgeuse ('the hand [of Orion]'), Aldebaran ('the follower [of the Pleiades]'), Fomalhaut ('the mouth [of the fish]'), Altair ('the flying eagle'), Vega, ('the falling eagle'), Spica ('the stem'), and Deneb ('the tail'). These elements hint at the influence of Arabic mathematicians and astronomers on later scholars in Europe; as we shall see, the effect has been far more than purely linguistic.

The religious calendar and the time and direction of prayer

Arabic scholars were motivated, in part, by obligations to Islam, as set out in the Qur'an. There was a need for a calendar, a reliable means of accurately determining the local times for prayer, and establishing the direction of prayer. These three requirements were the same across the whole of the expanding caliphate, and the mathematics and astronomy involved were far from trivial.

The Islamic calendar consists of a *Hijri* year of twelve lunar months (i.e., 354 or 355 days; 354.37 days on average), with no intercalary months. Hence there is no correspondence with the solar year, and religious observances such as Ramadan shift forward by ten or eleven days each year relative to the Gregorian calendar. Each day begins at sunset, and each month begins with the first observation of the evening crescent Moon (*hilal*) after sunset. If the *hilal* is not observed at sunset on the 29th day of the month (due to either weather conditions or the sky still being too bright at moonset), then the next day is deemed to be the 30th day of the same month. The next month will begin at sunset the next day regardless of whether or not the *hilal* is observed.

This traditional practice for determining the start of the month is still followed in the majority of Muslim countries. Each conducts its own monthly observations before declaring the beginning of a new month. The Moon does not typically become visible to the naked eye until a day after conjunction with the Sun, although it has been seen as early as twelve hours after (the actual moment of conjunction, or new Moon, has only recently been captured telescopically). As one moves from east to west, so the Moon moves further away from the Sun, and hence sets progressively later after sunset. This means that the *hilal* might be observed in a western country several hours after it was unobservable further east. The beginning of each month is therefore liable to differ from one Muslim country to another.

Muslims pray at five set times of the day and must face towards the holy sanctuary in the city of Makkah ('Mecca'). The prayers are *fajr* (onset of dawn to sunrise), *zuhr* (just after local noon until approximately twenty minutes before the start of the *asr* prayer), *asr* (late afternoon), *maghrib* (onset of evening twilight until onset of night), and *isha* (onset of night until onset of dawn, preferably before midnight). The prayers must not be performed outside their prescribed limits, or they will be invalid. The timings of the two daytime prayers, especially the *asr* prayer, are non-trivial. For Muslims living far from Makkah, determining the direction of prayer (*qibla*) is also problematic.

Determining the time for the *zuhr* is reasonably straightforward: the shadow cast by a gnomon will decrease in length up until local noon and increase thereafter. The *zuhr* time will therefore begin just as soon as the shadow is seen to increase from its noontime minimum. The time for the *asr* prayer was originally deemed to begin when the shadow of a gnomon is equal to its length, and its end was set at the time when the shadow length was equal to twice the height of the gnomon. In higher latitudes during winter, however, these conditions are never met because the Sun never rises high enough for a gnomon to cast a shadow shorter than itself. Accordingly, new definitions had to be adopted: the time for the *asr* prayer was deemed to begin when the shadow of a gnomon was equal to its length at noon, plus the length of the gnomon itself, and the end was when the shadow became equal its length at noon plus twice the length of the gnomon[5,6].

Such schemes assumed that instead of an actual gnomon, an individual relied upon their own shadow, and that the length of the shadow was measured in terms of the individual's feet by pacing it out. The ratio between an individual's height and foot length was typically assumed to be 7, although values of 6, 6½, or 6⅔ were sometimes used[5].

If the height of the gnomon (actual or a person) is h and the length of the shadow at noon is Sn, then the shadow length at the beginning of the *asr* prayer, S_1, will be $Sn + h$ and the shadow length at its end, S_2, will be $Sn + 2h$.

In terms of the **altitude** (a) of the Sun at noon on a particular day:

$Sn = h \cot (a)$.

If θ is the latitude of a particular location is and δ is the **declination** of the Sun at noon on a particular day, then:

$a = 90 - \theta + \delta$.

The geographic latitude θ for the specific location will be known or it can be determined. If λ is the **celestial longitude** of the Sun at noon on that day and ε is maximum declination of the Sun (i.e., the **obliquity of the ecliptic**), then:

$\sin (\delta) = \sin (\lambda) \sin (\varepsilon)$.

The problem thus reduces to one of obtaining a table of values for λ at noon on each day of the year; from these the shadow length of a gnomon at noon, and hence the limits for the *asr* prayer, can be determined. Provided that the obliquity of the ecliptic, solar eccentricity, and solar **apogee** are known, values for λ can be derived from the **mean longitude** ($\bar{\lambda}$) and **equation of centre** (q) of the Sun (see Chapter 12). Arabic astronomers did not simply use the values reported in the *Almagest* and instead chose to redetermine them from scratch. While it is probable that they recognised the importance of accurate values for determining limits for the *asr* prayer, it is also likely that they realised that they were fundamental to all astronomical work[6].

In addition to performing the prayers at the right times, Muslims are required to do so while facing in the direction of Makkah. This direction is known as the *qibla*. The obligation affects individuals at prayer and the construction of prayer niches in local mosques where the whole congregation faces the same direction during the communal Friday prayers. The required angle θ along the local horizon of any city C can readily be calculated by use of spherical trigonometry and is given by the relation:

$\cot(\theta) = [\sin(\Psi_C)\cos(\Delta L) - \cos(\Psi_C)\ \tan(\Psi_M)] \ / \ \sin(\Delta L)$

where Ψ_C is the latitude of C;

Ψ_M is the latitude of Makkah;

ΔL is the difference in longitude between Makkah and C.

It was possible to solve the problem using the mathematics of the ancient Greeks, but in the absence of the modern trigonometry functions (see Chapter 12), the solution would have been extremely clumsy[6].

Thus, astronomy and mathematics was able to provide tables and techniques for determining the times and direction for prayers, and the timing of Ramadan. This service to Islam provided a social legitimacy to astronomers and mathematicians, in addition to giving them interesting problems to solve. But the service to Islam went further in that science could reveal the glory of God's creation. Islam, in return, gradually freed astronomy from its long-standing ties to astrology. Although political considerations led the early Abbasid caliphs to take in interest in astrology, in the long term it was always going to be on shaky ground by ascribing powers to the stars and planets that should be reserved for God[7].

The House of Wisdom

The Round City of Baghdad was the original core of Baghdad, built by Caliph al-Mansūr between AD 762 and 766 as the seat of the Abbasid court. Unfortunately, no part of it has survived, and our knowledge of it is based entirely on literary sources. The site was carefully chosen by al-Mansūr, who is reputed to have sailed up and down the Tigris in search of a suitable site for his new capital, deep in what had been the heartland of the Sasanian Persian Empire. Located on the west bank of the Tigris and about 70 km (43 miles) east of the Euphrates, the city was described by ninth century geographer Ya'qubi as "*the crossroads of the universe*". By this time, it was already home to mathematicians, astronomers, poets, musicians, philosophers, and historians. The original Round City was 2 km (1.25 miles) in diameter, surrounded by a heavily fortified 24 m (80 ft) double outer wall and a deep moat. Within the city were two further concentric walls. The eleventh century historian al-Khatib al-Baghdadi (AD 1002-1071) noted that each wall consisted of 162,000 bricks for the lower third of its height, 150,000 for the middle third, and 140,000 for the upper third.

Four equidistant gates were set into the outer wall, from where straight roads led to the centre of the city. At the centre was the city's mosque and the caliph's official residence, the Qasr Bab al-Thahab ('Palace of the Golden Gate'). Also located within the centre, though at a distance from the mosque and the palace, were other royal accommodation, royal kitchens, barracks for the caliph's horse guards, and accommodation for the caliph's servants and officials. The two outer circles were occupied by residential and commercial buildings.

A sprawling conurbation rapidly grew up around the original Round City as pre-existing market districts were subsumed into the new city and other districts and commercial facilities were newly built. Baghdad was not just an administrative hub: it became a centre of art, culture, and trade. Caliph al-Mansūr soon built a larger palace, the Qasr al-Khuld, ('Palace of Eternity') on the eastern side of the Tigris. The new palace was on higher ground than its surroundings, and accordingly less affected by mosquitos during the summer. The old palace continued to be used for administration[2].

Islam has always encouraged learning and a devotion to scholarship, but this alone did not trigger the age of Arabic science. There was only limited scientific activity during the Rashidun and Umayyad caliphates. Two developments triggered the revolution. The first was a renewed interest in academic enquiry inherited by the Abbasid caliphs from the Persians. The second was the translation movement, which began in Baghdad and lasted for two centuries. During this period, much of the accumulated wisdom of the Greeks, Persians, and Indians was translated into Arabic.

From the start, the Abbasid caliphs sponsored and encouraged the large-scale translation of these ancient texts. The translation movement was in the first instance driven by the Abbasids interest in Persian culture, in particular astrology which played a fundamental role in Persian daily life. Caliph al-Mansūr developed a personal interest in astrology, but he also had a practical reason for his interest: he needed the support of the Persian aristocracy, many of whom were still Zoroastrian and had not converted to Islam. The second factor driving the translation movement was the invention of the paper mill, which meant that paper could be produced far more cheaply than papyrus or parchment[2].

For all its influence and achievement, much about the House of Wisdom remains uncertain. Not a trace of its structure has survived, so we have no definite knowledge about its location or architecture. There is no certainty regarding the range of activities carried out there. Some scholars claim that its scope and purpose have become embellished over time. When and by whom it was founded also remains uncertain. Caliph Abū al-'Abbās 'Abd Allah ibn Hārūn ar-Rashīd (reigned AD 813-833), usually known by his regnal name al-Ma'mūn, is often credited, but claims have also been made for al-Mansūr and Hārūn al-Rashīd (reigned AD 786-809). A possible scenario is that it started as the private collection of Caliph al-Mansūr. He collected books on medicine, astronomy, engineering, and literature that had been translated in his reign; this collection likely became the nucleus of the House of Wisdom. Al-Mansūr was the first caliph to encourage the study of the sciences and the translation of Persian, Greek, and Indian works[2,8].

Astronomer and astrologer Muhammad ibn Ibrahim al-Fazari (died AD 777), an advisor of al-Mansūr during the construction of the Round City, is associated with translating several Indian astronomical texts into Arabic including Brahmagupta's *Brahmasphuta siddhānta*[2]. The translated work was known as the *Zīj al-sindhind* ('Great astronomical tables of the *Siddhānta*'). *Zīj* is an Arabic word used for astronomical ephemerides, *sindhind* is an Arabized form of the Sanskrit *siddhānta*. Al-Fazari is also supposed to have constructed the first astrolabe in the Arab world. However, there is a degree of uncertainty about the extent to which his father Ibrahim ibn Habib al-Fazari was involved in the translation work, and whether it was father or son who constructed the astrolabe. Matters are further confused by the existence of several contemporaries who were also named al-Fazari. In the reign of Caliph al-Ma'mūn, the *Zīj al-sindhind* was abridged and updated by astronomer and mathematician Muhammad ibn Musa al-Khwārizmī (*c.*AD 780-850)[9,10].

The revised work comprises material on the Islamic, Seleucid, and Sasanian calendars; tables of sines and cotangents; spherical astronomy functions; a table for the **equation of time**; mean longitudes ($\bar{\lambda}$) of the Sun, Moon, and planets; planetary theories; **celestial latitudes** of planets; **stationary points and retrogradations of planets**; **parallax** theory; eclipse theory; and finally, astrological tables (see Chapters 10, 12, and 13). Much of the material is non-Ptolemaic and is based on pre-Islamic Sasanian Persian equations, which in turn were probably modified from Indian theories. The *Zīj al-sindhind* exerted a considerable influence on Arabic astronomy, and continued to do so even after translations of the *Almagest* became available. Al-Khwārizmī's version has survived more or less complete, though only as a Latin translation[9]. Many other astronomers would later produce *zījes* of their own during the Islamic Golden Age. Around 200 are known. While early *zījes* drew on Indian sources, later ones followed the tradition of Ptolemy's *Almagest* and *Handy Tables*[11].

Al-Khwārizmī was an important figure at the court of al-Ma'mūn. In addition to the revised *Zīj al-sindhind*, his astronomical works include a treatise on the sundial and two books about the astrolabe[12]. The *Kitāb Surat al-Ard* ('Book of the Description of the Earth') was a revised and more accurate version of Ptolemy's *Geographia*. The *al-Kitāb al-mukhtasar fi hisab al-jabr wa'l-muqabala* ('The Compendious Book of Calculation by Restoration and Balancing') was a treatise on algebra and the first work to treat it as a mathematical discipline in its own right. The word 'algebra' is derived from '*al-jebra*', which means 'rejoining of broken parts'[2]; and the word 'algorithm' is derived from the Latinized form of his name, Algorithmi. The 55 km (34 mile) diameter lunar crater Al-Khwarismi and the asteroid 11156 Al-Khwarismi are named for al-Khwārizmī.

Caliph Hārūn al-Rashīd is said to have founded a formal library at the court, with a director and several collaborators. Although its scope has been the subject of much debate, one of its directors was involved with the production of an early translation of the *Almagest*. There was a considerable increase in translation activity during Hārūn al-Rashīd's reign. However, there is no evidence that any of the translations actually took place in the library itself. The Hārūn al-Rashīd translation of the *Almagest* was superseded by later translations, and very little is known about it[2,12].

Hārūn al-Rashīd's successor, Caliph al-Amin (reigned AD 809-813) was deposed by his half-brother, al-Ma'mūn. In contrast to the conservative al-Amin, al-Ma'mūn was open to new philosophical movements and outside influences. In particular, he was a supporter of a rationalist movement known as Mu'tazila, which opposed literal interpretations of the Qur'an. Caliph al-Ma'mūn is usually acclaimed as the founder of the House of Wisdom, although he might in reality have done no more than extend a pre-existing library[12]. Regardless, as a state institution, the House of Wisdom was now managed by a high-ranking official known as a sahib. The full job title was *sahib al-Khazin*; to be considered for the role an impressive C.V. was required including a mastery of various sciences. In addition to its function as a public library and translation house, the House of Wisdom employed scribes and book binders to produce finished books from the manuscripts of authors and translators. Final copies would be distributed not just inhouse but to libraries outside Baghdad[13].

The collections of the House of Wisdom included books both newly authored and translated, manuscripts, and maps. It was considered a great honour for an author to have their books deposited in the House of Wisdom. Al-Ma'mūn also assigned a group to purchase books from Roman and Greek libraries. The House of Wisdom was both a reference library and a lending library. The literature was divided into sections, each of which was dedicated to a specific subject. Each section was stored on partitioned shelves, with books indexed as in modern libraries. There were large numbers of reading rooms, and in some of these a servant was on hand. Borrowers were required to pay a refundable deposit to make good possible damage or loss. But in addition to its functions as a library and a translation house, the House of Wisdom was a university where authors worked, and lecturers taught subjects including philosophy, astronomy, history, geography, mathematics, medical sciences, and music. We can assume that large sums of money were required to pay the translators, authors, map makers, book binders, lecturers, debaters, and servants. The budget must also have included habitation, food, and stationery[2,13].

By the middle of the ninth century AD, the House of Wisdom had become the largest repository of books in the world[2]. It became a model for similar institutions elsewhere in the Islamic world, as well as many private libraries. These libraries played a major role in translating and transmitting works of Greek, Persian, Indian, and Assyrian philosophers, scientists, and physicians; these work later became the basic textbooks in European schools of Bologna, Naples, and Paris, and made a major contribution to the revival of learning in Europe[13].

The *al-Zīj al-Mumtahan*

Caliph al-Ma'mūn believed that there was a need to verify the astronomical observations quoted in the *Almagest*, a project he pursued with great energy during the last six years of his reign. He began in AD 827 by ordering a new determination of the diameter of the Earth. He dispatched a team of surveyors led by astronomer Khalid ibn Abd al-Malik al-Marwarrūdhī (fl. AD 828) and mathematician Ali ibn Isa al-Asturlabi (fl. AD 832) to make an accurate measurement of one degree of latitude. The measurements were carried out on the Plain of Sinjar in what is now northwest Iraq. The team split into two groups, one of which headed north and the other south. Both groups continued until based on the positions of the stars, they had travelled through one degree of latitude. They then returned to base and compared measurements. The result came out at 56.6 Arabic miles which is about 68 statute miles or 110 km. This gives a circumference for the Earth of 39,400 km (24,480 miles) and hence a diameter of 12,540 km (7,790 miles). The correct value is 12,742 km (7,917 miles)[2,12].

Al-Ma'mūn later commissioned a second survey in the Syrian desert. This time, his team measured the distance from Palmyra to Raqqa to the north. They found that the two cities were one degree of latitude and 66 Arabic miles (80 statute miles; 130 km) apart, giving a circumference of 46,350 km (28,800 miles) and hence a diameter of 14,760 km (9,170 miles). The second result was rather less accurate than the first; Raqqa is actually 1.5 degrees of latitude and 100 statute miles (160 km) north of Palmyra, as well as being further east. The problem, as with Eratosthenes' measurements a millennium previously (see Chapter 8), is that there was no reliable method of measuring the distance between two points on the ground. Another astronomer of the period, Sanad ibn Ali (died AD 864), reputedly suggested to al-Ma'mūn around AD 832 that it would be better to measure the diameter of the Earth directly by measuring the dip of the horizon from the top of a mountain of known height. This method is perfectly sound, but there is no evidence that Sanad ibn Ali ever carried out the experiment[2,12].

In AD 828, al-Ma'mūn launched his next major scientific project, which was to establish an observatory at Shammasiyya, near Baghdad. The task fell to the young Sanad ibn Ali and Yahya ibn Abi Mansūr (died AD 830), senior astronomer and astrologer at the court.

Astronomer al-Abbas ibn Said al-Jawhari (*c.*AD 800-860) was put in charge of procurement of the requisite instruments for the observatory, who in turn selected al-Marwarrūdhī to construct them. Astronomer and engineer Ahmad ibn Muhammad ibn Kathīr al-Farghānī (*c.*AD 800-870) was brought on board for his expertise with astronomical instruments. Finally, al-Khwārizmī was called upon for his overall knowledge and expertise. The type of instruments built by al-Marwarrūdhī is uncertain, but they would have included a sundial with a gnomon to determine the height of the Sun from the length of the shadow cast, astrolabes (see chapter 12), and mural quadrants. The latter is a sighting instrument attached to a wall aligned on the meridian[2].

The work continued for about eighteen months between AD 828-29. Observations were made of the Sun, and Moon; and the positions of 24 stars were recorded. The solar observations were carried out around the **solstices** and the **equinoxes**. According to al-Bīrūnī, al-Ma'mūn took an active interest in the observations, and he rejected the first set of observations carried out during AD 828 because the values for the altitude of the Sun at the summer and winter solstices did not agree with later observations. A second set of observations was therefore carried out using a specially devised new method known as the *fusūl* ('seasons') method. A **maximum solar equation** of 1°59' was obtained[12].

Hipparchus and Ptolemy had determined the solar eccentricity and apogee by measuring the solar declination at the equinoxes and solstices. The problem is that at the time of the solstices, changes in the Sun's declination are barely discernible, and fixing the maximum (or minimum) value is difficult. An alternative was to obtain the measurements in the middle of the seasons rather than the beginning. The methods reported in the *Almagest* could still be used, provided the distance between the observations was 90°. Thus, the observations of the Sun's declination were made at the mid-points of Taurus, Leo, Scorpius, and Aquarius instead of the start-points of Aries, Cancer, Libra, and Capricornus[6].

Presumably still not convinced by the results, al-Ma'mūn ordered further observations. A second observatory was constructed at Dayr Murran, a monastery on the slopes of Mount Qasyun outside Damascus. Yahya ibn Abi Mansūr had died in AD 830, so al-Marwarrūdhī was chosen to head up the new project. He designed several new instruments to be used at the site, including a 5 m (16 ft) mural quadrant to measure solar angles. Despite problems caused by copper and iron instruments expanding and warping in the summer heat, his team carried out two sets of solar and lunar observations during the period AD 831-33[12].

An astronomical handbook entitled *al-Zīj al-Mumtahan* ('The verified tables') was produced on the basis of the observations carried out at the two sites. This work is commonly attributed to Yahya ibn Abi Mansūr, but as he had died before the Syrian observations were made, he could not have been its sole author. The only existent version is the manuscript Escorial II Codex árabe 927, although only a small part of this represents the original *al-Zīj al-Mumtahan.* The majority of the material is from earlier or later sources that are mostly unacknowledged. The contents, which were typical of most later *zījes*, include Hijri, Seleucid, Yazdegerd, Coptic, and Jewish calendars, trigonometric and spherical trigonometric tables, values of the equation of time for each degree of solar longitude, mean motions (mean longitudes of the Sun and planets), planetary equations for computing corrections to mean longitudes due to **solar anomaly** and **zodiacal anomaly**, celestial latitudes of planets, stationary points and retrogradations of the planets, tables of lunar parallax for various latitudes, a table of solar eclipses, and tables of the positions of 24 stars in both ecliptic and equatorial coordinates. The rate of **precession of the equinoxes** is given as 0; 0,54,44,20° (54;44,20 = 54.74 arcsec) per Hijri year, i.e., 65.77 Hijri years or 63.8 Gregorian years per degree, a considerable improvement on the one degree per century claimed by Hipparchus and Ptolemy[9]. In his *The Elements of Astronomy*, a compendium of the *Almagest* published after al-Ma'mūn's death, al-Farghānī listed a value of 23°35' for the obliquity of the ecliptic, which he claimed was determined by Yahya ibn Abi Mansūr and agreed upon by other scholars. The published result was very close to the correct value for the epoch (23°35'31")[14].

The 49 km (30 mile) diameter lunar crater Almanon is named for Caliph al-Ma'mūn. Al-Farghānī is commemorated by the small lunar crater Alfraganus.

Trepidation

The value for the rate of precession obtained at Shammasiyya – around 1° in 65 years – was far closer to the accepted value (one degree in 72 years) than the 1° per 100 years reported by Hipparchus. At the same time, Arabic astronomers realised that their values of around 23°35' for the obliquity of the ecliptic were much lower than Ptolemy's value of 23°51'. These figures led some astronomers to draw two erroneous conclusions: firstly, the rate of precession had increased since ancient Greek times; and secondly, that there was a link between this increase and the decrease in the obliquity. They failed to consider the possibility that the Greek observations might simply have been inaccurate. The rate of precession does not vary, and although the obliquity has been decreasing since 8700 BC, the rate of decrease is far less than that implied by the differences between the Shammasiyya results and those of Ptolemy. The actual value for the obliquity in Ptolemy's day was 23°41'.

The supposed combined motion became known to medieval astronomers as 'trepidation'. In an attempt to explain it, mathematician Thābit ibn Qurra (AD 830-901) picked up on a previously ignored proposal by Greek scholar Theon of Alexandria (AD 335-405). Theon suggested that precession reversed direction every 640 years after completing an 8° arc, with one such reversal happening in 155 BC (128 years before the start of the reign of the first Roman emperor, Augustus, in 27 BC)[15]. Thābit described his proposal in a treatise which exists only as a Latin translation *De motu octave spere* ('On the Motion of the Eighth Sphere'). As the title implies, the model involved the motion of the eighth sphere, i.e., the sphere of the fixed stars (see Chapter 12). The attribution of this work to Thābit has been questioned; one suggestion is that it was written by Thābit's grandson Ibrahim ibn Sinān (AD 909-946)[12].

Thābit (if it was Thābit) invoked a ninth sphere, surrounding the sphere of the fixed stars. In the Thābit model, this ninth sphere is responsible for all diurnal motion, which it communicates to the eight lower-lying encompassing spheres (see Chapter 12). The ninth sphere carries the **celestial equator**, and hence is known as the sphere of the equator. It also holds an unobservable 'fixed ecliptic'; an imaginary circle that makes an angle of 23°33' with the celestial equator. Two circles, each with a radius of 4°18'43", are centred on the two points of intersection between the equator and the 'fixed ecliptic', in Aries and Libra, respectively. Two points, known in medieval Latin astronomy texts as *caput Arietis* ('head of Aries') and *caput Librae* ('head of Libra'), move around the two circles at a uniform rate, and take 4,057 years to complete a circuit. They are always opposite one another, i.e., when one is north of the celestial equator, the other is always south of the celestial equator[15,16].

The eighth sphere, known as the sphere of the ecliptic, carries the 'real ecliptic' as well as the stars. The 'real ecliptic' passes through *caput Arietis* and *caput Librae*; the actual spring and autumn equinoctial points are where it intersects the celestial equator (i.e., the **First point of Aries** and the **First point of Libra**). *Caput Cancri* ('head of Cancer') is a point on the 'real ecliptic' located 90° from *caput Arietis*. At the same time, it always remains on the 'fixed' ecliptic, moving along it in an oscillatory motion. However, it does not represent the true summer solstitial point, which lies 90° from the First point of Aries[15,16].

The celestial latitudes of the stars (with respect to the real ecliptic) do not change, but the circular motion of *caput Arietis* ('head of Aries') and *caput Librae* ('head of Libra') causes the sphere of the ecliptic to wobble with respect to the sphere of the equator; hence the stars advance ('accession') and retreat ('recession') with respect to the equinoxes. The obliquity of the ecliptic also changes, going through two complete cycles every 4,057 years. The model gives a good agreement with the erroneous results it is trying to explain. Trepidation received a mixed reception in the Arab world, although it was widely accepted in Islamic Andalucía. As the effects of precession accumulated over the centuries, it became clear that Thābit's model was unsatisfactory, but trepidation was slow to die. Some believed in an oscillation that was superimposed onto a steady forward motion. One version of this model was based on a 49,000-year steady forward motion cycle with a superimposed 7,000-year cycle of ± 9° oscillations. Not until the time of Tycho Brahe (AD 1546-1601) (see Chapter 17) was trepidation finally abandoned[16]. The 57 km (35 mile) diameter lunar crater Thebit is named for Thābit ibn Qurra.

Refining the *Zīj*

Caliph al-Ma'mūn died in AD 833, supposedly after eating fresh dates that caused him and others with him to fall ill. The fruits of what is one of the world's first state-sponsored scientific endeavours, the *al-Zīj al-Mumtahan*, was his legacy. However, the quest to improve upon the accuracy of the astronomical parameters reported in the *Almagest*, and the production of updated *zījes*, continued throughout the history of the Arabic astronomical tradition. In addition to the public observation programs such as those conducted at Shammasiyya and Damascus, there were many observation programs carried out by private individuals in different parts of the Islamic world.

After al-Ma'mūn's death, mathematician Habash al-Hāsib (died after AD 869) revised the observational data gathered by the *Mumtahan* astronomers at Shammasiyya and Damascus. He had been in close contact with the group, although he was not a member himself. Prior to this period, he had written two *zījes*; the first based around al-Khwārizmī's updated *Zīj al-sindhind* and the second on the work of al-Fazari. His third and most important *zīj*, the '*zīj* of Habash' (it is known under four different names) saw a shift to Ptolemaic methodology, although some of the computational methods used were almost certainly of Indian origin, or at least inspired by Indian methods. In this work, he records his own observations of the Sun, Moon, and planets over a period that coincides with the reigns of Caliph al-Mutawakkil (reigned AD 847-861) and Caliph al-Mu'tazz (reigned AD 866-869). The work is thought to have been finalised after al-Mu'tazz's death during the decade of political instability that followed the assassination of al-Mutawakkil. The *zīj* of Habash is the earliest independently compiled Ptolemaic astronomical handbook in the Arabic language that has survived in its entirety, and it was one of the most influential *zijes* of its time[12].

Born towards the end of al-Hāsib's life, Abū Abd Allah Muhammad ibn Jābir ibn Sinān al-Battānī al-Harrānī al-Sābi (AD 858-929), Latinized as Albategnius, was one of the most notable astronomers of this period. He was probably born in Harran, southern Anatolia, and his father is thought to have been Jābir ibn Sinān al-Harrānī, a well-known instrument maker. Al-Battānī moved to Raqqa and established a private observatory from where he carried out regular observations between AD 877 and 918. He made use of instruments including a gnomon, sundials, a triquetrum (used for measuring stellar altitudes), an astrolabe, an armillary sphere, and a mural quadrant. To obtain greater accuracy, he used large-sized instruments[12].

At Raqqa, al-Battānī redetermined standard astronomical parameters including the planetary motions and the parameters of the lunar model; the latter were in agreement with Ptolemy. He obtained a value of 23°35' for the obliquity of the ecliptic (the value in AD 880 was 23°35'6"), measured the apparent diameters of the Sun and the Moon and their variations over the course of a year and an **anomalistic month** respectively, and considered the possibility of annular solar eclipses[11,12].

The earliest surviving redetermination of Ptolemy's value of 1,210 Earth radii for the mean solar distance was made by al-Battānī. Although it is unclear how he arrived at his result of 1,146 Earth radii, it became as widely accepted as Ptolemy's figure had been[17]. It was, of course, still hopelessly underestimated. He obtained values of 2;4,45 / 60 (= 0.0347) for the solar eccentricity, corresponding to a maximum solar equation of 1°59'; and 82°17' for the longitude of solar apogee[17]. The actual value in AD 900 was 83°39'. Al-Battānī realised that the solar apogee is not fixed, as Ptolemy had believed. He established a rate of 1° in 66 years (the true rate is 1° in 310 years) for both it and for the precession of the equinoxes. For his tables, he assumed that the latter was fixed, although in line with contemporary thinking on trepidation, he had some doubt that the rate had remained unchanged since Ptolemy's time[16,12].

Al-Battānī estimated the mean length of the tropical year by comparing the time of the autumn equinox in AD 880 with one recorded by Ptolemy in AD 139. His value of 365.24056 days (365 days 5 hrs 46 min 24 sec) was 2½ minutes too short, but it was an improvement on the 365¼ - 1/300 days (365 days 5 hrs 55 min 12 sec) of Hipparchus (6½ min too long). The date for the equinox as recorded by Ptolemy is now known to be one day too late; if this is taken into account then al-Battānī's estimate is only too short by half a minute[1].

Al-Battānī's *zij*, entitled *al-Zij al-Sābi*, survives in its entirety. It was the first to be written entirely in the Ptolemaic tradition, with very few Indian or Sasanian influences. It went on the become very influential in the Islamic world, and later in medieval and Renaissance Europe[12]. The 129 km (80 mile) diameter lunar crater Albategnius is named for al-Battānī.

Another important figure is Abū al-Hasan Alī ibn Abd al-Rahmān ibn Ahmad ibn Yūnus al-Sadafī (AD 950-1009). He was born in Egypt, and in his teens lived through the Fātimid conquest of Egypt in AD 969. The Fātimids were a Shi'a dynasty claiming descent from the Prophet Muhammad through his daughter Fātima. They came to power in AD 909 and established an empire across North Africa. In AD 973, the fourth Fātimid caliph, al-Mu'izz li-Din Allah (reigned AD 953-973) established Cairo as his capital. Between AD 977 and 1003, ibn Yūnus worked for al-Mu'izz's successors al-Azīz (reigned AD 975-996) and al-

Hākim (reigned AD 996-1021), and he dedicated his major work, the *al-Zīj al-Kabir al-Hakimi* ('the large *Zīj* of Hākim') to the latter. This *Zīj* was intended to replace the *al-Zīj al-Mumtahan*, which was now around 180 years old[12]. The 58 km (36 mile) diameter crater Ibn Yunus, located on the Moon's far side, is named for ibn Yūnus.

The early fifteenth century marked the latter days of the Islamic Golden Age, but it was still a period of considerable achievement. Ulugh Bēg (AD 1394-1449) was a Timurid sultan and a grandson of the all-conquering Timur. In 1409, aged sixteen, he was appointed governor of Turkestan and Transoxiana. He had a longstanding interest in mathematics and astronomy, and in 1417 he established a school in Samarqand where these were among the most important subjects taught. Two of the most prominent teachers were mathematician Jamshīd al-Kāshī (died AD 1429) and astronomer Qādīzāde al-Rūmī (*c.*AD 1359-1440). Al-Kāshī was one of the most significant mathematicians of the fifteenth century. He is best known for deriving the cosine rule, a trigonometric identity that allows the length of any side of a triangle to be determined provided the other two lengths and one angle are known. He was also an early pioneer of the use of decimal fractions and calculated a value for *pi* that was accurate to sixteen decimal places.

In AD 1424, Ulugh Bēg established an observatory on a hill outside Samarqand. The observatory consisted of a circular building 50 m (165 ft) in diameter and 35 m (115 ft) high, decorated with glazed tiles and marble plates. The main instrument was a quadrant 40.4 m (132 ft 6 in) in radius in the vertical north-south plane. It was sited partly beneath ground level and consisted of two parallel walls just sufficiently far apart to accommodate an astronomer seated between them. The walls were marked in degrees with grooves. The backsight was probably a wooden rectangle with handles, ridged on the underside, which could be set down accurately between a pair of degree markings. In the middle of this was a rectangle with a slot along which the observer could move a pinhole up and down and read off fractions of a degree on a scale. The front sight was an opening at the top corner of the building and would be provided with cross-wires marking the centre of the circle of which the quadrant formed part. The vertical cross-wire marked due south. The Sun's position could be read off with a precision of 5 arcsecs. Smaller instruments were placed on the flat roof of the main building, including an armillary sphere and a triquetrum. Other instruments known to have been used at the site include astrolabes and quadrants[1,12].

Al-Kāshī was invited to become the Director of the new observatory. He remained in the post until his death in AD 1429, when he was succeeded by al-Rūmī and then by Ali al-Qūshjī, a former student and assistant of his two predecessors. The results of the observation program include a value for the obliquity of the ecliptic of 23°30'17" (the actual value for that date was 23°30'48") and for the precession of the equinoxes of 51.4 arcsec per year. The tropical year was determined to be 365.24253 days (365 days 5 hrs 49 min 16.6 sec), about 30 seconds too long. Planetary eccentricities and epicyclic radii were redetermined, and the coordinates of the fixed stars in Ptolemy's star catalogue were verified and corrected. For all this, the observatory remained in use for little over thirty years, and it fell into ruins soon after Ulugh Bēg's assassination in 1449. In time, it became buried with earth and remained lost until 1908, when it was rediscovered by Russian archaeologist Vassily Vyatkin[1,12].

Ulugh Bēg, al-Kāshī, al-Rūmī, and al-Qūshjī co-authored the *Zīj-i jadīd-i Guragānī* ('New *Zīj* of Guragan'). The four-volume work covered calendars, spherical astronomy, spherical coordinate system on the celestial sphere, geographical coordinates, spherical distances between stars, the direction of *qibla*, motions of the Sun, Moon, and planets, the solar, lunar, and planetary distances, and astronomical calculations. Ulugh Bēg also was the author of the mathematical treatise *Risāla fī istikhrāj jayb daraja wāhida* ('On the Determination of Sine of One Degree')[11]. Ulugh Bēg is commemorated by the 54 km (33 mile) diameter lunar crater Ulugh Beigh and by the asteroid 2439 Ulugbek.

Al-Bīrūnī

Abū Rayhān Muhammad ibn Ahmad al-Bīrūnī (AD 973-1050) was one of the most important scientists of the medieval period. He was the author of 146 works, although only 22 have survived. Around half of these concern the exact sciences: in addition to mathematics and astronomy, al-Bīrūnī was also an authority on geography, pharmacology, and meteorology. He also wrote *The Chronology of the Ancient Nations*, which provides information about the calendars used by Persians, Sogdians, Kwārizmians, Jews, Syrians, Harrānians, Arabs, and Greeks[12].

Al-Bīrūnī's major astronomical work, the *al-Qānūn al-Masūdī*, comprises eleven treatises. Treatise 1 is an introduction, covering basic astronomical and cosmological concepts; Treatise 2 concerns calendars including the Hijri, Seleucid, and Persian; Treatise 3 concerns trigonometry; Treatise 4 concerns spherical astronomy; Treatise 5 concerns geodesy and mathematical geography; Treatise 6 concerns time differences, the solar motion, and the equation of time; Treatise 7 concerns lunar motion; Treatise 8 concerns eclipses and the visibility of the crescent Moon; Treatise 9 is a catalogue of 1,029 fixed stars; Treatise 10 concerns the planets; and Treatise 11 describes astrological operations. The work is primarily based on the *Almagest*, but with the addition of elements of Indian, Iranian, and Arabic origin. Al-Bīrūnī updated Ptolemy's astronomical parameters with observations made by his predecessors and by himself, and his catalogue of fixed stars was corrected for the effects of precession since Ptolemy's day. The 1954-56 translated edition of *al-Qānūn al-Masūdī* runs to more than 1,400 pages. The work is of enormous value not only for al-Bīrūnī's own contributions but for the great volume of information that he gathered from other sources, many of which are now lost[18,12].

Al-Bīrūnī is commemorated by a 77 km (48 mile) diameter crater on the Moon's far side, and by the asteroid 9936 Al-Biruni.

Doubts about Ptolemy

Theoretical astronomy is conventionally supposed to have been a later development in the Islamic world than the practical astronomy of the ninth century AD. The usual view is that it first emerged in the eleventh century AD with the work of mathematician, astronomer, and physicist Hasan Ibn al-Haytham (AD 965-1040). In fact, the theoretical basis of Ptolemaic astronomy was being challenged as early as the ninth century AD. A text from this period disputes Ptolemy's assertion that celestial physics is completely separate from terrestrial physics. It insists that celestial spheres must follow the same laws of physics as terrestrial objects, and that they must behave in a recognisable physical manner. Again, the distinction between celestial and terrestrial physics is traditionally not thought to have been abolished until later in medieval and Renaissance Europe[19].

The text in question is a treatise entitled "*Book on the Mathematical Proof by Geometry that there is not a Ninth Sphere outside the Sphere of the Fixed Stars*", although all we have of it is a long citation in a late thirteenth-century AD text written by Persian astronomer Qutb al-Din al-Shirazi (AD 1236-1311). Al-Shirazi attributes the work to Mohammad ibn Mūsā (died AD 873), one of three brothers known collectively as Banū Mūsā. Other sources from the time suggest that the author was Muhammad's younger brother Ahmad[19,12].

According to the Ptolemaic cosmological model outlined in *Planetary Hypotheses*, the 'ninth sphere' is a sphere intended to handle precession. The 41-sphere version of Ptolemy's model (see Chapter 12) envisages eight 'movers' that are responsible for the diurnal motion of the Sun, Moon, five planets, and fixed stars. Each 'mover' sphere encompasses an assemblage of subsidiary spheres to account for orbital motions; for example, Mercury requires seven, but the Sun requires only one. All these spheres, 'movers' and subsidiary spheres alike, are composed of an invisible and frictionless substance known as aether. Accordingly, the rotations of a 'mover' can only affect subsidiary spheres if the latter are off-centre. The problem arises with the need to account for precession of the equinoxes. The ninth sphere handles the diurnal westwards motion of the fixed stars. Within is a concentric eighth sphere, which is tasked with handling the slow eastwards movement of precession – but, as noted, the rotation of an exterior sphere cannot induce the rotation of an interior sphere if the two are concentric.

These considerations left Mohammad (or Ahmad) Mūsā to conclude that "*...it has become clear that it is not in any way possible that there be beyond the orb of the fixed stars a circular body which moves by its own particular motion, and moves through that motion the orb of the fixed stars around the centre of the world*"[19].

Another problem was the equant (see Chapter 12). The basic concept envisages two points, *D* and *Q*, on opposite sides of the Earth (*E*). *D* is the centre of a deferent circle, around which an epicycle moves. The latter is centred on a point (*C*), lying on the deferent circle. *C* moves at a uniform angular speed not in relation to *D* but in relation to *Q*, which is known as the equant point. As set out in the *Almagest*, this works very well as a means to explain planetary motions. The problems begin when you try to envisage a sphere that rotates at uniform speed, in place, about an axis that does not pass through its centre. If you accept that celestial spheres must obey the everyday laws of physics, then such a sphere is a physical impossibility.

During the tenth and eleventh centuries AD, critiques of the Greek astronomical tradition known as *shukūk* ('doubts') began to appear. The best known of these is a text by Ibn al-Haytham entitled *al-Shukūk 'alā Batalamyūs* ('Doubts about Ptolemy'), which is a systematic exposure of the contradictions and issues with the *Almagest*, *Planetary Hypotheses*, and *Optics*. In his pivotal *Kitāb al-Manāẓir* ('Book of Optics'), Ibn al-Haytham applied **Occam's Razor** to Ptolemy's theory of vision. He argued that if an eye emits rays, they must be reflected back for the eye to perceive an object the rays have fallen on; and that the eye must therefore see by reflected light. But if this so, then the light could come directly from the object if it is luminous, or be reflected from it if not. In either case, there is no need to postulate rays emitted by the eyes. On account of this work, Ibn al-Haytham is often referred to as 'the father of modern optics'. In 2015, he was made the focus of the International Year of Light and Light-based Technology. He is also commemorated by the 32 km (20 mile) diameter lunar crater Alhazen and by the asteroid 59239 Alhazen.

What Ibn al-Haytham failed to do in any of his works was to put forward solutions to the problems he had exposed with Ptolemy's astronomical work[2,6].

Reforming Ptolemy, influencing Copernicus

Over the next two centuries, many astronomers attempted to solve the problem of the equant, but without success. The breakthrough that enabled Ptolemaic astronomy to be reformed eventually came at the Marāgha observatory, in the far northwest of Iran. The observatory was founded in AD 1259 through the efforts of its first Director, astronomer Nasīr al-Dīn al-Tūsī (AD 1201-1274). As a young man, al-Tūsī fled the Mongols and joined the Ismailis in their fortress at Alamut, about 100 km (62 miles) north of present-day Tehran. After Alamut was seized by the Mongols in AD 1256, al-Tūsī became an adviser to Hūlāgū Khan. Although his forces utterly destroyed the House of Wisdom in Baghdad two years later, the Mongol leader was persuaded to support the construction of a new astronomical observatory in Alamut. Also comprising a library and a school, the Marāgha observatory was one of the most ambitious scientific institutions of its day, and remained the world's leading centre of astronomy for several hundred years[2,12].

Engineer and instrument maker Mu'ayyad al-Dīn al-Urdī (died AD 1266) was among those recruited by al-Tūsī. Like others of the Marāgha School, he was concerned about the inconsistencies with Ptolemaic astronomy, and he proposed a solution to the longstanding problem of the equant: the Urdī lemma. The lemma states that if two equal lines form equal angles (either internally or externally) with a third (base) line, then a line joining the other ends of these two lines will always be parallel to the base line[6].

Al-Urdī applied this theorem to the problem of equant. Instead of assuming that the epicycle is carried by a Ptolemaic deferent circle (D_P) rotating uniformly around an axis (Q) that does not pass through its centre (T), the centre of the new deferent (D_T) is located at a point K located halfway between the centre of the old Ptolemaic deferent (T) and the equant point (Q). The new deferent (D_T) carries a small epicycle (E_T) with a radius (R) equal to $TK = KQ$. The small epicycle (E_T) moves at the same speed and in the same direction as the old Ptolemaic deferent circle (D_P), and in turn carries the Ptolemaic epicycle (E_P). The combination of the equal motions allows the lines joining the extremities of E_T's radius to points K and Q respectively to be always parallel. This makes the centre of the Ptolemaic epicycle (C_P), now carried at the extremity of al-Urdī's small epicycle (E_T), appear to be moving uniformly around Ptolemy's equant point (Q). In fact, it moves around the centre (C_T) of its own epicycle (E_T); and C_T in turn moves around the centre (K) of the new deferent circle (D_T). All the spheres now move uniformly and in place, around axes that pass through their centres. The Urdī lemma thus avoids the physical impossibility of the equant, while retaining its utility for explaining astronomical observations[20].

In the meantime, al-Tūsī investigated the problem of a planet's celestial latitude. A mechanism was required to reproduce the latitudinal 'see-saw' motion around the plane of the ecliptic without affecting the celestial longitude. He proposed the theorem now known as the Tūsī couple, which describes a small interior circle rotating inside a larger exterior circle with a diameter twice that of the interior circle. The rotation causes a point (P) on the circumference of the interior circle to oscillate back and forth in linear motion along a diameter of the exterior circle, thus transforming two uniform circular motions into a linear motion. Applying this theorem to the Ptolemaic model, al-Tūsī replaced the circles with spheres, with an interior sphere placed tangentially inside an exterior sphere. The result was that the desired latitudinal oscillations could be reproduced without affecting longitudinal motion. The principle is used in mechanical engineering to translate a piston's reciprocating motion into the rotary motion of a wheel. With a name that recognises its astronomical origins, the so-called 'sun-and-planet' gear was invented by Scottish engineer William Murdock, and later patented by James Watt[20].

A hundred years later, Damascus-born astronomer Ibn al-Shātir (*c.*AD 1305-1375) took full advantage of the Urdī lemma and Tūsī couple to systematically reformulate Greek astronomical theories. His intention was to formulate mathematical theories of solar, lunar, and planetary motions that were consistent with the cosmological presuppositions of the ancient Greeks. He accepted that the Earth was the centre of the Solar System and the paradigm of regular circular motion[6].

Ibn al-Shātir published a work entitled *Nihāyat al-su'l fī tashīh al-usūl* ('A final inquiry concerning the rectification of planetary theory') in which he developed a set of theoretical models for describing planetary motions. These models were intended as alternatives to the Ptolemaic models. The observational data to be fed into these models was supposedly discussed in another work, *Tasliq al-Arsad* ('Discourse on Observations'), which unfortunately is lost. However, we know that he obtained an angular diameter of the Sun that varied from 0°29'5" at apogee to 0°35'55" at perigee (a much greater variation than predicted by Ptolemy), and that he confirmed that the solar apogee changes at a different rate to the precession of the equinoxes[21].

The lunar model eliminated the great variation in the Moon's distance that was a major issue with the Ptolemaic lunar theory. The solar model accounted for the variation in the angular diameter of the Sun, and for the movement of the solar apogee. It used four spheres to describe the Sun's behaviour: a 'parecliptic' concentric with the Earth (E), of radius 60 'parts', rotating in a prograde direction to account for the Sun's daily mean motion; a smaller deferent circle of radius $4\ ^{37}/_{60}$ 'parts' carried by the parecliptic and rotating at the same speed, but in a retrograde direction; and a 'director' of radius $2\ ^{30}/_{60}$, carried by the deferent circle and carrying in turn the Sun. The fourth sphere was the 'all-encompassing one', which rotates slowly in a prograde direction to account for the movement of the solar apogee. In accordance with the Urdī lemma, the Sun appears to move at a uniform speed around a point C, where EC represents the eccentricity $= (4\ ^{37}/_{60} - 2\ ^{30}/_{60})/60 = (2\ ^{7}/_{60})/60 = 0.0353$. This is close to Ptolemy's eccentricity of 0.0417 and thus the model predicts longitudes very similar to those predicted by the Ptolemaic model; but unlike the latter it allows for a variation in the angular diameter of the Sun[21]. So successful were these methods in reformulating Ptolemaic astronomy that over the last few decades it has become clear that they were utilised by Copernicus 150 years later. His lunar theory was identical to that of Ibn al-Shātir, and he made use of the Tūsī couple in his model to describe the motion of the planet Mercury[22] (see Chapter 17).

The 52 km (32 mile) diameter lunar crater Nasireddin and the asteroid 7058 Al-Tusi are named for al-Tūsī.

Legacy

It will be clear from the above that Islamic scholars deeply questioned the Greek scientific tradition; they then dismantled, overhauled, and extended it, before finally reconstructing the whole edifice on a much sounder basis. Yet the view persists that they somehow did little more than serve as scribes, translating and preserving the learned texts of the ancient Greeks until such time as Europe would emerge from eclipse and resume its leading role. This view fails to recognise a genuine rebellion against the limitations and inconsistencies of Greek astronomical theory and practice, and how for centuries Islamic scholars sought to establish their own ideas; these ideas would eventually play a significant part in the seismic paradigm shift now known as the Copernican revolution[23,6].

15
China

The early civilisations of China

Classical texts place the earliest Chinese civilisations on the central plains of the Yellow River. The semi-mythological Five Emperors and the Xia, Shang, and Zhou Dynasties are described by the historian Sima Qian (*c.*145-86 BC) in a work entitled the *Shiji* ('Records of the Grand Historian'), compiled between 109 and 91 BC. The Five Emperors reigned during a remote period characterised by walled cities and recurrent warfare among rival chiefdoms. This period was followed by the emergence of the Xia Dynasty, traditionally regarded as China's first dynasty. Archaeologists have associated the Five Emperors period and the Xia Dynasty with the archaeological cultures known respectively as the Longshan and the Erlitou. The Xia was followed by the Shang, then the Zhou. They are often referred to collectively as the Three Dynasties. The Longshan dates from around 2800 BC, and it is divided into Early (2800 to 2600 BC) and Late (2600 to 2000 BC); the Xia Dynasty dates from 2070 to 1600 BC; and the Shang Dynasty from 1600 to 1046 BC.

The Zhou Dynasty comprised two periods: the Western Zhou Dynasty from 1045 to 771 BC, and the Eastern Zhou Dynasty from 770 to 221 BC. The latter is divided into two periods: the Spring and Autumn Annals period from 770 to 481 BC, and the Warring States period from 481 to 221 BC. Both names originate from contemporary historical texts: *Chunqiu* ('Spring and Autumn Annals') by Confucius, and *Zhan Guo Ce* ('Strategies of the Warring States'), thought to be the work of several authors during the period in question. During the Warring States period, China fragmented into seven large competing states and many smaller ones, and a series of wars were fought over a period of 260 years. Eventually, Qin in the west prevailed over its rivals and re-established a united China under the Qin Dynasty. However, the new empire did not long survive the death of its founder, Ying Zheng, in 210 BC. Rebellions broke out, and there followed another eight years of fighting before the pivotal Han Dynasty emerged victorious in 202 BC. The era of the Han Dynasty (202 BC-AD 220) is regarded as a golden age in China, and to this day the main ethnic group in China identifies itself as 'Han Chinese'.

Mandate of Heaven

China can lay claim to one of the world's oldest astronomical traditions. Chinese astronomical records go back around four thousand years and throughout that time there was an intense interest in predicting the movements of the Sun, Moon, and planets, and an insatiable demand for accuracy. The driver for this tradition is commonly supposed to be a need to provide farmers with an accurate calendar so they would be able to carry out tasks such as sowing and harvesting at the right time: China is one of only a few places in the world where agriculture and state-level societies arose independently of developments elsewhere, and its economy remained predominantly agrarian until the overthrow of the final imperial dynasty in 1912. But such explanations fail to take into account the variability of weather and local conditions, and also presuppose that farmers could not judge for themselves when conditions were right for seasonal activities. Moreover, the degree of calendrical precision attained far exceeded any practical considerations[1].

The real motives behind this ancient astronomical tradition are undoubtedly related to the Chinese belief that society is a reflection of the heavens. The ancient and Imperial-era Chinese regarded celestial phenomena as *tian* ('Mandate of Heaven'): a sacred regulation of matters pertaining not only to an agrarian economy but to society as a whole. The emperor was viewed as a pivotal element linking the human microcosm to the natural macrocosm, and his role was to ensure the orderly functions of both realms. Any unexpected celestial phenomenon, such as a comet or an unexpected eclipse, would be seen as ominous and reflect badly upon him. But if an eclipse could be predicted, its significance as a portent would be greatly reduced. It was therefore essential that that as many astronomical phenomena as possible be reduced to predictable occurrences. Furthermore, it was necessary that various imperial rituals be carried out at the correct time: if, for example, the timing of the winter solstice was out by even an hour, the prediction could fall on the wrong side of midnight. The appropriate rituals would not be carried out on the correct day, with potentially harmful effects. Such considerations extended to all levels of society: the timing of activities ranging from marriage to business deals relied on choosing 'lucky' days and avoiding 'unlucky' ones. It could be argued that there was much state-sponsored scientific activity for reasons that have little to do with modern science[2].

The predictable, regular motions of celestial bodies were considered to be *lixiang* ('Calendrical Phenomena') and all other astronomical phenomena were *tianwen* ('Celestial Patterns'). To track Calendrical Phenomena, Chinese astronomers devised systems of *lifa* ('Calendrical Astronomy') for predicting the motions of the Sun, Moon, and planets. In the meantime, the Celestial Patterns of heavens were carefully watched for portents meaningful to rulers. Astronomical and astrological activities were wholly controlled by the Emperor, *tyanzi* ('the Son of Heaven'); this practice dates to no later than the Zhou Dynasty[3].

Only the Emperor was entitled to maintain a *lingtai* ('Platform for Heavenly Communication'), which was both an observatory and a site for the worship of heaven. Unofficial astronomers could well be rebels preparing their own calendars – and the results of the official astronomers were kept partly secret. To maintain a precise calendar was vital to the prestige of any ruler and one of their first priorities upon taking office was to commission a new calendar[4,3]. *Scientia potentia est.*

A Longshan astronomical observatory

Taosi is a Longshan site located in Shanxi Province, north-central China. The site is located 20 km (12 miles) from Pingyang, which is believed to have been the capital of the Emperor Yao, the penultimate of the proto-historical Five Emperors. Yao is supposed to have been a sage ruler who lived around the twenty-third to the twenty-second century BC. Three decades of excavations at Taosi have revealed the ruins of a walled city which was established in the Early Taosi period(*c.* twenty-third to twenty-second centuries BC), and considerably expanded during the Middle Taosi period (*c.* twenty-first century BC). In this second phase of development, the city attained an area of 2.8 sq. km (1 sq. mile), making it one of the largest known settlements of this era. Archaeologists have uncovered a palace enclosure, elite residential area, royal cemeteries, ceremonial centres, craft production districts, an exclusive storage area, and a commoners dwelling area. These features make Taosi a good example of an early urban centre. Of particular interest is a structure in the southeast of the site, which is believed to have been an observatory[5].

The structure comprises three levels. The innermost, top level is a semi-circular platform 21 m (69 ft) in radius. On its boundary is a curved wall 22.5 m (83 ft 8 in) long, 1.1 m (3 ft 8 in) wide, and 2.7 m (8 ft 9 in) in depth. The wall was originally made up of 11 rectangular pillars made of rammed-earth, erected at regular intervals from south to north along its arc to create a series of slots (E1-E10). Each slot has a width of about 150-200 mm (6-8 inches). Further to the north and on the middle level of the structure are two further pillars, making a total of 13 pillars and creating two further slots (E11 and E12). As viewed from the centre of the platform, the slots are oriented toward the eastern horizon. After taking into account changes in the **obliquity of the ecliptic**, it was found that at around 2100 BC the range in azimuth defined by the slots E2-E12 matched the arc defined by the azimuth of the Sun at sunrise between the summer and winter solstices to within a few minutes of an arc. This was a strong hint that the site was used to observe the rising sun in order to determine the seasons, with the slots serving as backsights. Sure enough, when archaeologists excavated the area at the centre of the arc, they found a rammed-earth core 250 mm (10 inches) in diameter, which probably served as the observation point[6,7].

As viewed from the observation point, the southernmost slot E1 lies about 6° further south than slot E2, and the Sun could never have risen in it. However, the Moon could rise in the slot, suggesting that it was used to mark the **lunar standstill limit** – relevant to the prediction of eclipses. Observing the sunrise in slot E7 gives the dates of the equinoxes, albeit with less accuracy than the solstices. The slots as a whole give twenty dates of the year in the form of a 'horizon calendar'. However, this number differs from the 24 dates of the traditional Chinese solar calendar, where the year was divided into roughly equal intervals[6,7].

A painted wooden stick with calibrated scales was found in a tomb at Taosi in 2002. The size and contents of the tomb suggested the person buried there must have been royalty. The wooden stick was decomposed, but its shape was restored from paint remains embedded in the soil. Estimated to have been 1.718 m (5 ft 8 in) long, the stick was painted in a repeating pattern of light green and black bands, delimited by narrower pink bands. There is one exception, where a pink band separates two light green bands. The stick appears to be a template for a gnomon, i.e., a graduated scale for measuring the length of shadow cast by the gnomon. If the height of a gnomon is 8 *chi* (approximately 2 m), the pink markings on the painted template will indicate lengths of the gnomon shadow at noon on specific days, producing a series of annual dates. The *chi*, sometimes known as the Chinese foot, is a traditional unit of length that has varied over time. Archaeologists assumed that the gnomon stick had originally measured 1.98 m, corresponding to a *chi* of 247.5 mm (9.74 inches)[8,9].

On the day of the summer solstice, the gnomon shadow extends almost exactly to the pink band between the two light green bands. The stick is not long enough to determine the winter solstice; it would have had to be 3.444 m in length. However, it could be flipped over to measure longer shadows, from which it was suggested that the original length was 1.722 m. That the gnomon was found in a royal tomb implies that it was a symbol of power. The dates derived from the pink markings are very consistent with those obtained from the observatory's twelve wall slots[8].

Gnomon shadow measurement was a fundamental part of ancient Chinese astronomical tradition. The earliest records of such measurements in China are found in the *Zhou li* ('Rituals of the Zhou Dynasty') and the *Zhou bi suan jing* ('Arithmetical Classic of the Gnomon and Circular Paths'), which date to the sixth century BC. The gnomon was crucial for determining the seasons, cardinal points, and geographic latitude. A traditional Chinese gnomon was 8 *chi* high and combined with a template to measure the shadow lengths at noon during the year[8].

Some early documents, such as the *Yao dian* ('Canon of Yao') in the *Shangshu* ('Book of Documents') mention astronomical practices in remote ancient times. The Canon of Yao was compiled in the late first millennium BC, but it is believed to contain historical information from the time of Emperor Yao. The Canon of Yao notes that Yao had commissioned two brothers named Xi and He to compile a calendar and "...*reverently to follow August heaven, calculating and delineating the sun, moon, and other celestial bodies in order respectfully to grant the seasons to the people. . .*". This could be a reference to the type of work being carried out at the Taosi observatory. The brothers were also instructed to obtain timings for the solstices and equinoxes, using the four groupings Bird (α Hydrae) for spring, Fire (Antares) for summer, Void (Altair) for autumn, and Hair (the Pleiades) for winter. Assuming the four groups marked the positions of the solstices and equinoxes at that time, then the date of the observation can be shown to have been around the twenty-third century BC. Scholars have always been sceptical about the reliability of accounts such as the Canon of Yao, but the Taosi observatory suggests that astronomical knowledge was already well advanced before the 21st century BC. The calendar was to be used in "*regulating the various officers to make all works in the year fully performed*", confirming that the importance Chinese rulers attached to calendrics goes back to late prehistoric times[2,3,7].

The Chinese calendar

Most Westerners will be familiar with the Chinese calendar, but there is much more to it than what most people would associate with a 'calendar'. The Chinese calendar was not just a means of arranging days, months, and years; it included mathematical techniques for predicting the movements of the Sun, Moon, and five planets, and it was the basis for producing astronomical ephemerides and annual almanacs. The calendar's ultimate purpose was to provide knowledge and technique for maintaining harmony between Heaven and the human world[1].

Even the familiar twelve animals of the traditional Chinese calendar are only part of a larger cycle. In order, they are *Zi* (Rat), *Chou* (Ox), *Yin* (Tiger), *Mao* (Rabbit), *Chen* (Dragon), *Si* (Snake), *Wu* (Horse), *Wei* (Sheep/Goat), *Shen* (Monkey), *You* (Rooster), *Xu* (Dog), and *Hai* (Pig). Thus, 2020 is the Year of the Rat, 2021 will be the year of the Ox, and so on. The cycle will begin again in 2032. The twelve animals are often referred to as the Chinese zodiac, although they are not related to **ecliptic** constellations.

But the animals are actually the second of two components assigned to each year in a sixty-year cycle, and they are known as the *Zhi* ('earthly branches'). The first component is known as the *Gan* ('heavenly stems'), of which there are ten: *Jia* (growing wood), *Yi* (cut timber), *Bing* (natural fire), *Ding* (artificial fire), *Wu* (earth), *Ji* (earthenware), *Geng* (metal), *Xin* (wrought metal), *Ren* (running water), and *Gui* (standing water). The first year of the cycle is *Jia-Zi*, the second *Yi-Chou*, and the tenth is *Gui-You*. At this point, the *Gan* cycle restarts; the eleventh year is Jia-Xu and the twelfth is *Yi-Hai*. The *Zhi* cycle then restarts, so the thirteenth year is *Bing-Zi*. The cycle continues until the sixtieth year, which is *Gui-Hai*, then it restarts with *Jia-Zi*. The current cycle began in 1984. The *Gan-Zhi* system was originally used as a sixty-day cycle for naming days. It was not until the Han period that its use was extended to naming years[2].

The traditional Chinese calendar is lunisolar, requiring periodic intercalation of a month. Although the Gregorian calendar is used for administrative purposes in the Peoples Republic of China, the traditional calendar is still used for festivals and for timing agricultural activities, and by Chinese communities around the world. The Chongzhen calendar was the final version of the lunisolar Chinese calendar. It was developed by the Jesuit scholars Johann Schreck and Johann Adam Schall von Bell with Chinese mathematician Xu Guangqi from 1624 to 1644 for Chongzhen, final emperor of the Ming Dynasty. After Chongzhen's death, the calendar was adopted by new Qing Dynasty emperor, Shunzhi, and renamed the Shíxiàn calendar.

The Shíxiàn calendar's workings are as follows[10]: in an ordinary year of twelve months there are 353-355 days, and in a leap year of thirteen months there are 383-385 days. Months begin on the day of the new Moon, but this is taken to be actual **syzygy** rather than the first sighting of the evening crescent. There is no consistent system of alternation between 29-day and 30-day months. A day runs from midnight to midnight. A day was divided into twelve *shi*; hence one *shi* is two hours. The *shi* are numbered using the *Zhi* 'earthly branches' system.

In addition, the **tropical year** is divided into 24 'terms', each corresponding to a movement of the Sun through 15 degrees of longitude. Starting from winter solstice, the odd-numbered solar terms are known as *zhōngqì* ('major solar terms'), and the even-numbered solar terms are called *jiéqì* ('minor solar terms'). The tropical year thus contains 12 major terms and twelve minor terms. Like the season, and for the same reason (the non-uniform movement of the Sun along the ecliptic), terms are not all the same length.

Between successive winter solstices there will be either twelve new Moons or thirteen new Moons. In the latter case, an intercalary month known as a *jian* is required. Despite the extra month, there will still be only twelve major terms between the two winter solstices. It follows that one of the next thirteen months will not contain a major term, and the intercalary month is inserted after that month. The Chinese New Year begins on the first day of the second regular (i.e., non-intercalary) month after the month containing the most recent winter solstice. The twelve regular months are either numbered or use the *Zhi* 'earthly branches' naming convention.

The above system was the endpoint of many centuries of refinement. There are no calendrical treatises from prior to the Qin period, but oracle bone inscriptions from the Shang period suggest that dates were recorded using the *Gan-Zhi* ('stems and branches') system. In the 1940s, archaeologist Dong Zuobin reconstructed the Shang calendar from the inscriptions and showed that it was lunisolar. Intercalary months were inserted every two or three years to keep the calendar in step with the solar year, although there was no fixed rule for their placement. During the Zhou period, calendar-making became more formalised and state officials were assigned to the task[1,11].

By the fifth century BC the Chinese were aware of the Metonic cycle of 235 **synodic months** (*zhang yue*) equals 19 tropical years (*zhang fa*). These do not correspond to a whole number of days, but when multiplied by four they give 76 years (*bu fa*) = 4 × 235 = 940 months (*bu yue*) or 27,759 days (*bu ri*); and hence one synodic month = 29 $^{499}/_{940}$ days. The 76-year cycle was known as a *bu*. The number of intercalary months in a *bu* is 940 - 76 × 12 = 28, i.e., seven intercalary months to be added in every 19 tropical years. If the number of years elapsed since the start of the *bu* is *n*, and if 235 × *n* / 19 gives a remainder greater than 12, then an intercalary month is required. It is possible that Chinese astronomers learned of the Metonic cycle from the Babylonians (see Chapter 10)[4,12].

The Quarter Remainder calendar was introduced at this time, and it was based on a year of 365¼ days. The system was used by all the rival states during the Warring States period, but they did not all use the same month for the beginning of the year. One of the first three months of the *Zhi* 'earthly branches' sequence was commonly used. However, the victorious Qin dynasty used the twelfth month. Although the theory of the Mandate of Heaven required a new ruling dynasty to establish its own calendrical system, the Han Dynasty simply adopted the same calendar and continued to use it despite the consternation of state officials. It was not until the time of Emperor Wu (reigned 141-87 BC) that a calendrical reform known as the *tai chu* ('Great Inception') was instigated. The new calendar took its start point to be midnight on the day of the winter solstice of 105 BC, coinciding with the new Moon and the start of the month. The new system was later revised by Emperor Liu Xin (reigned 7-1 BC) to become the *san tong* ('Triple Concordance') calendar[1]. In addition to Sun, Moon, and planet data, the Triple Concordance calendar contained a list of historical dates, the lengths of tubes for producing musical notes, units of length and volume, and standards for temple buildings and ceremonial dress[4,2].

Although it might seem strange, the Triple Concordance calendar replaced the 365¼ day year with a less accurate year of 365 $^{385}/_{1539}$ days. This fraction was obtained by combining the Metonic relation of 235 months equals 19 years with a new value for the synodic month of 29 $^{43}/_{81}$ days, i.e., 81 synodic months = 2,392 days, to give the number of days in the year as 235/19 × 2,392/81 = 365 $^{385}/_{1,539}$ days. Over the next seven hundred years, other fractions were used: *da ming* (AD 462) 9,859/3,9491 (= 0.24965 days); *tian bao* (AD 550) 5,787/23,660 = 0.24459 days; and *da yen* (AD 724) 743/3,040 = 0.24441 days. The denominators of these rather unwieldy fractions were known as *ri fa* ('day denominator'). Various sources list a total of 47 different *ri fa*[4].

Chinese calendrical theory had a tendency towards interlocking cycles, so that the first day of a large cycle would also be the first day of each of the smaller cycles it contains; for example, the first day of the year must coincide with the new Moon. This was a problem in as much as it resulted in attempts to find the *shang yuan* ('Grand Origin Point'). It meant that rather than use parameters obtained by direct measurement, Han period astronomers preferred to derive them from cycles thought to have important cosmic meaning. For example, the *ri fa* for the Triple Concordance calendar, 1,539, is 81 × 19 (9 × 9 × 19). Nine was regarded as the largest Heavenly number, and nineteen is the sum of the largest Heavenly number and the largest Earthly number (ten), as well as being the number of years in the intercalary cycle. Similarly, the new value for the synodic month reflects a desire to use the number 81 as a denominator, albeit the older value was more accurate[2,1].

The precise prediction of lunar and solar motions was essential for the prediction of lunar and solar eclipses, which were regarded as a key function of the calendar. Until the discovery of the **solar anomaly** and the **lunar anomaly**, predictions were not very accurate and relied on eclipse cycles similar to the saros. However, Chinese astronomers obtained eclipse cycles from their own observations and different calendars used different cycles; for example, the Three Concordance Calendar relied on an eclipse cycle of 135 synodic months[1]. The problem was compounded by the use of values for the tropical year and the synodic month that were too high, with the result that the calendar gradually lagged behind the heavens. By the time that the Triple Concordance Calendar had been in use for a century, it was out by almost a day[2].

Over time, calendars became more refined; for example, the lunar anomaly was incorporated into the *qian xiang* ('Supernal Emblem') calendar by Liu Hong in AD 179 and precession into the *yuan jia* ('Epochal Excellence') calendar by He Chengtian in AD 443. Planetary motions were computed using methods similar to the Babylonian System A whereby a planet's synodic period was divided into up to six zones, with the planet assumed to be moving at a constant speed in each. Around AD 560, Zhang Zixin of the Northern Qi dynasty discovered the solar and **zodiacal anomalies**, which were first incorporated into the *huang ji* ('Sovereign Pole') calendar by Liu Zuo around AD 600. The model used now resembled the Babylonian System B, with the speed of a planet increasing and decreasing in each zone. The number of zones was increased to a maximum of ten (see Chapter 10)[1].

A weakness of Chinese astronomical calendars was that revisions and refinements typically only happened when a new emperor came to power rather than in response to more accurate data. As noted above, during the Warring States period between the third and sixth centuries AD, China was fragmented into a series of independent states. Ironically, this political instability resulted in many of the advances in Chinese astronomy, as each state supported its own astronomical establishment. Over the next four hundred years, the accuracy of planetary motion calculations progressively improved. The emperors of the Song Dynasty (AD 960-1279) invested considerable resources in astronomy, with an emphasis on verification by observation. By the eleventh century AD, Chinese calendars had attained a level of accuracy that would not be achieved in Europe until the sixteenth century AD[1].

The Yuan Dynasty (AD 1271-1368) is familiar to us as the dynasty of Kublai Khan and the period when Marco Polo visited China. Early on in this period, astronomer Guo Shoujing (1231-1316) made a series of precise astronomical measurements using large instruments of his own design. The *shou shi* ('Season Granting') calendar was based on these results[1], which included accurate results for the length of the tropical year and the obliquity of the ecliptic. The year length was given as a decimal rather than a fraction: 365.2425 days, which is within 20 seconds of the modern value of 365.242189 days. A value of 23.903 *du* (23°33'34") is given for the obliquity of the ecliptic, close to the correct figure (23°32'01") for that epoch. Unfortunately, it is not known how either result was obtained. The Season Granting calendar also makes use of sophisticated mathematics for describing planetary motions[4].

A unique sky system

The ancient Chinese developed a system of mapping the heavens that was unlike any that were adopted in western countries. Instead of 360 degrees in a circle, there were 365¼ *du*; i.e., one *du* for every day of the year. The sky was divided into 28 orange-like segments of unequal width, known as *xiu*, each of which was named for a lunar 'mansion' or 'lodge'. These were equatorial constellations that served as reference points for the measurement of the position of celestial bodies. For each *xiu*, one star was designated *ju xing* ('determinative'). A star was deemed to be in a particular *xiu* if it reaches upper **culmination** after the determinative of the *xiu* and before that of the next *xiu*. The longitudinal width of each *xiu* was the longitudinal distance between its determinative star and that of the next *xiu*; the sum of the widths of the 28 mansions was 365¼ *du*. The position of a celestial body was defined by its distance from the celestial pole (*qu ji du*) and distance into a certain *xiu* (*ru xiu du*), both being given in *du*. The *qu ji du* is thus equivalent to the **declination** and *ru xiu du* is equivalent to **right ascension** of the equatorial coordinate system. It has been suggested that the system is related to the Indian *nakshatras* (see Chapter 13), of which there were also 28. But the widths of these were all very similar, whereas the *xiu* varied in width from 33° to 2°, suggesting different origins[4,2,13].

Despite the apparent equivalence of this sky system to a celestial sphere, Han period astronomers and their predecessors had not yet begun to think of the heavens as spherical; the concept was not widely accepted until the first century BC. The *du*, which is readily interpreted as a measure of angular distance, was to them a measure of time. The *du* can be thought of as the progress along the ecliptic of the **mean Sun** in one day. If two stars are separated by a certain number of *du* (say n), they will reach upper culmination at a particular time (say t) n days apart[2].

The first names of stars date to the Shang period. There are references to the 'fire star' (Antares) and the Canon of Yao names the groupings Bird (α Hydrae), Fire (Antares), Void (Altair), and Hair (the Pleiades) in connection with the four seasons. The *Xia xiao zheng* ('Small Calendar of the Xia Dynasty') makes reference to the Warrior (Orion), South Gate (α Centauri), Northern Dipper (Ursa Major), and Weaving Damsel (Vega), along with their culminations at dusk and dawn, their **achronycal rising**, **cosmical setting**, **heliacal rising**, and **heliacal setting**. All told, just under forty names of stars are found in pre-Qin period literature. Most of these names pertain to stars of the 28 mansions[13].

The earliest star catalogues were compiled by astronomers Shi Shen and Gan De in the fourth century BC. Some 93 constellations plus the 28 *xiu* are attributed to Shi Shen and a further 118 are attributed to Gan De. 44 constellations were also attributed to an astronomer named Wuxian, who supposedly lived in the ninth century BC. None of their maps have survived and it is believed that many of the constellations attributed to them were actually named by astronomers during the Han period. The earliest surviving work to describe the night skies is *Tian guan shu* ('Book on the Celestial Officials'), which is a chapter of Sima Qian's *Shiji*. The title suggests that Sima Qian regarded the constellations as 'officials of the sky'. He divided the sky into five palaces (*gong*), with a central palace around the north celestial pole and four others on the ecliptic at the cardinal points. The central palace held constellations with names such as emperor, queen, princes, and various court officials. The other palaces held the most prominent stars of the four seasons, with the 28 *xiu* divided into four groups of seven, one for each ecliptic palace. The *Tian guan shu* lists a total of around 90 constellations, but it appears to be based heavily on Shi Shen's list[13].

The definitive list of Chinese constellations was established by astronomer Chen Zhuo, who lived during the Three Kingdoms period (AD 220-280). He consolidated the constellations attributed Shi Shen, Gan De, and Wuxian into a single work with 283 constellations comprising 1,464 individual stars. The constellations were based on the Chinese concept of a correlation between Heaven and humanity. In addition to the royal court of the central palace, the other 'palaces' held various public buildings, military installations, materiel, and personnel, common folk, scenes of rituals, ceremonies, and social life, philosophical and religious concepts, mythological figures, and geographical features. Thus, an entire society with all its hierarchical complexity was projected upon the sky, with higher-ranking people placed closer to the north celestial pole. The constellations visible at different times of the year reflected the seasons; for example, harvest scenes in autumn[13].

The earliest known Chinese star maps are to be found on the ceilings of tombs. The purpose appears to have been to afford a view of the night sky to the deceased. The oldest such maps date to the Han period. One such tomb was excavated in 1990 on the campus of Xi'an Jiaotong University in Shaanxi Province. The 28 *xiu* are depicted in a circular belt, organised into four groups that represent the 'Four Auspicious Beasts' of Chinese tradition: the Azure Dragon of the East, the Vermilion Bird of the South, the White Tiger of the West, and the Black Tortoise of the North. These animals are also depicted. The 'fire star' Antares is depicted in red. The Sun is portrayed as a round disk with a three-legged crow, and the Moon as a round disk with a toad. All are superimposed onto a colourful background of swirling clouds[13].

The earliest star maps depicting the whole sky are those in the Dunhuang Collection of manuscripts from the Buddhist Archives in Dunhuang, Gansu Province, which date to the eighth to ninth century AD. There are twelve equatorial maps and one showing the northern circumpolar stars. The positions of the constellations are reasonably accurate, although the maps are not intended as a proper star atlas. Other early star maps include the *Tianwen tu* ('Astronomical Planisphere') of the Song Dynasty, which is engraved on a stone tablet and is preserved in Suzhou in Jiangsu Province and the *Xin yixiang fa yao* ('Essentials of the New Instruments') by astronomer Su Song (AD 1020-1101). The *Tianwen tu* depicts the circumpolar circle, ecliptic, equator, Milky Way, and longitudinal lines corresponding to the 28 *xiu*, with star positions accurate to within 1°30'. The *Xin yixiang fa yao* is a set of maps that were used in the Water-Powered Astronomical Clock Tower, a large automated clock combining an armillary sphere with a celestial globe. Measuring 12 m (39 ft 4 in) in height and 8 m (26 ft 3 in) in width, it was powered by scoops which received water from a water clock[13,14].

Cosmology

Remarkably, Chinese astronomers believed that the Earth was flat, and they continued to do so until Jesuits reached China at the end of the sixteenth century. Even then, the Jesuit astronomer-priests were still wedded to the geocentric Tychonic system (see Chapter 17) rather than the heliocentric model of Copernicus[4]. Despite the considerable achievements of Chinese astronomers during the period from 500 to 200 BC (roughly from the time of Confucius to the rise of the Han Dynasty), little thought seems to have been given to the shape, size, arrangements, and motions of the Earth and the heavens prior to the first century BC[2].

However, around AD 180 the scholar Cai Yong described three schools of cosmological thought. The first of these was the *gai tian* model, which held that the Earth was a flat square surface, overlain by a rotating hemispherical dome. The term *gai* refers to a large umbrella-like canopy used to cover an ancient Chinese chariot. The celestial canopy was pivoted around the north celestial pole. The Sun, lying some way from the pole, is carried over the various regions of the Earth to give day in each region; when it moves away from a region, night falls. The Sun is closest to the celestial pole in summer and furthest away in winter. A second model, *hun tian*, proposed a celestial sphere rotating on a central axis, but the Earth was viewed as a flat surface floating on water. Essentially, the *hun tian* universe was a planetarium, which reproduced the celestial phenomena seen by an observer in the latitude of the central plains of the Yellow River. Finally, the *xuen ye* model envisaged a flat Earth surrounded by infinite space in which the Sun, Moon, planets, and stars were composed of freely floating condensed vapour. Cai Yong noted that the *xuen ye* model had been abandoned, and while the *gai tian* model was still in use, its computational methods gave poor results in comparison to the *hun tian* model[2].

As noted above, Han period astronomers and their predecessors lacked the concept of a celestial sphere. This would not necessarily have been a handicap as the Babylonians also managed perfectly well without it (see Chapter 10). However, there was a significant difference between Babylonian and Chinese astronomy. The Babylonians were primarily interested in horizon phenomena, whereas Chinese astronomers took a good deal of interest in meridional phenomena, i.e., the timing of culminations[15,16,2]. The importance attached to these is underlined by the primary role of the gnomon in Chinese astronomy, which was required to find the direction of the meridian. Actual timings of meridian transits were made using a water clock[2,17].

The first century BC saw the *hun tian* model and the concept of the celestial sphere adopted. Gnomons gradually fell out of use and were replaced by instruments such as the armillary sphere (*hun yi*), with the view that astronomical instruments should reflect the structure of the heavens. At the same time, there was a shift in emphasis from meridional observations to measuring the angular separation of celestial bodies[2,18].

An important feature of Chinese cosmology was the belief that the Earth had a centre. The *Zhou li* claimed that the founders of the Zhou Dynasty had built their capital at the centre, having used a gnomon to determine its location. They believed that this lay at the latitude where a standard 8-*chi* gnomon cast a 1.5 *chi* shadow at noon on the day of the summer solstice. By the first century BC, astronomers were identifying the location with Yangcheng (now Gaocheng, near Dengfeng, in Henan Province). Thereafter, Yangcheng became the location of choice for making calendrical measurements of the Sun's shadow length. It was incorrectly believed that the length of the Sun's noontime shadow length on any given day of the year increased in proportion to the distance north of Yangcheng and decreased similarly in proportion to the distance south. The shadow cast by a standard gnomon was believed to shorten by one 1/10 *chi* (1 *cun*) for every thousand *li* of southward travel (1 *li* = 1,800 *chi*). That this was incorrect (due to the Earth being spherical rather than flat) was not appreciated until experiments were carried out in the fifth and sixth centuries. Eventually, between AD 721 and 725, a monk-astronomer named Yi Xing carried out a survey at thirteen points between 52° N in the north and 17°24' N in the south, including Yangcheng. He obtained a complete set of values for the altitude of the north celestial pole and the length of the Sun's shadow at noon at the solstices and equinoxes, and from this data he correctly concluded that it is the altitude of the north celestial pole rather than the noon shadow length that changes in proportion to the distance along the meridian. To mark the 'prime meridian' at Yangcheng, a special stele was erected there in AD 723[4,9].

Observational astronomy

China has the longest continuous tradition of recording astronomical phenomena anywhere in the world. The 28 official histories of the Chinese dynasties contain a wealth of astronomical information. Unusual events in the heavens were of particular significance because of their astrological significance. Records cover phenomena including solar and lunar eclipses, sunspots, 'guest stars' (novae and supernovae, and possibly variable stars), comets and meteors, and a wide range of planetary phenomena. Other phenomena recorded include earthshine on the crescent Moon, auroras, oddly shaped clouds, rainbows, parhelion ('sun dogs'), and lunar and solar halos. Thirteen oracle bone inscriptions from the Shang period refer to solar and lunar eclipses, and the *Spring and Autumn Annals* contain records of 37 solar eclipses. Nearly 100 'guest stars' are recorded in ancient Chinese literature, including the supernovae of 1006 and 1054. Chinese astronomers noted that the tails of comets always point away from the Sun, and there are records of sunspot observation from as long ago as 28 BC. The earliest western reports of sunspots are from Russia in the fourteenth century AD[4,19].

The *Shangshu* contains one of the earliest records of a solar eclipse, occurring in the reign of King Zhong Kang of the Xia Dynasty. According to later accounts, two court astronomers named Hsi and Ho failed to give a warning of the eclipse allegedly due to being drunk. Zhong Kang was distinctly unimpressed and ordered Prince Yin to command an expeditionary force to punish Hsi and Ho. The eclipse has now been dated to 16 October 1876 BC[20], although the rest of the account is almost certainly apocryphal. The Chinese did not learn how to predict eclipses until much later.

Prior to the invention of paper during the Han period, Chinese astronomical observations were recorded on oracle bones, strips of bamboo, and silk manuscripts[21]. Oracle bones are ox shoulder blades and the undersides of turtle shells upon which questions were carved with a sharp instrument prior to heating until the bone cracked. The crack patterns were then interpreted, and the prognostication written on the piece. The questions typically concerned warfare, hunting, rainfall, agriculture, or the health of members of the royal family. Over 100,000 have been found, mainly from Anyang in Henan Province. Most date to the Shang period, though examples dating back to the Longshan Culture have been found at Chengziyai in Liaoning Province. They were first noted in 1899, when scholars identified archaic texts inscribed upon bones that were intended for use in traditional Chinese medicine. Unfortunately, this means that many have been lost[22,23].

Astronomical records were first identified from the inscriptions by the archaeologist Guo

Moruo in the 1930s. Working from Japanese collections of ink rubbings of the inscriptions, he listed star names and other astronomical phenomena. Subsequent work, especially since the 1980s, has revealed extensive records concerning the Sun, Moon, planets, comets, stars, and solar and lunar eclipses[11].

The earliest silk manuscripts date back to 300 BC. The Mawangdui Silk Manuscripts were found in the tomb of the son of a prefecture chief in 1973, and they date to 168 BC. There are three texts in the collection that relate to astronomy or celestial divination: *Wuxing zhan* ('Divination on the five planets'), *Tianwen qixiang zazhan* ('Miscellaneous divination on the astrological and meteorological phenomena'), and *Riyue fengyu yunqi zhan* ('Divination on the Sun, Moon, wind, rain, clouds, and vapours'). The *Wuxing zhan* text contains chapters concerning divination rules for each of the five visible planets, together with records of their motions in the period 246-177 BC. The planetary data are actually predictions based on observations made in 246 BC, suggesting that even at this date Chinese astronomers understood how to predict the motions of planets from their sidereal and synodic periods; and that they recognised the need for more precise observation, documentation, and study of the motions of the celestial bodies. Similarly, *Tianwen qixiang zazhan* and *Riyue fengyu yunqi zhan* concern celestial divination and make predictions by means of miscellaneous astronomical and meteorological phenomena. *Tianwen qixiang zazhan* also describes occultations of stars by the Moon, and there is a section depicting 29 types of comet categorised by differences in their tails; for example, 'broom comet'[21,3].

These meticulously collected records have found practical applications for present-day scholars. Records of occultations of stars and planets by the Moon and of stars by the planets are useful for studying secular variation of the Earth's rotation[24,19], and records of gnomon shadow length observations have been used to show the variations in the obliquity of the ecliptic over the course of three thousand years[25]. Conversely, all thirteen Shang period oracle bone eclipse records have been matched to solar or lunar eclipses. Events and rulers mentioned in the texts can thus be accurately dated, and thus used as a foundation for building a detailed chronology of the Three Dynasties period of ancient Chinese history[26].

China now has the world's second largest economy, surpassed only by the United States. In 1970, she followed the Soviet Union, the United States, and France in placing a satellite in Earth orbit, and in 2003 became the third nation to develop an indigenous crew-carrying spacecraft, *Shenzhou*. Current Chinese plans include sending orbiters and landers to the Moon, with the ultimate goal of landing humans there by 2030.

16
The Maya

Civilisation emerges in the New World

The Maya, along with the Aztec and the Inca, were one of the three great civilisations of the New World at the time of the Spanish conquest, and they are as familiar to us as the ancient Egyptians, Greeks, and Romans. However, a protracted transition to agriculture meant that complex societies arose later in the New World than they did in Eurasia. Early farmers were hampered by a lack of animal species suitable for domestication: there was nothing comparable to the cattle, pigs, sheep, and goats that were herded and domesticated in Eurasia. The New World also lacked beasts of burden such as the horse, the ass, or the ox. Another difficulty lay in the geography of the Americas. Unlike Eurasia, the principal axis of the Americas lies north-south. Plants and animals tend to be adapted to particular latitudes, and do not thrive when farmers attempt to spread north or south rather than east or west. Hampered by these issues, agriculture spread far more slowly in the New World than it did in Eurasia[1,2].

Another feature of New World civilisation that puzzled the Conquistadores was the contrast between its rich, dynamic culture and its effectively Stone Age technology. There were no animals for transport, and no technology to augment manpower. Pulleys, wheeled vehicles, and large sailing ships were all lacking. There was a complete absence of any form of complex machinery, and tools were largely made from wood and stone[3]. The wheel was known, but it was used only for toys and clay idols. Lacking draft animals, the peoples of the New World failed to appreciate the full possibilities of the wheel[4]. Gold, copper, and even meteoric iron was used to make hammered artefacts from as early as 2000 BC[5,6,7], but smelting of metal ore was a later development. It did not come into use in South America until 500 BC[8], and in Mesoamerica until as late as AD 650[9]. Even then, metals were almost entirely used to make ritual objects and jewellery rather than tools or weapons[3].

The timeline of Mesoamerica is as follows: the Late Archaic/Preceramic ended around 2500 BC and was followed by the Preclassic (or Formative) Period (2500 BC to AD 250). This period is subdivided into an Early (2500 to 900 BC), a Middle (900 to 400 BC), a Late (400 BC to AD 100), and a Terminal (AD 100 to 250). The first settlements that could reasonably be described as 'cities' did not appear until the Middle Preclassic, and the earliest states did not emerge until the Late Preclassic. The Preclassic was followed by the Classic (AD 250 to 925), comprising an Early (AD 250 to 600), Late (AD 600 to 750), and Terminal (or Epiclassic) (AD 750 to 925). During the Classic Period, large expansive states flourished in the Valley of Oaxaca and Valley of Mexico. This period also saw the zenith of Maya civilisation. The Classic was followed by Postclassic (AD 925 to 1519), which ended with the Spanish conquest.

The first complex pre-state culture in the New World were the Olmec, who flourished during the Early and Middle Preclassic periods between 1400 and 400 BC on the Gulf Coast of southern Mexico. To the east of the Olmec heartland, the Maya inhabited the southeast corner of Mesoamerica, from the Yucatán Peninsula to the Pacific Coast. Indigenous Maya live in the region to this day, and Maya languages are spoken by over six million[10]. Around 1000 BC, the region was populated with Maya-speaking farming societies, with social organisation at the tribal level. In comparison to the neighbouring Olmec territory, the Maya region was still a backwater. However, by 900 BC, more complex societies were starting to emerge, and Middle Preclassic Mayan culture took its first steps towards the great civilisation that would flourish during the Classic period from AD 250 to 925[11,12]. The Maya entered a period of decline thereafter, and many of the Classic period cities were abandoned and never reoccupied. Despite this, Maya civilisation survived into the Postclassic period, and their astronomical tradition continued up until the arrival of the Conquistadores.

Writing in Mesoamerica

The origins of Mesoamerican writing are uncertain, but they predate the Maya. From around 400 BC, related glyphic scripts and calendrical systems were in use in the Isthmus of Tehuantepec and the Valley of Oaxaca, in addition to the Mayan region. The similarities between the three writing and calendrical systems suggest that they all probably developed from a common ancestral script during the preceding Middle Preclassic[13]. However, the Maya are the only Mesoamerican civilisation that has left us a significant body of texts. Around 15,000 texts have been recovered, although they mostly date to Late Classic period after around AD 600. They are mainly recorded on pottery found in tombs or elite residences, although some are carved on monuments or buildings.

Unfortunately, only a very few Mayan books have survived. These include the Dresden, Madrid, and Paris Codices, named for the cities in which they are now kept. They are made from the flattened bark of wild *Ficus* (fig) trees, surfaced with lime to provide gloss. They are unbound, each comprising a single folded sheet, with writing on both sides. The glyphs are painted with a fine brush in either black or red, with other characters additionally in yellow, green, and blue[14]. The codices are thought to date to the Late Postclassic period (AD 1250-1521) and hold updated versions of tables first composed during the Late Classic period (AD 600-900)[15].

The Dresden Codex probably dates to the later part of the Late Postclassic period, so it was fairly new when the Conquistadores arrived. We do not know how it reached Europe, although it was probably within a few decades of the conquest. It was acquired for the Royal Library in Dresden from a private collector in the eighteenth century. Unfortunately, it was severely damaged by water during the Allied bombing of Dresden in World War II, and much of the original text was irretrievably lost. Fortunately, a facsimile edition had been published in the nineteenth century. Of 74 pages, 48 contain astronomical content, much of which is in the form of astronomical tables concerned with solstices and equinoxes, the movements of Venus and Mars, and the prediction of eclipses[15]. The term 'page' refers to the folded sections of the document, and does not always reflect the reading order. Pages are typically subdivided; for example, the upper and lower halves of page 55 are numbered 55a and 55b, respectively. The reading order precedes from 55a to 52a through to 58a before moving down to 55b[14].

Most Mayan books were destroyed on the orders of Diego de Landa (1524-1579), a Franciscan friar sent to the Yucatán to convert to Maya to Roman Catholicism. De Landa was considered over-zealous even for the Spanish Inquisition, and he was eventually recalled to Spain. On the credit side, he was appalled by the practice of human sacrifice, and on one occasion he intervened to prevent the sacrifice of a young boy. His *Relación De Las Cosas De Yucatán*, a study of Mayan culture, religious practices, and writing systems, is considered to be an "*ethnographic masterpiece of the Maya*"[16]. With the help of Maya interlocutors, De Landa attempted to match Maya glyphs with the letters of the alphabet. His findings were included in *Relación de las cosas de Yucatán*, but nineteenth century attempts to translate Maya texts with his 'alphabet' were not successful. He failed to recognise that Mesoamerican writing systems are not alphabetic, and the glyphs he had recorded represented syllables and not letters. For example, the glyph he had recorded for the letter *B* had the phonetic value of */bay/*, corresponding to the Spanish pronunciation of *B*.

A breakthrough came in the 1950s, when the Soviet scholar Yuri Knorozov realised the need to treat de Landa's 'letters' as syllables. He also suggested that Maya script followed a rule known as synharmony whereby a word ending in a consonant will be written with a glyph representing a syllable whose vowel sound is the same as that of the preceding glyph. When spoken, the vowel sound is silent. For example, the Maya word for *turkey* is *kutz*. Knorozov suggested that it would be made up of glyphs with the values */ku/* and */tzu/*. De Landa's book contained a glyph for *C*, pronounced *cu* in Spanish, and the same glyph appears in the Madrid Codex accompanied by a picture of a turkey. The synharmony theory predicts that the next glyph should have the value */tzu/*. The Dresden Codex contains an example of this second glyph, associated with a picture of a dog. The Maya word for *dog* is *tzul*, and this not only confirmed the value of the glyph, but also suggested that the one following it should have the value */lu/*. By this approach, Knorozov had obtained the phonetic value of two previously unknown signs. The same principle has since been used to decipher a large number of signs, although many remain unknown to this day.

That astronomical data is encoded among the Mayan texts has been accepted since the pioneering work of American chemical engineer and Mayan researcher John Teeple (1874-1931) and astronomer Robert Willson (1853-1922) in the 1920s. Between the 1920s and 1970s, British archaeologist Sir Eric Thompson (1898-1975) published extensively about Mayan writing systems, calendrics, and astronomy. More recently, much important work is due to American anthropologists Floyd Lounsbury (1914-1998), Harvey Bricker (1940-2017), Victoria Bricker (born 1940), and Anthony Aveni (born 1938).

The Mesoamerican Calendar

Mesoamerican calendrical systems have become well-known to the general public in recent years as a result of the Long Count, which ended on 21 December 2012. It led to a flurry of doomsday predictions, although there is absolutely no reason to suppose that the Maya expected anything untoward to occur on that day. The Long Count is one of three calendars that were in use in pre-Columbian Mesoamerica. The others were a 'vague year' or *haab* cycle of 365 days and a ritual calendar of 260 days known as the *tzolkin* or sacred almanac. From the Maya point of view, it was this third count that was the most important[17]. All three calendars used a vigesimal or base-20 system of counting (units 0-19). This might have come about through counting the digits on the feet as well as on the hands. Notably, the Maya system employed a place-value notation and a zero, long before these concepts were introduced into the Hindu-Arabic numeral system (see Chapter 13). Numbers were represented by combinations of ones (dots), fives (bars), and zeros (often represented by a seashell or a grasping hand) stacked vertically, with place value increasing from bottom to top[14]. The other number that featured prominently in Mesoamerican calendrical systems was 13, possibly representing the number of levels of heaven in Mesoamerican cosmology (*cf.* the seven levels of heaven in the Jewish, Islamic, and Hindu traditions)[14].

Like the Gregorian calendar, the Long Count calendar generated dates from a fixed start point, and that were to all intents and purposes unique. The basic unit of time was the *tun* of 360 days, which was subdivided into 18 *uinals* each comprising 20 *kins* (days) numbered 0-19. There were 20 *tuns* to the *katun* (360 × 20 = 7,200 days) and 20 *katuns* to the *baktun* (360 × 20 × 20 = 144,000 days or just over 394 years); again, these were numbered 0-19. The *katun* and the *baktun* may be seen as the vigesimal equivalent of decades and centuries. The Maya did not invent the Long Count, but by AD 250 nobody else was using it[18].

The Maya implementation of the Long Count is generally accepted to have begun on 11 August 3114 BC in the Proleptic Gregorian calendar. Correlating the two calendars is a highly complex problem, but it is now accepted that the Modified Thompson 2 correlation of 584,283 is correct[12]. This correlation establishes that the Maya Long Count began on Julian Day Number 584,283. The Julian Day Number is a count that began at noon on 1 January 4714 BC, and is frequently used by astronomers and software developers. Sir Eric Thompson[19] originally obtained a value of 584,285, based in part on earlier work by American researcher Joseph Goodman (1838-1917) and Spanish archaeologist Juan Martinez Hernandez (1866-1959); hence the correlation is often referred to as the Goodman-Martinez-Thompson or GMT correlation. Thompson later revised his value to 584,283.

The date 11 August 3114 BC is written as 13.0.0.0.0 in Maya script, although the 13 represents a zero. The thirteenth *baktun* from that date ended on 21 December 2012. It is uncertain what was supposed to follow. The usual view is that 13 *baktuns* (a period of just over 5,125 years) represents a creation epoch, and that the count then returns to zero[14]. However, there are claims that the Maya intended the count to continue. There may be higher-order units beyond the *baktun*, which modern scholars have named the *piktun*, *kalabtun*, *kinchiltun*, and *alautun*.

The words *tzolkin*, *haab*, *tun*, and the multiples of the *tun* are conventions derived from present-day Maya languages. *Tzolkin* (or *tzolk'in*) means 'count of days' and *haab* means 'year'. '*Tun*' refers to the end of a year, but not necessarily the 360-day civil year. We do not know what the original pre-Columbian terms were, although it has been suggested that the Calendar Round was known as the *hunab*[20].

Days in the 260-day *tzolkin* cycle were denoted by a number known as a day coefficient from 1 to 13 and one of 20 names (*Imix, Ik, Akbal, Kan, Chicchan, Cimi, Manik, Lamat, Muluc, Oc, Chuen, Eb, Ben, Ix, Men, Cib, Caban, Edznab, Cauac, Ahau*), for a total of 260 days. The cycle worked on a similar principle to the Chinese sexagesimal *Gan-Zhi* system (see Chapter 15), beginning with 1 *Imix*, then 2 *Ik*, up to up to 13 *Ben*, then 1 *Ix*, 2 *Men*… 8 *Imix*, and so on. The 365-day *haab* cycle comprised 18 regular 'months' of 20 days each (*Pop, Uo, Zip, Zodz,*

Tzec, Xul, Yaxkin, Mol, Chen, Yax, Zac, Ceh, Mac, Kankin, Muan, Pax, Kayab, Cumku), plus a short intercalary 'month' of 5 days (*Uayeb*). Each date was denoted by a day coefficient of 0 to 19 paired with one of 18 month-names, or 0 to 4 paired with the short month. Like the pre-Ptolemaic Egyptian calendar, it did not take the quarter day into consideration. There was no extra intercalary day every fourth year, and hence the *haab* cycle did not accurately track the tropical year.

The *tzolkin* and *haab* cycles were combined into the Calendar Round; for example, a date would be written as 4 *Ahau* 8 *Cumku* and the next day would be 5 *Imix* 9 *Cumku*. The Calendar Round date was in turn combined with the Long Count date. Perhaps surprisingly, the two were not synchronised: the Calendar Round date at the start of the Long Count was 4 *Ahau* 8 *Cumku*. The Calendar Round cycle repeats every 52 *haab* 'years' or 18,980 days, which is the lowest common multiple of 260 and 365. Note that this means that only one-fifth of the 94,000 possible combinations of *tzolkin* and *haab* can appear. The 52-year cycle was a period of great significance throughout Mesoamerica, and each completion was celebrated by elaborate ceremonies[14].

The reason for a 260-day ritual count is not known. One suggestion is that it originated at a location between 14°42' and 15°00' N., where the Sun crosses the zenith at 260 and 105-day intervals, possibly at Izapa on the Pacific Coast of Mexico[21]. The problem is that the 260-day cycle then simply repeats and does not factor in the 105-day interval[22]. There is also evidence that the 260-day cycle was in use at the site of Monte Albán in the Valley of Oaxaca before its use at Izapa[22], and Olmec inscriptions suggest that it may date to as early as 650 BC[13,23], well before any adoption by the Maya.

Other possibilities are the average human gestation period (266 days), or various astronomical cycles. Two *tzolkin* cycles (520 days) correspond closely to three eclipse half-years ($3 \times 173.31 = 519.93$ days), i.e., the period between successive **eclipse seasons**[24]. The *tzolkin* cycle is also close to the average of 263 days that Venus remains visible as either a morning or evening object before disappearing into the dawn or twilight skies. Links to Mars have been suggested: the Martian **synodic period** is almost exactly three *tzolkin* cycles, or 780 days[14].

The Maya were proficient astronomers who were able to predict the movements of Venus and Mars in some detail. The earliest known Maya astronomical tables are murals painted and incised on the walls and ceiling of a small room at the site of Xultun in Petén, Guatemala. At least two of the tables concern the movements of the Moon, and possibly Mars and Venus. They date to the early ninth century AD and resemble the later calendrical and astronomical tables in the Dresden Codex[25]. Unlike astronomy in the Old World, the primary focus of Maya astronomy seems to have been to identify commensurate relationships between celestial cycles and their calendrical systems[26]. However, it does not necessarily follow that astronomical phenomena were responsible for the 260-day period in the first place. Given the importance of the numbers 13 and 20 in Mesoamerican culture, then 260 could represent nothing more significant than 13 multiplied by 20[14].

As far as is known, a lunar calendar was never developed in ancient Mesoamerica, but after around AD 200 the Maya began to include lunar and other information as a supplement to their Long Count and Calendar Round dates[27]. This additional information is referred to as the Supplementary Series or (as it was largely restricted to lunar information) Lunar Series. The Supplementary Series often began with the non-lunar Glyph G, identifying the Lord of the Night. According to Maya tradition, each night was ruled in turn by one of the nine Lords of the Night in a nine-night cycle. Some of these Lords brought good fortune to the night they ruled; others bad. Only nine variants of Glyph G are known, designated G1-G9[14].

Lunar data is recorded in a series of glyphs designated A to F and X, Y, Z. The meanings of Glyphs F, Y, and Z are as yet unknown. Glyph B is usually a sky sign with an elbow shape or animal head. It may refer to the house or constellation in which the Moon is residing. Glyphs D and E indicate the age of the Moon counted from the previous new Moon for the indicated Long Count date. Glyph D is a compound comprising a hand with a forefinger pointing to the right and/or a half-Moon with three vertical dots. This glyph has a coefficient of up to 19 and appears on its own when the Moon is less than 20 days old; for example, 8D means an 8-day-old Moon. Glyph E is usually a single element of a half-Moon with a coefficient of zero to nine, and it indicates the age of the Moon is 20 or more days by the amount of the coefficient; for example, 3E followed by a D with no coefficient means a 23-day-old Moon. Both D and E without coefficients indicates a new Moon. Glyph A is a Moon glyph invariably accompanied by a coefficient of 9 or 10; it has always been interpreted as an indicator of a 29 or 30-day month[20,14].

Glyph C counts the number of the **synodic months** in a completed cycle of six (i.e., 177 days), a period known as a 'lunar semester'. It is similar in form to D, often comprising a hand and a half-Moon, combined with two variable elements: a numerical coefficient and three head variants: skull, young female head, or male or jaguar deity head. During the Classic period, the sequence of Glyph C head variants was fixed and followed the order: (1) skull, (2) female head, (3) male or jaguar deity head. The coefficient is always 2-6; no coefficient implies the first synodic month of the semester. Thus, 4C 8D means 8 days after the fourth new Moon out of six. Depending on the convention in use, the number could refer to either current or elapsed synodic months. Different conventions were followed in different cities; for example, the fourth Moon in the Usumacinta area was counted as the fifth Moon in Tikal[20,14,27]. Finally, Glyph X records the position within an 18-month lunar calendar consisting of three semesters of six months each[28].

Despite the lack of a formal lunar calendar, the Maya obtained accurate values for the length of the synodic month. At Palenque, southern Mexico, several dates have been connected by the relationship 81 synodic months = 6.11.12 days, corresponding to a synodic month of 2,392/81 = 29.53086 days. The average length of the synodic month is 29.530589 days, so this figure is accurate to within 24 seconds. The Xultun murals indicate the same relationship, expressed as 4,784 days = 162 synodic months. At Copan, western Honduras, a formula of 149 synodic months = 12.4.0 days was used, corresponding to a synodic month of 4,400/149 = 29.53020 days. Despite the use of a longer time-base, the result is less accurate than the Palenque formula, though still out by only 34 seconds[14,25,27].

Tropical year vs *tzolkin* year

On the Stela A at Copan are recorded three dates (days after epoch commencing 13.0.0.0.0 4 *Ahau* 8 *Cumku* given in brackets):

(a) 9.14.19.8.0 12 *Ahau* 18 *Cumko* (1,403,800 days)

(b) 9.15.0.0.0 4 *Ahau* 13 *Yax* (1,404,000 days)

(c) 9.14.19.5.0 4 *Ahau* 18 *Muan* (1,403,740 days)

Date (b) marks the end of a *katun* and as such was the most important date on the stela, as the ending of a *katun* was an important event for Maya diviners. Date (c) occurs 19.5.0 (= 6,940) days after the start of that *katun*. Date (a) occurs 200 days before the *katun*-ending date (b), but what is the significance of these other dates? The 6,940-day interval recalls the Metonic cycle of 235 synodic months = 19 tropical years = 6,939.60 days. This interval is also found within the Dresden Codex. It suggests that the Stela A dates have a lunisolar significance, but what was it?

The Maya would have been aware that the 365-day *haab* calendar soon falls out of step with the tropical year: even over a single lifetime, the missing 0.2422 days per year would cause the seasons to fall two or three weeks behind the *haab* calendar. If an individual's age were reckoned in *haab* years, they would reach their 70th birthday 25,550 days after birth; but in tropical years that particular milestone is not reached for another 17 days, 25,567 days after birth.

The date (b) 9.15.0.0.0 4 *Ahau* 13 *Yax* is fractionally more than 3,844 tropical years after 13.0.0.0.0 4 *Ahau* 8 *Cumku*, and at that point the discrepancy between the tropical and *haab* years was 3,844 × (365.2422 - 365) = 931 days. This corresponds to two completed *haab* cycles and 200 days (actually 201 days). It suggests that the Maya wanted to work back from the 13 *Yax haab* date of date (b) to what the *haab* date would have been after allowing for the slippage of the tropical year against the *haab* cycle, but disregarding the two completed cycles. Going back by 200 days from date (b) takes us to date (a)[14].

Teeple[24] suggests that the calculations were made as follows: the period of 3,844 tropical years can be broken down into 202 completed Metonic cycles of 19 years each plus 6 tropical years, i.e., 202 × 235 = 47,470 synodic months + 6 tropical years. Applying the Copan Moon formula of 149 synodic periods of the Moon = 4,400 days (accurate to within better than 1½ hours in 12 years), we obtain 47,470 synodic months = 47,470 × 4,400 / 149 = 1,401,799 days. Finally, add the number of days in six *haab* (as opposed to tropical) years to obtain 1,401,799 + 6 × 365 = 1,403,990 days. This corresponds to a date of 9.14.19.17.10 7 *Oc* 3 *Yax*. Thus by 9.15.0.0.0 the Calendar Round date had moved from 4 *Ahau* 8 *Cumku* to 4 *Ahau* 13 *Yax*, but the computed Calendar Round date was 7 *Oc* 3 *Yax*. The *haab* date of 3 *Yax* precedes 8 *Cumku* by 200 days, i.e., the remainder portion of the slippage of the tropical year against the *haab* year. Now we take our *katun*-ending date (b) of 9.15.0.0.0 4 *Ahau* 13 *Yax* and go back by 200 days to reach 9.15.19.8.0 12 *Ahau* 18 *Cumku*, i.e., date (a).

If the Maya were indeed aware that the tropical year had slipped by 930 days against the *haab* year by Long Count date 9.15.0.0.0, then this can be recast as follows: in 1,404,000/365 = 3,846.575 *haab* years the tropical year had fallen behind by 930/365 = 2.548 *haab* years;

i.e., 3,846.575 *haab* years = (3,846.575 - 2.548) tropical years;

or (3,846.575 + 2.548) *haab* years = 3,846.575 tropical years;

Hence, 1 tropical year = (1 + 2.548/3,846.575) = 1.000662 *haab* years = 365.2418 days.

This is just 35 seconds shorter than the accepted average length of the tropical year, but similar calculations from other inscriptions at Copan reveal a Maya value for the tropical year that was even closer to the recognised length. At 365.2420 days, it was out by just 16 seconds.

Mayan Astronomical tables

The four Mayan codices may be regarded as collections of almanacs, analogous to present-day almanacs such as the *Old Farmer's Almanac*. A modern almanac is an annual calendar containing an eclectic compilation of holiday dates, weather forecasts, tide tables, and astronomical data. The latter includes sunrise/sunset times, Moon phases, forthcoming eclipses, and planetary positions. The Mayan almanacs were based around the 260-day ritual *tzolkin* calendar, the most important of their three calendrical systems. Many concerned matters such as the weather, agriculture, medical matters, and ceremonies, as well as astronomical tables[14].

The 260-day *tzolkin* cycle was most commonly divided into 5 rounds of 52-day time periods. Other common divisions were 4 × 65 and 10 × 26 – though, surprisingly, not 20 × 13. Larger almanacs used multiples of 260; for example, 9 × 260. Pages 17c-18c of the Dresden Codex contains Almanac 48, a 5 × 52-day almanac, with each round beginning at one of the five numbered and named entry-points known as *lubs*. The *lubs* are *tzolkin* dates, which appear in a vertical column of glyphs on the left-hand side of the page with four red dots above. The glyphs denote the *tzolkin* day names *Ahau*, *Eb*, *Kan*, *Cib*, and *Lamat*; the dots denote day coefficients, hence 4 *Ahau*, 4 *Eb*, 4 *Kan*, 4 *Cib*, and 4 *Lamat*. To the right of the *Ahau* glyph are six numbers in dot-and-bar notation: three black and three red. The three black numbers (15, 33, and 4, adding up to 52 or the number of days in a round) are known as distance numbers. The three red numbers (6, 13, and 4) refer to *tzolkin* day coefficients. Above each pair of black and red numbers is a 2 × 2 block of glyphs and below is a larger illustration. The almanac is thus divided into three subunits known as *t'ols*. The glyphs and illustrations represent prognostications. In the first *t'ol*, a woman is depicted carrying a skeletal death god on her back. The upper left glyph in the block above is a verb that states the action of the deity depicted below; the upper right glyph gives the object toward which the action is directed; the lower left glyph is the subject or name of the deity; and the lower right glyph states the result of the action of the deity or the appropriate offering for their action. In this case, the whole reads, "*Death is the burden of the Moon Goddess, bad winds; Two-Blue-Yellow is the burden of Ix Chel, the Moon Goddess; Muy is her burden, the divination of the Moon Goddess*"[14].

We enter the table at the point defined by the first *lub*, 4 *Ahau*, and moving 15 days on from 4 *Ahau*, we reach 6 *Men*, the point identified by the first red number (albeit the *tzolkin* day name is understood). From there, 33 days takes us to 13 *Lamat* and 4 days takes us to 4 *Eb*, which is the second entry point *lub*. From there we commence a second run: 4 *Eb* + 15 days ⇒ 6 *Manik*; + 33 days ⇒ 13 *Ahau*; + 4 days ⇒ 4 *Kan*, which is the third entry point, and so on. In this way, we cycle through all five rounds of the almanac before returning to 4 *Ahau*, which is where the 5 × 52 = 260-day sequence commenced. At each stopping point, the prognostication for that *t'ol* will apply[14]. A similar principle applied to astronomical tables, where the primary concern was to link celestial cycles to the 260-day ritual *tzolkin* calendar[15].

Astronomical tables all followed a similar format. An introductory table provided calendrical data that was used to determine base dates and *lubs* for the table. The data begins with a 'ring number' (so called because it is circled by an image of a tied red band) and a Long Count date. We begin by using the ring number to move back from 13.0.0.0.0 4 *Ahau* 8 *Cumku* to a date in the previous creation epoch, i.e., earlier in time than the Long Count zero date (11 August 3114 BC). From this point, we jump forward by the Long Count date to reach the base date. If, for example, the ring date is 17.12 (i.e., 352 days) and the Long Count date is 9.19.8.15.0 (i.e., 1,435,980 days), we move back from 13.0.0.0.0 4 *Ahau* 8 *Cumku* by 17.12 days to the previous epoch date 12.0.19.0.8 3 *Lamat* 1 *Uayeb*. Next, we add 9.19.8.15.0 to give a base date of 9.19.7.15.8 3 *Lamat* 6 *Zotz*. The base date is used to position the table in time, though it is not typically used as an entry point to the table. Instead, a further interval is added to the base date to obtain the *lub*. This interval is selected from a table of multiples and submultiples of a planet's synodic period; for example, 78 or 780 for Mars. The table often also includes correction factors to make up for differences between the working values and true values of a particular cycle; for example, the Maya used a working value of 584 days for the synodic period of Venus, which is slightly more than the true value of 583.92 days. These preliminary items are followed by the actual table, which is organised into discrete units known as frames. The frames contain glyph captions, dates referring to the events described, and pictures[29,15].

Eclipses

The Maya, like other peoples around the world, attached great significance to eclipses. That eclipses were something the Maya feared is evident from the Dresden Codex. A picture on page 53 features a dead woman hanging by her hair and the death god above her, and prognostication glyphs give ominous warnings. The text on page 53 has been translated as "*... damage the earth sky, the lord ... eclipse of the sun, moon by One Sky Bearded Lord bad omen or wind damage to the seats.*" Given such potentially dire consequences, there is little doubt that the Maya would have sought to predict eclipses[14].

The picture is part of a table spanning pages 51a-58b of the Dresden Codex. The table is often referred to as a lunar table, but it is thought to contain predictions of eclipses over a period of about 33 years[30,31,14]. Each page contains up to nine blocks, labelled as follows: Block I Long Count dates; Block II multiples of the length of the table; Block III *lubs*; Block IV augural and omen glyphs; Block V a column comprising 13 occurrences of the number thirteen; Block VI eclipse pictures; Block VII intervals of 177, 178, and 148 days; Block VIII running totals; and Block IX *tzolkin* calendar dates[32]. The upper halves of these pages are by convention labelled 'a' and the lower halves are labelled 'b'. Reading begins on page 51a, then moves from left to right to page 52a, continuing to 58a. It then shifts to the lower segments, beginning at page 51b and continuing to page 58b. The user manual is on pages 52a and 53a, and contains dates, a table of multiples, and other introductory material. There are five Long Count dates and corresponding *tzolkin* dates on page 52a (Block I):

(a) 9.16.4.10.8 12 *Lamat* 1 *Muan* 10 November AD 755 (new Moon)
(b) 9.16.4.11.3 1 *Akbal* 16 *Muan* 25 November AD 755 (full Moon)
(c) 9.16.4.11.18 3 *Edznab* 11 *Pax* 10 December AD 755 (new Moon)
(d) 9.19.8.7.8 7 *Lamat* (?)
(e) 10.19.6.1.8 12 *Lamat* 6 *Cumku* 25 September AD 1210

Date (d) should have been 8 *Lamat* 6 *Kankin*, so either the Long Count date or the *tzolkin* date was incorrect. Dates(a), (b), and (c) are each 15 days apart.

The table proper runs from page 53a to 58b. It comprises 69 columns with three levels each (Blocks VIII, IX, and VII). In each column, the upper and lower level (Blocks VIII and VII) contain numbers in dot-bar notation, and the middle level (Block IX) contains three rows of successive *tzolkin* dates. The numbers on the lower level (Block VII) are mostly 177s (53 in all), but there are seven 178s and nine 148s. They suggest an eclipse table; the intervals are close to six (177.16 days) and five (147.65 days) synodic months, respectively. The 178-day interval represents the periodic use of an intercalary day to prevent the table from drifting against the synodic month. The numbers on the upper level (Block VIII) are running totals of the numbers from the lower level (Block VII). These Block VIII/IX/VII groups run from left to right in sequences ranging from as few as three to as many as ten. Each sequence is terminated by a 148 and is followed by a picture (Block IV). There are nine pictures delimiting eight sequences.

The Block IX *tzolkin* dates in column 1 on page 53a are 6 *Kan*, 7 *Chicchan*, and 8 *Cimi*. The middle date, 7 *Chicchan*, is the *tzolkin* date reached by adding the numerical increment in column 1, 8.17 (i.e., 177) days, to a date of 12 *Lamat*, the *tzolkin* component of date (a) in the introduction to the table. The middle *tzolkin* date in column 2, 2 *Ik*, is reached by adding the numerical increment in column 2 (again 177) days to the previous *tzolkin* date, 7 *Chicchan*. The adjacent *tzolkin* dates, differing by ± 1 day, were probably intended for one-day corrections. The process continues through the table until the Block VIII running total reaches 11,959 days (as a result of an error in the table, the tabulated cumulative value is actually 11,958, but the sum of the individually listed increments is 11,959). The *tzolkin* component of date (a) and the 69 middle Block IX *tzolkin* dates cluster into three small, evenly spaced 'arcs' on a 260-day 'wheel'. This is a natural consequence of the two-to-three relationship between the *tzolkin* cycle and the eclipse half-year, as noted above: two *tzolkin* cycles (520 days) corresponds closely to three eclipse half-years (3 × 173.31 = 519.93 days). The Maya were obviously aware that only during these three portions of the *tzolkin* cycle was there the danger of an eclipse. The arcs could be considered to be eclipse windows, with an eclipse possibility occurring on every second passage through an arc. The discrepancy between the 177-day six-lunation interval and 173.31 eclipse half-year causes a gradual advance of the table against the *tzolkin* cycle; this was accounted for by the periodic substitution of a 148-day five lunation interval, enabling the 'recovery' of the original *tzolkin* dates.

Across the middle of pages 51a and 52a are seven columns comprising the table of multiples (Block II), obtained (with some exceptions) by multiplying 1.13.4.0 (i.e., 11,960) by integers ranging from 2 to 39. Below these are repeated columns (Block III) of five *tzolkin* dates: 12 *Lamat*, 1 *Akbal*, 3 *Edznab*, 5 *Ben*, and 7 *Lamat*. The first three of these dates are the *tzolkin* components of the Block I Long Count dates (a), (b), and (c). These are the *lubs*, or entry points into the table. If date (a) is taken to be the base date and the first day of the cycle, and it is incremented by 11,959 (i.e., the total of the 69 increments), then the total length of the cycle is 1.13.4.0 (i.e., 11,960) days or 46 *tzolkin* cycles. This is just 2 hours 38 minutes in excess of 405 synodic months (11,959.89 days). Dividing by five, we obtain the 2,392 days = 81 synodic months relation in use at Palenque. The *lub* and first *tzolkin* day of the cycle is 12 *Lamat*, the 11,960th and last day is 11 *Manik*, and the 11,961st day or the first *tzolkin* day of the next cycle is once again 12 *Lamat*. The table satisfies the Maya desire for cyclicity, and it was evidently intended to be reused; the other *lubs* listed could be used as later or earlier entry points.

The interval totals between the nine pictures are 1,742, 1,034, 1,210, 1,742, 1,034, 1,210, 1,565, and 1,211 days. These correspond to recognised eclipse cycles of 1,033.57, 1,210.75, 1,565.12, and 1,742.30 days (35, 41, 53, and 59 synodic months)[14]. There is little doubt that the pages 51a-58b of the Dresden Codex represent an eclipse table consisting of groups of five and six synodic months, with eclipses occurring at the positions of the pictures. The question, though, is the table a record of past eclipses or is it intended to warn of future eclipses – and are the eclipses lunar, or solar, or both?

Anthony Aveni[33,14] considered solar and lunar eclipses visible in the Yucatan during the fifth century AD. He noted intervals between consecutive lunar eclipses could be composed from sequences of 177s and 148s, but this could not be done for any solar-to-lunar or lunar-to-solar eclipse interval. Consequently, as read, the table could neither record nor predict both solar *and* lunar eclipses. However, consecutive full and new Moons fifteen days apart are often associated with eclipses. A 'dual-purpose' table would require a 15-day interval in addition to those of 177 and 148 days. It is possible that that is the significance of the 15-day spacing between the Block I Long Count dates (a), (b), and (c) above.

Regardless of whether the table was used for solar or lunar eclipses, predictions must have been based on recorded observations of actual eclipses visible from the Yucatan. During the fifth century AD, there were 73 lunar eclipses visible from the Yucatan. By contrast, there were only ten solar eclipses exceeding 50 percent totality. For this reason, Aveni[14] believes that lunar data was used to construct the table. It would soon become clear that once a lunar eclipse had occurred, another was reasonably likely to follow after either 177 by ± 1 days, or 325 (177 + 148) days, or 354 (2 × 177) days. By contrast, it would take centuries to build up a comparable database for solar eclipses. Of the 72 eclipses noted above, 42 could have been successfully predicted by this rule. Furthermore, all 72 were separated by intervals of either multiples of 177 ± 1 days or multiples of 177 ± 1 + 148 days.

Notwithstanding Aveni's reservations about solar eclipse prediction and the need for a lengthy period of observation to establish the underlying cycles, some do believe that this was in fact the purpose of the Dresden Codex lunar table. Harvey and Victoria Bricker[30] have put forward a model by which the table could have been used to predict solar eclipses right up until the 1990s. By applying the Modified Thompson 2 correlation to the Block I Long Count dates, it can be seen that dates (a) and (c) and their increments correspond to new Moons, and date (b) and its increments correspond to full Moons. During the approximately 33 years between the Block I Long Count date (a) on 10 Nov AD 755 and the 69th increment of date (c) on 6 Sept AD 788, a total of 77 solar eclipses occurred somewhere in the world. All occurred within three days of an eclipse warning; 32 of them with (a) as the base date and the remaining 45 with (c) as the base date. This leaves no eclipse occurring on 63 of the 140 predicted dates (a) and (c).

In most cases, each increment of the two dates takes the user to one valid prediction and one null prediction, but on seven occasions (the 7th, 16th, 30th, 31st, 45th, 54th, and 69th increments), both the date (a) and date (c) predictions are valid. The mean error of the valid predictions is 1.92 days early. As we have seen, the Block IX *tzolkin* dates were listed in columns of three. It is generally understood that this was to allow for a range of ± 1 day on either side of the day reached by the increment to date (a). The Brickers accept this interpretation and propose that it was also applied to the date (c) group; they suggest that the *tzolkin* dates for this group were left unwritten to save space. If an uncertainty of ± 1 day was built into the predictions, then the 'excess' mean error is less than a day. However, it was necessary for the user to know in advance whether the valid prediction was in the (a) group, the (c) group, or both.

As noted above, the base date (a) and its 69 increments may be thought of as clustering into three evenly spaced 'arcs' on the 260-day *tzolkin* 'wheel'. If base date (c) and its increments are added, the valid predictions cluster into three 'eclipse windows' running from (1) 10 *Caban* to 13 *Cimi* (30 days centred on 11 *Chuen* to 12 *Eb*), (2) 8 *Kan* to 4 *Cauac* (36 days centred on 7 *Cib* to 8 *Caban*), and (3) 1 *Cauac* to 10 *Ix* (36 days centred on 13 *Chuen* to 1 *Eb*). All but three of the null predictions lie outside these limits. In summary, nearly all the predictions lying within 15 to 18 days of the three centres are valid, and all of those lying further out are null. Such an outcome is only to be expected: a solar eclipse can only occur during an eclipse season, which ranges in duration from 31 days 8 hours to 37 days 13 hours, i.e., when the Sun is within 18 days of one of the **nodes of the Moon's orbit**.

In practice, the width of the eclipse windows is reduced for observers in the Yucatan. Eclipses occurring on the extremities of the range are invariably partial in the sense that the zone of totality misses the Earth entirely. Such eclipses are seen only in higher latitudes, and they would never be visible in the Maya lands. When they are excluded, window (2) is reduced to a length of 31 days. It starts on 12 *Lamat*, is centred on 1 *Akbal*, and ends shortly before 3 *Edznab*. We can see that these three *tzolkin* dates are the same as those for base dates (a), (b), and (c). The Brickers suggest these base dates served a secondary function of defining the limits and midpoint of the eclipse window. Although date (b) is a date located close to the full Moon, its *tzolkin* component defines the centre of the window. If we now assume that the midpoints of the other two windows lie one half-eclipse year before and one half-eclipse year after 1 *Akbal*, and that the windows are likewise 31 days in duration, then window (1) starts on 8 *Men*, is centred on 10 *Oc*, and ends on 12 *Chicchan*; and window (3) starts on 3 *Imix*, is centred on 5 *Cib*, and ends on 7 *Chuen*.

What should be noted is that while the table successfully predicted the occurrence of all 77 solar eclipses occurring between November AD 755 and September AD 788, only eight would have been visible in the Maya lands, and of these only four would have been particularly conspicuous. While this paucity of eclipses does underline the view that the table was not a record of eclipses actually observed, it does raise the question of how the table was viewed by the Maya when, seemingly, most of its predictions failed to materialise. The clue is in the images and texts foretelling of the deleterious effects of eclipses. If we replace the neutral term 'eclipse window' with 'eclipse danger zone', then the explanation becomes clear. The table was an early warning system around which appropriate ceremonial precautions were planned. These precautions were seemingly able to ward off all but a small number of the predicted eclipses. If there were any questions to be asked, they would have concerned the eclipses that were seen rather than those that were not.

Although most believe that the table was intended to be reused, problems would arise from simply 'recycling' it with the multiples of 11,960 listed in Block II. The Brickers note that the lunar cycle of 405 synodic months (405 × 29.530589 = 11,959.89 days) falls short of 46 *tzolkin* cycles (11,960 days) by 0.11 days. Furthermore, the table's conceptual length of 69 eclipse half-years (69 × 173.31 = 11,958.39 days) is short by 1.61 days. On the first reuse of the table, the midpoint of the eclipse danger zone centred on 1 *Akbal* would have slipped to 13 *Ik* or 12 *Imix*.

To permit reuse of the table while allowing for periodic correction to deal with both eclipse half-year shortfall and synodic month shortfall, the Brickers propose that the Maya used two types of base date, which were corrected by different means. The 'long-term base dates' remained unchanged for 63,431 days, i.e., five 11,960-day runs through the table plus an intercalary period of 3,631 days. The 'window-defining base dates' changed after every 11,960 days, i.e., a single run through the table. During the first 11,959 days of each 63,431-day period, the two sets of base dates were identical, as was the case when the table's original 12 *Lamat* commencement. At the start of each successive reuse of the table, the values of the base dates (a), (b), and (c) used to define the eclipse danger zones were set back by at least one day in the *tzolkin* cycle. The 'long-term' values of these dates were left unchanged, and they continued to be used in conjunction with the requisite multiples to re-enter the table.

This partially corrected the eclipse half-year shortfall, but further action was periodically required to achieve a full correction, and also to cater for the synodic month shortfall. Accordingly, after every fifth completed run, the next run was ended after 3,631 days rather than the usual 11,960 days, and the date reached was used as the new 'long-term' base date for the next sequence of five full and one shortened, intercalary run. After every thirty full runs, a count of 2,071 days instead of 3,631 days was used for the intercalary run. With these corrections applied, the table remained extremely accurate even into the twentieth century. It predicted that an eclipse would occur between the 8th and 10th of July 1991. The eclipse actually took place on 11 July.

Impressive as this is, there are issues with the Brickers' interpretation of the table. The correction constants of 3,631 and 2,071 are derived from a complex interpretation of the user manual, the distance number 5.1.8, the Block V 'thirteen thirteens', certain entries in the table of multiples, and an inferred subtraction of 30 days. Canadian mathematician Michael Closs[34] notes that despite their importance, the correction constants do not themselves appear in the text, neither does the negative count of 30 days. The explanation for their absence, along with that of Block IX *tzolkin* date ranges for the base date (c) is to 'save space' – yet Block V contains nothing but thirteen repetitions of the same numerical constant.

American anthropologist John Justeson[35], while also favouring the solar eclipse interpretation, has proposed a very different model to the Brickers. Justeson believes that the table was in use during a 405-month interval that began on 19 April AD 1083 and that it makes use of an eclipse cycle of 88 synodic months. This period (88 × 29.530589 = 2,598.69 days) is within an hour of 15 eclipse half-years (15 × 173.31 = 2,599.65 days), and like the saros gives rise to 'families' of eclipses that eventually die out because the relationship is not exact.

Researcher Frederick Martin[36] has shown how the Dresden Codex table could be used to predict both lunar and solar eclipses using separate runs through the table. He took a series of eclipses that occurred between 1970 and 1992 and found he could 'overlay' the two sets of eclipse predictions with a good degree of accuracy. A match for lunar eclipses occurred with the Calendar Round day name 12 *Lamat* 1 *Muan* set to 26 May 1975, which was the day after a lunar eclipse. This was 'paired' with a solar eclipse occurring 14 days earlier on 10 *Ben* 6 *Kankin*. All the subsequent solar eclipses fell on Calendar Round days consistent with a parallel progression from the 42nd entry of the eclipse table. The eclipses alternate between lunar and solar, with periodic reversals (i.e., lunar followed by lunar or solar followed by solar). These reversals always occurred after the interval of advance was 148 days rather than

177 or 178 days. Martin[37] also suggests that a Venus glyph on page 53a in the fourth entry of Block IX refers to a lunar eclipse coinciding with the last visibility of the planet as an evening object before superior conjunction.

The problem, as Anthony Aveni[38] notes, is that interpreting the table is not unlike interpreting Stonehenge. There is little doubt that pages 51a-58b of the Dresden Codex contain an eclipse table, but beyond that it is difficult to come to any definite conclusions. While favouring a lunar explanation, Aveni notes that "*some models are better than others*", but he concedes that the Brickers' model is "*a very good one*".

Venus

There is no doubting the pre-eminence of Venus in Maya art and cosmology. The Maya referred to it as *Noh ek* ('great star'), *sastal ek* ('bright star'), and *xux ek* ('wasp star'). The significance of Venus lies in its uniqueness: a highly conspicuous object, bright enough to cast shadows, that is obviously linked to the Sun. On some occasions it blazes forth in the west as dusk falls; at others, it heralds the dawn in the east. More sacrifices were made to Venus than to any other celestial object barring the Sun. In sharp contrast to the Romano-Greek tradition, the 'Star Wars' hypothesis proposes that Venus was the patron star of war, and its appearance in the sky was seen as a propitious date for military action. Decisive victories were marked with a glyph representing the planet above one representing the vanquished[14,12].

However, the most significant manifestation of Venus in Maya culture is undoubtedly the Venus table on pages 24 and 46 to 50 of the Dresden Codex. That the Maya were familiar with the synodic period of Venus and the Dresden Codex texts represented a Venus calendar were generally accepted as long ago as the 1920s[39,40]. The synodic period of Venus is 583.92 days, which may be approximated to 584 days or one canonical Venus Round. The Venus Round begins with the **heliacal rising** of Venus after inferior **conjunction** and is split into four stations 236, 90, 250, and 8 days apart. The intervals correspond to visibility as a morning star, invisibility around superior conjunction, visibility as an evening star, and invisibility around inferior conjunction. The accuracy of this four-part division is poor: the true figures are 263, 50, 263, and 8 days, so only the last interval is correct. The 584-day approximation gives five Venus Rounds = 1 Venus Almanac cycle = 2,920 days = 8 *haab* cycles. To commensurate the Venus Almanac cycle with the ritual *tzolkin* cycle requires a much longer period of 37,960 days (slightly less than 104 years), or one Great Venus cycle. This number is the lowest common multiple of 260, 365, and 584; it corresponds to 13 Venus Almanac cycles, or 65 Venus Rounds, or 104 *haab* cycles, or 146 *tzolkin* cycles[31,20].

The Dresden Codex Venus table comprises a user manual on page 24 and a main body on pages 46 to 50. The non-contiguous numbering reflects an earlier numbering convention, now known to be incorrect but retained due to longstanding use[20]. The main body tabulates the Venus Almanac cycle on thirteen lines for a Great Venus cycle of 37,960 days. The user manual comprises four sections. Block I is a list of Long Count dates; Blocks II-1 to II-3 are lists of multiples of 2,920 for the thirteen cycles through the table; Block II-4 gives corrections to keep the table in line with the actual Venus synodic cycle; and Block III is a list of ring numbers and *lubs*. From Blocks I-III, we can determine entry points into the main body of the table. Block IV contains augural and omen glyphs[32,14].

The five pages of the main body each comprise five sections. Block VI contains three Venus images with explanatory texts: Block VI-1 is an augury pertaining to Venus apparition; Block VI-2 is Venus at apparition spearing a victim; and Block VI-3 is the unfortunate victim being speared. Above each are further groups of augural and omen glyphs. Venus was believed to be especially dangerous when it emerged from inferior conjunction, having just emerged from spending eight days in the underworld. Its bright rays were likened to arrows raining down death, destruction, and pestilence upon the Earth. It was therefore essential to perform the appropriate rituals to ward off these evil effects. Block VII gives the appearance and disappearance intervals for Venus; Block VIII lists cumulative figures; Block IX gives *tzolkin* dates for the various stations of Venus; and Block X gives directional glyphs (north, south, east, west). Successive apparitions of Venus were evidently associated with different quarters of the horizon[32,14].

Up to eight base dates appear to have been in use at various times, all positioning the heliacal rise of Venus at 1 *Ahau*[20]. To the Maya, it was imperative that the two should coincide. 1 *Ahau* is associated with the Maya deity God L, patron of warriors and traders, who was personified by Venus. The ancient Maya pantheon is poorly understood; in 1904 German researcher Paul Schellhas identified many Maya gods, which in the absence of the original names, he designated with letters – God A, God B, etc.

On page 24 are a pair of Long Count dates; 9.9.16.0.0 (1,366,560) and 9.9.9.16.0 (1,364,360) and the ring number 6.2.0 (2,230). 1,366,560 is divisible by 260 (*tzolkin* cycle), 365 (*haab* cycle), 584 (Venus Round), 780 (Mars Round/triple *tzolkin*), 2,920 (Venus Almanac cycle), 18,980 (Calendar Round), and 37,960 (Great Venus cycle), meaning that all these cycles were positioned at the same point as they were at the start of the epoch on 13.0.0.0.0 4 *Ahau* 8 *Cumku*. It could have been for this reason that the Maya chose 9.9.16.0.0 as a base date. Subtracting the ring number 6.2.0 from 13.0.0.0.0 4 *Ahau* 8 *Cumku* gives us a date in the previous epoch: 12.19.13.16.0 1 *Ahau* 18 *Kayab*. Adding the Long Count number 9.9.16.0.0 takes us to 9.9.9.16.0 1 *Ahau* 18 *Kayab*. The corresponding Gregorian date is 4 Feb AD 623, a date when Venus was still disappearing into the evening twilight some 16 days before its heliacal rising. Possibly the date was based on calculation rather than observation, and that at that time the Maya still believed that the synodic period was exactly 584 days. It has been suggested that the table was actually inaugurated three Great Venus cycles later, on 10.5.6.4.0 1 *Ahau* 18 *Kayab* (20 November AD 934), when a heliacal rising of Venus did occur[14].

Taking the 1 *Ahau* 18 *Kayab* base date, we add 236 days to enter the table on page 46 line 1 at 3 *Cib* 8 Zac (morning star disappears into the dawn). Next, we move forward 90 days to 2 *Cimi* 18 *Muan* (Venus reappears as an evening star); from there 250 days forward to 5 *Cib* 4 *Yax* (evening star disappears into the twilight); and finally, 8 days forward to 13 *Kan* 12 *Yax* (Venus reappears as a morning star) and the completion of the first Venus Round. Similarly, the second Venus round is outlined on page 47, line 1; the third on page 48, line 1, the fourth on page 49, line 1, and the fifth on page 50, line 1. At the completion of five Venus Rounds (i.e., one Venus Almanac cycle of 2,920 days) the *tzolkin* date will be 9 *Ahau*. At this point, we move to page 46, line 2 to begin the second Venus Almanac cycle, and so on until 13 Venus Almanac cycles (i.e., one Great Venus cycle) have been completed. Note that because 2,920 is divisible by 365, the *haab* portion of the Calendar Round date for each of the twenty points in the Venus Almanac cycle is unchanged for each repetition. 2,920 is also divisible by 20 but not 260, hence the *tzolkin* day names will also remain unchanged, but not the *tzolkin* day coefficients. The latter do not repeat until the completion of a whole Great Venus cycle, at which point we return to the 1 *Ahau* 18 *Kayab* that commenced the sequence[20].

At this point, we should note that the 37,960-day Great Venus cycle has fallen out of step with the actual Venus cycle by (584 - 583.92) × 5 x 13 days = 5.2 days. There is a need to apply a correction, but at the same time the heliacal rising of Venus must still occur on the ritual *tzolkin* day 1 *Ahau*[41]. This means that any correction must displace the table by a multiple of 260 days. The solution was a shortened, intercalary run: stop the count at the end of 61 Venus Rounds, subtract four days and go back to the beginning. At the end of 61 Venus Rounds = 35,624 days, the *tzolkin* day is 5 *Kan*. Subtracting four days before restarting the cycle sets the start date to the all-important 1 *Ahau*. However, the actual Venus cycle at this point would be out by (584 - 583.92) × 61 = 4.88 days, so this is an under-correction of 0.88 days. After four Great Venus cycles, the error would still amount to 3.52 days. Accordingly, every fifth Great Venus cycle, there was a modified intercalary run. The count was stopped after just 57 Venus Rounds, and eight days rather than four days were subtracted before the restart. After 57 Venus Rounds = 33,288 days, the *tzolkin* day is 9 *Muluc*; the subtraction of eight days will once again ensure that the restart day is 1 *Ahau*. The nett result is that 24 days are subtracted from 61 × 4 + 57 = 301 Venus Rounds. Remarkably, the error is just (301 × 583.92) - (584 x 301 - 24) = 0.08 days, or less than two hours in 481 years[17].

Victoria Bricker[42] believes that pages 30c-33c of the Dresden Codex may connect the motions of Venus with those of Mercury. The table spans a period of 2,340 days, equal to nine *tzolkin* cycles or four synodic periods of Venus. The count begins on 11 *Ahau* and there are nine stopping points, each associated with intervals of 13 days for a total of 117 days. The synodic period of Mercury is 115.88 days. The first column of day signs on page 30c mandates five runs through this table for a total of 585 days (one synodic period of Venus). Three adjacent columns indicate three more runs, for a total of 4 × 585 = 2,340 days, or nine *tzolkin* cycles. The slightly high value of 117 days for the synodic period of Mercury enabled the Maya to equate five synodic periods of Mercury to one synodic period of Venus. The interval of 117 days is also close to four synodic months (118.12 days), albeit this time a little low.

Mars

Mars is less prominent than Venus in Maya mythology, and no deity has been identified as the 'Mars God'. However, an entity known as the Mars Beast (or Sky Beast or Lightning Beast) is depicted in both zoomorphic (with cloven hooves) and anthropomorphic (with human hands) guises in the Dresden and Madrid Codices[29]. It is widely believed that Pages 43b-45b of the Dresden Codex represent a Mars almanac, although the case is less clear cut than for the Venus table. The synodic period of Mars averages 779.94 days = 2 × 260 or three *tzolkin* cycles, but it has been suggested that the tables involving this number are nothing more than triple-tzolkin almanacs that in all probability refer to weather forecasts[32]. However, on the basis of the Maya interest in commensurable cycles, a Martian connection seems likely[14].

The introduction, in the leftmost column of page 43b, provides a ring number of 17.12 (352) days and a Long Count date of 9.19.8.15.0 (1,435,980) days. Going backwards from the zero date at 13.0.0.0.0 4 *Ahau* 8 *Cumku*, the ring number takes us to the last 3 *Lamat* before the zero date; adding the Long Count date gives a base date of 9.19.7.15.8 3 *Lamat* 6 *Zotz*, which falls in March AD 818. 3 *Lamat* is the day of Mars, equivalent to 1 *Ahau* for Venus. 352 days is close to the average number of days between the superior conjunction of Mars and the beginning of **retrogression**; possibly this is why 3 *Lamat* was important. These dates are followed by a series of multiples of 78 and 780. The table proper, comprising the rightmost part of page 44b plus page 45b, divides a 78-day period into four intervals: 19, 19, 19, and 21 days. Information about each interval appears in a column containing glyphs, a picture of the head of a 'Mars beast', and two numbers in dot-and-bar notation alternating between red numbers and black numbers. The red numbers (9, 2, 8, and 3) are the *tzolkin* day coefficients, and the black numbers (19, 19, 19, 21) are the distance numbers that must be added to them in order to proceed to the next. We begin at the bottom right of 44b, entering the table at 3 *Cimi* (i.e., 78 days later than 3 *Lamat*), and proceeding leftwards. As 78 is divisible by 13, the day coefficients of subsequent dates will also be 3[29,14].

In contrast to Venus, the heliacal rising of Mars occurs when the planet is on the far side of the Sun and closest to its maximum distance from Earth. The maximum brightness of Mars occurs at **opposition**, and the Maya probably considered this phenomenon to be of greater importance than the heliacal rising. Due to the eccentric orbit of Mars, opposition brightness is very variable, ranging from magnitude -2.9 to -1.2, but it is always far brighter than at heliacal rising. The retrograde period is centred on opposition, and although the duration varies from 60 days to over 80 days, the average is around 75 days. This is close to the tenth-triple-tzolkin period of 78 days, possibly explaining why the table is built around this interval[29].

There is also some reason to suppose that the Maya were also interested in the sidereal motion of Mars. If the empirical sidereal interval of Mars is defined as the interval between successive returns to a given **celestial latitude** while not in retrogression, then there will be two typical intervals: a short interval of around 543 days that does not include a retrograde loop and a long interval of around 707 days that does include a retrograde loop. A short interval is followed by either seven long intervals (543 + 7 × 707 = 5,492 days; just over 15 years) or eight long intervals (543 + 8 × 707 = 6,199 days; just under 17 years). Long and short intervals follow the repeating sequence 7L 1S 7S 1S 8L 1S… repeat. This gives an average interval of (22 × 707 + 3 x 543)/25 = 687.32 days, which is just under 16 hours shorter than the **sidereal period** of 686.98 days (what is actually being measured is the **tropical period**, but for Mars this is only ten minutes shorter than the sidereal period).

On pages 69-74 of the Dresden Codex is a complex table related to rainfall. It comprises upper and lower portions known respectively as the Upper and Lower Water Tables (UWT and LWT). The UWT is a 702-day interval table. It contains nine Long Count base dates, seven of which appear to mark the beginning of a long interval that immediately follows a short interval. The chance of seven out of nine dates confirming to this pattern by pure chance are extremely low, suggesting that Maya astronomers were aware of the sequence. One question is, why was a value of 702 days chosen for the long interval over the more accurate value of 707 days? The answer was almost certainly the Maya desire for commensuration. The synodic period of Mars may be factorised as 10 × 78; 702 may be factorised as 9 × 78; and if the short interval is taken to be 546 rather than 543, it may be factorised to 7 × 78. Another sequence possibly known to the Maya is that 17 sidereal periods of Mars is approximately equal to 20 synodic periods of Venus. Using the Maya approximations and the sequence 7L 1S 8L 1S gives 15 × 702 + 2 × 546 = 11,622 days for Mars and 20 × 584 = 11,680 days for Venus; a difference of 58 days. However, using modern values gives 17 × 686.98 = 11,678 days 16 hours for Mars and 11,678 days 10 hours for Venus; a difference of just six hours[26].

The Maya zodiac

The Paris Codex came to light in the Bibliothèque Nationale in Paris in 1859, and it is in poor condition. It was found amongst a pile of soot-covered papers in the corner of a chimney. Like the Dresden Codex, it is thought to date to the Postclassic period. The fragmentary pages 22 and 23 depict several animals hanging from sun symbols below a continuous band. The latter is thought to represent the body of the two-headed sky serpent (*caan*). The animals then continue across a lower band, which is undecorated. A total of thirteen animals are evident – four on top of page 24, three on top of page 23, and three on the bottom half of each page. The animals are bird (1), rattlesnake (2), turtle (3), scorpion (4), vulture (5), serpent (6), bird (7), frog (8), possible deer or bat (9), possible peccary (10), unidentified (11), death head (12), and peccary or ocelot (13)[33].

There are seven columns of *tzolkin* dates on page 24 followed by six columns on page 23. The reading order is right to left. Five rows span both pages. From right to left, each *tzolkin* entry is separated by 28 days. Each row thus comprises 28 × 13 = 364 days, a period known as a computing year. Five rows are 5 × 364 = 1,820 days = 7 *tzolkin* cycles, so the two line up at the end of this interval. However, the 'computing year' upon which the table is based is 1.25 days less than the sidereal year. It would be necessary to periodically move the *tzolkin* cycle forward to bring the table back into alignment with the constellations, although it is not clear how this was accomplished[20].

The animals featured in the Paris Codex are very similar to a sequence of animals carved on a lintel on the façade of the east wing of the Nunnery (Las Monjas) at Chichén Itza. Here there are 24 symbols. These include seven animals, four of which appear in the Paris Codex and in the same order: the turtle (18), scorpion (20), vulture (22), and serpent (24). Other possible matches, albeit out of sequence, are bird (11), death head (9), and peccary (4). It is unlikely that this correspondence is a coincidence: the Paris Codex and the lintel are almost certainly referring to the same sky phenomena. That the animals represent a zodiac is suggested by the repeated appearance of Venus glyphs in the Nunnery panels – but what constellations do they depict[14]?

While most scholars accept that the figures represent a zodiac, there is no agreement on the identification of the constituent constellations. One possible scheme is as follows: bird (Libra), rattlesnake (Pleiades), turtle (Orion), scorpion (Scorpius), vulture (Gemini), serpent (Sagittarius), bird (Capricorn), frog (Cancer), possible deer or bat (Aquarius), possible peccary (Leo), unidentified (Virgo), death head (Pisces), and peccary or ocelot (Aries)[20].

E-Group architecture

Uaxactun is a Maya site located in the Petén Department of northern Guatemala (part of the larger Petén basin, which extends into southern Mexico). The site is noted for a cluster of buildings known as the E-Group that is widely believed to have functioned as a solar observatory. The complex comprises an open plaza with a single pyramid designated E-VIII sub on its western side and a north-south aligned platform on its eastern side. Upon the latter stand three small equally spaced buildings designated E-I, E-II, and E-III. From the point of view of an observer standing on the pyramid E-VIII sub, the small buildings line up with the rising of the sun over the horizon at the summer solstice (just to the left of E-I), equinoxes (above E-II) and winter solstice (just to the right of E-III). What became known as 'E-Group' architecture was subsequently identified at over sixty other sites, mostly but not entirely located within a 100 km (62 mile) radius of Uaxactun. They generally consist of a pyramidal structure on the west side of a plaza and a tripartite structure on the eastern side. The plazas are commonly the setting for stela, and for other structures lining the north and south sides. The term 'E-Group' or 'Group E arrangement' simply refers to the alphabetic labels that were assigned to the various architectural groups at Uaxactun when the site was first mapped in the 1920s, and it does not imply any resemblance of the complex to the letter "E"[14,43].

The astronomical alignment was first noted by Danish archaeologist Frans Blum in 1924, and has been generally accepted ever since. But the alignment seems to have worked better in the earlier phases of the complex's history. Maya buildings typically underwent phases of extensive modification over the centuries, and the Uaxactun E-Group was no exception. Firstly, the central E-II structure was built up, so it blocked the view of the equinoctial sunrise from the pyramid E-VII sub. Subsequently, the height of the pyramid was increased from around 3.3-3.8 m (10 ft 9 in – 12 ft 6 in) to 8 m (26 ft 3 in). To an observer at its summit, the horizon now lay above the level of E-I and E-III, so these buildings were no longer able to pinpoint solstitial or sunrises. It is likely that by this time the function of E-Group had changed, and that it no longer served any astronomical function. It should be remembered that the Sun's day-to-day movement is at its slowest around the solstices (see Chapter 14). Therefore, even before these later phases, the Uaxactun E-Group complex could not have been used to accurately designate solstitial dates[14].

Further problems arise with E-Group architecture at other sites, where the Uaxactun alignment is not duplicated. It was suggested that Uaxactun was the original and the later sites were non-functional copies, where the focus was on the architecture and the ceremonies rather than on any actual observations. This view finds some support in that the astronomical purpose of Uaxactun itself was seemingly lost in the later phases of construction; perhaps these later E-Group sites dated to such a time. However, later work demolished this theory. While Uaxactun's E-Group complex was the first to be identified, it is now known that it was not the first to be constructed. Furthermore, earlier examples have long platforms on the eastern side lacking clear evidence for the tripartite architecture. These were commonly only added during later phases of construction. One possibility is that the E-Group complexes were intended to delimit the zodiacal band of the sky rather than mark specific alignments. On this view, they were 'planetariums' that provided a dramatic setting against which the Sun, Moon, planets, zodiacal constellations, and zodiacal light could be viewed[43].

Anthony Aveni and colleagues[44] have suggested that while early E-Group complexes functioned as solstice and equinox markers as at Uaxactun, later examples were concerned with zenith passages of the Sun (i.e., days on which the Sun is directly overhead at noon). At the latitude of Petén (17°30' N), this occurs on 10 May and on 3 August each year. Specifically, what was of interest was a 'countdown' to the May passage in intervals of twenty days; 19 February (zero minus 80 days), 11 March (zero minus 60 days), 31 March (zero minus 40 days), etc. Aveni and his colleagues identified up to three alignments each on the northern and southern structures on the eastern platforms as viewed from the west. They note that given the significance of the number twenty, the Maya would also have surely been aware that at 17°30' N the winter solstice precedes the May zenith passage by 140 (i.e., 7 × 20) days; the May zenith passage precedes the summer solstice by 40 (i.e., 2 × 20) days; and the summer solstice precedes the August zenith passage by 60 (i.e., 3 × 20) days. Aveni and colleagues believe that before the introduction of the written calendar seen in the Codices and on monumental inscriptions, the Maya used a solar orientation calendar. During the Middle Preclassic to Late Preclassic periods, calendars were mainly based upon solstices and equinoxes; but later during the Classic period there was a shift to the May zenith passage and preceding intervals, and the summer solstice. Such a calendar would have been more useful from an agricultural point of view. The E-Group complexes might have been the settings for rituals anticipating the onset of rainfall and the planting season.

Chichén Itzá

Chichén Itzá on the northern tip of the Yucatán is one of the most famous of all Maya sites, although it only rose to prominence during the Terminal Classic. The site is dominated by an iconic 30 m (98 ft) high step pyramid known as El Castillo ('the castle'), aligned on the cardinal points. It is ascended by stairways on each face, and surmounted by a temple to the Mayan feathered serpent deity Kukulkan. An hour before sunset at the equinoxes, the nine platforms that make up the pyramid of El Castillo casts shadows on the balustrade of the north stairway in a way that forms an undulating line that meets a large sculpture of a snake's head at the foot of the stairway; a dramatic visual effect that would seemingly bring the serpentine form to life.

A few hundred metres to the southwest of El Castello is a structure known as El Caracol ('the snail'): a round, domed tower superficially resembling a present-day astronomical observatory standing atop a trapezoidal upper platform, which in turn stands on a lower, rectangular platform. Dating to around AD 800, the lower platform measures 52 × 67 m (170 × 220 ft) and is elevated 6 m (20 ft) above the flat terrain. A stairway provides access to the two levels, which incorporates a niche containing a pair of columns. This so-called stylobate is aligned asymmetrically with respect to the upper platform. The 15 m (49 ft) high tower, which was a later addition, contains a solid lower body, a central portion with two circular galleries, and a spiral staircase leading to an observation chamber on the top. The building is named for the stone spiral staircase.

In contrast to its neighbour, El Caracol is a rather awkward-looking structure. Sir Eric Thompson[45] was one of the building's critics: "*Every city sooner or later erects some atrocious building that turns the stomach: London has its Albert Hall; New York its Grant's Tomb; and Harvard its*

Memorial Hall. If one can free oneself from the enchantment which antiquity is likely to induce and contemplate this building in all its horror from an aesthetic point of view, one will find that none of these is quite so hideous as the Caracol of Chichén Itzá. *It stands like a 2-decker wedding cake on the square carton in which it came. Something was pretty wrong with the taste of the architects who built it.*" One wonders what Thompson would have made of some of the 'carbuncles' that in more recent decades have attracted the ire of HRH The Prince of Wales.

The overall inelegant appearance of El Caracol has led many to supposed that purely utilitarian motives were behind its design such as, for example, a military watchtower. However, the most widely accepted explanation is that it served an astronomical function. Anthony Aveni and colleagues[46], also Aveni[14] proposed that a number of horizon alignments were built into the structure for an assumed construction in phases between AD 850 and 1000. The alignments were viewed looking along the corner-to-corner diagonals of the upper and lower platforms in the directions southwest-to-northeast; or southeast-to-northwest; or perpendicular to the base of the stylobate platform; or through windows on the upper observation level, using diagonal sightings inside-left to outside-right across the window jambs. The alignments include sunrise at the summer solstice, sunset at the equinoxes and zenith passage dates, the most northerly and southerly setting points of Venus, the **heliacal setting** of the Pleiades, and the setting point of the bright star Achernar (α Eridani).

Although the heyday of the Maya lay in the past at the time of the Spanish conquest, their astronomical tradition continued. We can but speculate on how astronomy might have developed in the New World had the Aztec and the Inca empires remained isolated for a few more centuries. That so much astronomical knowledge was seemingly encoded in the Dresden Codex; that the lengths of the synodic month and tropical year were known with such accuracy hints at an astronomical tradition of observation and prediction as sophisticated as anything in the Old World at that time. We shall probably never know just how much Maya astronomical knowledge was lost as a result of De Landa's book burnings, and what we have is almost certainly but a fraction of the whole. However, the recent discovery of astronomical murals at Xultun gives hope that at least some of it still remains to be discovered, and that ongoing archaeological investigations of Maya sites will one day bring it to light.

17
The Copernican revolution

The Renaissance

The European Renaissance followed the Middle Ages, and was a period of intellectual and cultural flowering that saw a resurgence of learning based on Classical philosophy, literature, and art. It was characterised by the flourishing of some of the greatest scientists, artists, and statesmen in human history. It is unfair to characterise the European Middle Ages as a grim period of war, famine, pestilence, and religious superstition, but the fact remains that throughout the medieval period the important advances in science and mathematics were made in India, the Islamic world, and China, rather than in Europe.

By the twelfth century, however, trade was opening up between Europe and the East, and Eastern knowledge began to spread to the West. The English scholar Robert of Chester translated Muhammad ibn Musa al-Khwārizmī's algebra treatise *al-Kitāb al-mukhtasar fi hisab al-jabr wa'l-muqabala* into Latin. In AD 1202, mathematician Leonardo Bonacci of Pisa popularised the Hindu-Arabic numeral system in Europe. Usually known by the Latinized form of his name, Fibonacci ('son of Bonacci'), he also introduced the Fibonacci number sequence to European mathematicians, although it was originally described by Indian mathematicians as early as 200 BC.

The term 'Renaissance' is means 'rebirth', but it was not used to describe the period until the eighteenth century. It began in Italy during the late fourteenth century AD and gradually spread across Europe over the next hundred years. The Renaissance was underpinned by a movement now known as Renaissance humanism, which was more a method of learning than a philosophy. Renaissance scholars studied and appraised ancient texts through a combination of reasoning and empirical evidence. Humanist education was based on the five humanities: poetry, grammar, history, moral philosophy, and rhetoric. Renaissance humanism may be viewed as a movement to recover the literature and knowledge of Classical antiquity, but most of all to assert the genius and extraordinary ability of the human mind.

The invention of the movable type printing press in the middle of the fifteenth century started a revolution of mass communication. Instead of being laboriously handwritten, books could now be mass-produced. By the end of the century, printing presses were widespread across Europe, and books were being produced by the million. In Renaissance Europe, the unrestricted circulation of information and new ideas threatened the power of religious and political elites. At the same time, literacy increased, breaking the elite's monopoly on education and learning, and further undermining their grip on power.

The study of natural philosophy was still dominated by Aristotle's works, and Ptolemy's geocentric cosmology was backed by the might of the Catholic Church. All of this was about to change, beginning with a fifteenth century Polish mathematician and astronomer who began a revolution that would transform our understanding of the universe in little over 150 years.

Copernicus

Nicolaus Copernicus (1473-1543) was born in Toruń, northern central Poland, the youngest of four children. His father, Nicolaus Copernicus, Sr., was a well-to-do copper merchant who had moved to Toruń from Krakow; and his mother, Barbara Watzenrode, was from a leading merchant family in Toruń. Until 1411, the town had been part of the Hanseatic League, after which it fell under the control of the Teutonic Order. In 1466, Toruń was ceded to Poland. After Nicolaus Copernicus, Sr. died in 1483, the four children were looked after by their maternal uncle, Lucas Watzenrode, a senior cleric who became bishop of Warmia in 1489. Watzenrode took charge of his nephew's education with a view to a career for him in the church[1,2].

Copernicus enrolled at the University of Krakow in 1491. The university offered courses in mathematics, astronomy, and astrology, which appear to have sparked his interest in science. During his four years in Krakow, Copernicus became interested in the contradiction between Aristotle's regular circular motion and Ptolemy's equants (see Chapter 12). Despite his new interest, Copernicus was still set for an ecclesiastical career. In 1495, Watzenrode obtained Copernicus's election as canon at the cathedral of Frombork. This would make him financially secure for life, although it would be two years before he eventually took up the position. In the meantime, in 1496, he left Krakow without completing his degree and went to Italy to study canon law at the University of Bologna. This was supposedly in preparation for the Frombork role, but his real interests lay elsewhere. He sought out Professor of Astronomy Domenico Maria Novara da Ferrara, from whom he undoubtedly gained knowledge. He also made his first known astronomical observation, on 9 March 1497, of a close approach of the Moon to Aldebaran (α Tauri)[1,2].

In 1501, Copernicus returned to Frombork, after spending time in Rome the previous year, and having again failed to complete his studies. As canon, he sought leave of absence for two years and a grant to study medicine at the University of Padua – although he knew that the medical degree required three years of study. Two years later, with his leave about to expire, and no doubt feeling that he should not return a third time without a degree, he successfully applied to the University of Ferrara to sit his examinations for Doctor of Canon Law[1,2].

On returning to Poland, he went to live with his uncle at the episcopal palace in Lidzbark-Warminski. Astronomy had to take second place to church matters and his uncle's declining health, but Copernicus decided against advancement in the church despite the entreaties of his fellow canons. In 1510, he returned to Frombork, where he remained for the rest of his life. He was still involved with administrative matters and other church responsibilities, but he was able to build an observatory and devote more time to astronomy[1,2].

In making observations, he was hampered by fogs rising from the nearby Vistula River, and there is a longstanding claim that he never managed to see the planet Mercury[3]. The story is almost certainly apocryphal[4], but it is a fact that most people go through life without ever seeing the Solar System's innermost planet. Although Mercury is brighter than any star, it is extremely elusive, and is typically only visible for two or three weeks in a year. I have only seen it on a handful of occasions, and only once when I was not actively looking for it.

Between 1510 and 1514, Copernicus put forward an initial outline of a **heliocentric theory** in an essay known from later transcripts as *Nicolai Copernici de hypothesibus motuum coelestium a se constitutis commentariolus* ('The little commentary of Nicolaus Copernicus on the hypothesis of the motions of the heavenly bodies'), usually referred to simply as the *Commentariolus*. The work was not intended for publication (it was eventually published, but not until the nineteenth century), and Copernicus saw it as nothing more than a study for a proposed book. He circulated it to friends and colleagues, including a group of astronomers from Krakow with whom he had collaborated in observing eclipses.

In the *Commentariolus*, Copernicus listed seven postulates:

1. Celestial orbs and spheres do not all revolve around a single point;
2. The Earth is only the centre of the heavy bodies (i.e., earth, air, fire, and water) and the lunar orb;
3. All the spheres encircle the Sun, which is near the centre of the Universe;
4. The distance between the Earth and the Sun is infinitesimal in comparison to the distance from the Earth and the Sun to the stars, and hence **parallax** is not observed in the stars;
5. The stars do not move, and their apparent daily motion is caused by the daily rotation of the Earth;
6. The apparent motion of the Sun is caused by the several motions (i.e., rotation around the Sun, axial rotation, and **precession**) of the Earth;
7. **Retrogression** of planets is caused by the Earth's motion around the Sun[5].

Copernicus did not entirely break with Classical cosmology. He continued to think in terms of solid spheres rather than planets moving through space, albeit the Earth was now carried around the Sun by an ethereal orb like the other planets. But the Earth had not been entirely dethroned. Rather than placing the Sun at the centre of the Universe, Copernicus placed the latter at the centre of the Earth's orbit, which is slightly eccentric to the Sun. The rest of *Commentariolus* explains how the motion of the Earth affects the apparently complex motions of other planets. *Commentariolus* was certainly not the finished article. The work lacked the detailed mathematical treatment that he intended to put forward in his book[5].

Copernicus then began working on his book *De revolutionibus orbium coelestium* ('On the Revolutions of the Heavenly Spheres'), but it would be a lengthy process. The work was not published until just before his death in 1543. The main reason for the delay was that the full-blown work required observational data and complex mathematical proofs. A Latin edition of the *Almagest* had come out in 1515, and Copernicus would have realised any treatise hoping to compete with Ptolemy's work would have to be just as comprehensive. As noted, fog often prevented him from making observations. He used 26 of his own observations, but he borrowed a further 45 from the *Almagest*[1,2].

In addition, Copernicus still had the encumbrance of his day job. His duties included organising a defence against the Teutonic Order, who had occupied the Prussian territory to the east; and proposing currency reform. Here, he anticipated English financier Thomas Gresham (1519-1579) in stating the law that bad money (i.e., debased coinage) drives out the good[1]. Around 1537 or 1538, Copernicus is reputed to have had an affair with a wealthy woman named Anna Schilling, who he met in Danzig (now Gdansk). As a canon, Copernicus had taken a vow of celibacy, and Anna was already married. She was also some fifteen years younger than Copernicus. After she moved to Frombork and became Copernicus's housekeeper, local sensibilities were offended. Although Copernicus insisted that there was nothing else to the relationship, Johannes Dantiscus, a senior bishop, demanded Anna's dismissal. She eventually returned to Danzig in March 1539, and as far as is known she and Copernicus never saw each other again.

Despite these issues, the six-volume work was largely complete by 1539. By now, word of Copernicus's theory had spread, and he was encouraged to publish by his friend Tiedemann Giese (1480-1550), Bishop of Chelmo, and by Nicholas Schönberg (1472-1537), Cardinal of Capua. Despite this, he held off publishing. It is not clear if he was concerned about how a heliocentric system might be received by the religious authorities, or simply about objections from astronomers and philosophers. It is also likely that Copernicus was hampered by his location in Frombork, which was a long way from the any of the major international centres of printing capable of handling a book as large and technical as *De revolutionibus.* Furthermore, he was far from the major centres of academe, where he could have discussed his work with technically qualified peers[6,2].

Eventually, in 1539, Georg Joachim Rheticus (1514-1574), a Lutherian professor of mathematics at the University of Wittenberg, arrived in Frombork. The university was a major centre for mathematics, but it was also at the heart of the Reformation. Martin Luther (1483-1546) held the position of Professor of Theology there from 1512 until his death. Rheticus was on a two-year sabbatical to visit Europe's leading mathematicians and astronomers. In the course of his travels, he heard of Copernicus and his theories, and decided to pay him a visit. Rheticus brought with him some mathematical and astronomical volumes, which enabled Copernicus to make some revisions to his work. These volumes also showed Copernicus the quality of printing available in Germany for mathematical texts. Rheticus was particularly keen to promote Nuremberg publisher Johann Petreius (1497-1550) as a possible publisher of Copernicus's work. In 1540, Rheticus published the *Narratio prima* ('First account'), which was a non-mathematical introduction to the theories of Copernicus. That its publication did not lead to a backlash against heliocentrism was the deciding factor that finally persuaded Copernicus to publish[5,2].

The work was dedicated to Pope Paul III (reigned 1534-1549), no doubt to keep the Catholic Church onside. He thanked Schönberg and Giese for encouraging him to publish; but he made no mention of Rheticus due to the latter's Protestant connections. In 1542, Rheticus took the manuscripts of *De revolutionibus* to Nuremberg for publishing by Petreius. He oversaw most of the printing, but later that year he was then forced to leave in order to take up the position of Professor of Mathematics at the University of Leipzig. He handed over the task to Andreas Osiander (1498-1552), a Lutheran minister. Without the knowledge or consent of Copernicus, over the objections of both Rheticus and Giese, and in contradiction to its actual text, Osiander added an anonymous preface to the work claiming that it was a hypothesis rather than a true account of the working of the heavens. By now, Copernicus was in poor health, having suffered a stroke. He died in May 1543, just after the publication of *De revolutionibus*[5,2].

It should be noted that much of the text is a reworking of Ptolemy's *Almagest* and only around five percent concerns heliocentricity. In common with the *Almagest* and the Arabic *zijes*, it included trigonometric rules and tables, a lengthy star catalogue (adapted from Ptolemy), a detailed determination of planetary parameters from both ancient and modern observations, and tables from which predictions could be made. Volume II describes the principles of spherical astronomy; volumes III-V are devoted to solar, lunar, and planetary theory; and volume VI deals with the digression in latitude from the ecliptic of the five planets[1].

Copernicus's Earth orbit was essentially the same as Ptolemy's solar theory: for the purpose of calculation, it makes no difference whether it is the Earth or the Sun that moves. The orbit of the Earth is centred not on the actual Sun but the **mean Sun** $\bar{S}$. In his planetary theory, Copernicus explained retrogression in terms of Earth's circular motion around the Sun rather than with large epicycles, but he did not eliminate epicycles altogether. For all the planets except Mercury, he replaced Ptolemy's equant with a minor epicycle or epicyclet carrying the planet P. The epicyclet, centre C, moves on a deferent, whose centre D is offset from $\bar{S}$. Copernicus usually took the radius of the epicyclet CP to be one-third of the eccentricity of the orbit, i.e., $\frac{1}{3}D\bar{S}$. The centre of the epicyclet C moves uniformly in a prograde direction around the deferent, and the planet P moves uniformly in a prograde direction around the epicycle; both C and P revolve at the same speed, each completing a single revolution in one sidereal period of the planet P. The whole is equivalent to the approach taken by Mu'ayyad al-Dīn al-Urdī to eliminate the equant[7,5] (see Chapter 14).

Copernicus's lunar theory was the same as that of Ibn al-Shātir; and his Mercury model was very similar. Like Ibn al-Shātir, he used a Tūsī couple to cater for celestial latitude in his Mercury model. Copernicus used and derived a proof for the Tūsī couple that was identical to that used by al-Tūsī, even labelling his diagrams in the same way. Copernicus could not read Arabic, and al-Tūsī's work had not been translated into Latin. However, some of Copernicus's contemporaries could read Arabic and it is possible that they communicated their knowledge to him. It has been suggested that the contents of many Arabic works were common knowledge in Italy by 1500. Although Copernicus makes no reference to the Islamic astronomers, there seems little doubt that he was in some way strongly influenced by their work[8,9].

The question also remains as to what extent Copernicus was influenced by Aristarchus, who had proposed a heliocentric system 1,800 years previously (see Chapter 12). *On the Sizes and Distances of the Sun and Moon* is Aristarchus's sole surviving work, but Copernicus would have known of him through the *Almagest* and writings of other later Greeks. The question is, how much did he know about Aristarchus's heliocentric theory? The original manuscript of *De revolutionibus*, preserved at the library of the Jagiellonian University in Krakow, mentions Aristarchus six times. Four of the citations are misattributions of results obtained by others; a fifth includes Aristarchus in a list of those who believed that the year was exactly 365¼ days long. The sixth is more interesting: it refers to Aristarchus's cosmology, but what Copernicus wrote says very little about it. The implication is that he actually knew very little about it, or he would surely have used it to support his theory. Furthermore, the passage was deleted from the published work during the editing that followed the arrival of Rheticus[10].

Copernicus quotes from *Opinions of the Philosophers*, a work misattributed to Plutarch, which mentions Aristarchus, but only in the context of eclipses. Two works genuinely by Plutarch do reference Aristarchus's heliocentricity: *Platonic questions*, and *On the face in the orb of the Moon*. *Opinions of the Philosophers* was published in 1509 in Venice, in a Greek language volume entitled *Plutarchi opuscula* ('Plutarch's work'). This compilation included the genuine Plutarchian references to Aristarchus's heliocentricity. It is possible that Copernicus consulted the work and simply failed to find the relevant citations in what was a dense, bulky volume. However, there is no evidence that he either owned or had access to it, or indeed if he was even fluent in Greek. It is more likely that he used a Latin edition of *Opinions of the Philosophers*, which was available in the cathedral library. If so, Copernicus never saw the genuine Plutarchian references to Aristarchus; and it is therefore probable that he arrived at his heliocentric theory independently[10].

Reaction to *De revolutionibus* was mixed. It read by nearly all the leading mathematicians and astronomers of the day, but most withheld judgment. Many were still caught up in the Aristotelean view that the Earth could not be in motion. Conversely, the replacement of the equant with the epicyclet was received with great enthusiasm. Based on annotations he made on his copy, astronomer Erasmus Reinhold (1511-1553) was more interested in the technical details of the solar, lunar, and planetary theories than he was in the heliocentric hypothesis. Reinhold later produced a book of tables known as the Prutenic Tables ('Prussian Tables') based upon the planetary theory. The theological reaction was mild. Martin Luther was unhappy, as was Philip Melanchthon (1497-1560). The latter was a prime mover in building the German school system and a key figure in the Protestant Reformation. But their complaints were not pursued, and indeed Melanchthon was and remained on good terms with Rheticus. Within the Catholic hierarchy, there was even less reaction: Copernicus had almost certainly sought permission to dedicate the work to Pope Paul III. In 1616, the Catholic Church did belatedly decree heliocentricity to be 'erroneous' and *De revolutionibus* was put on the prohibited list pending 'correction'. Four years later, a list of corrections was issued, intended to give the impression that heliocentrism was merely a convenient hypothesis to make mathematical calculations and observational predictions rather than a physical truth. This censorship had very little effect outside of Italy[5,1].

In the mid-twentieth century, British cosmologist Sir Hermann Bondi coined the term 'Copernican principle' for the view that the Universe is much the same from any vantage point, and that there is nothing unique about the terrestrial or solar perspectives. Copernicus has been honoured with the 93 km (58 mile) diameter lunar crater Copernicus, one of the largest on the Moon and clearly visible to the naked eye from Earth. He has an even larger 300 km (185 mile) diameter crater on Mars. The asteroid 1322 Coppernicus takes the variant spelling of his name that Copernicus actually used until later life. In 2015, the star 55 Cancri A, which has five known planets, was named Copernicus. In 2009, the synthetic chemical element 112 ununbium was officially named copernicium (Cn). Rounding out this list are the palm tree genus *Copernicia* and Copernicus Airport Wroclaw. Although uncredited in *De revolutionibus*, Rheticus has the 45 km (28 mile) diameter lunar crater Rhaeticus and the asteroid 15949 Rhaeticus named for him.

Tycho Brahe

Copernicus had started a revolution, but it would take some decades to gather momentum. Among those sceptical of the heliocentric system was Danish astronomer Tycho Brahe (1546-1601), who eventually proposed a rival Tychonic system (he is typically referred to by either his first name or his full name). Both Tycho's parents were prominent members of the Danish nobility. His father, Otte Brahe, was a royal Privy Councillor, and his mother, Beate

Clausdatter Bille, was a member of the royal court. He was born at the family's ancestral home of Knutstorp Castle, Scania (which was then part of Denmark), the oldest of twelve children. At the age of two, young Tycho was sent away to be raised by his uncle Jørgen Thygesen Brahe (1515-1565). It is unclear why this arrangement was reached, as all his other siblings were raised at Knutstorp. It was nevertheless a happy arrangement, as the childless Jørgen treated his nephew as his own son, and made him his heir.

At the age of just 12, Tycho began studying law at the University of Copenhagen, but he became interested in a variety of other subjects including astronomy and Aristotelian physics and cosmology. He witnessed the solar eclipse of 21 August 1560, which was only partial from Denmark, and took great interest in the ability to predict eclipses. Early in 1562, now aged 15, Tycho set off on an educational tour of Europe. His uncle wanted him to become a civil servant and assigned him the future historian Anders Sørensen Vedel (1542-1616), then aged 19, as a mentor. In March 1562, the pair arrived in Leipzig, where they enrolled at the University of Leipzig. In August 1563, Tycho observed a conjunction between Jupiter and Saturn, but the date as predicted by 200-year-old Alphonsine Tables was out by a month, and even Reinhold's Prutenic Tables were several days out. This made Tycho realise that the key to more accurate predictions was systematic, rigorous observation, with the most accurate instruments obtainable. He began maintaining detailed journals of his astronomical observations[1].

By the time Tycho and Vedel returned to Denmark in 1565, the Seven Year War was in progress between Denmark and its allies and Sweden. Tycho's uncle, serving in the navy, had already distinguished himself in action, but he fell ill and died in June that year. The next year, Tycho left to study medicine at the University of Rostock. While there, he lost a part of his nose while fighting a sword duel with a third cousin, Manderup Parsberg, over who was the best mathematician. Although the two later reconciled and indeed became good friends, Tycho had to wear a prosthetic for the rest of his life. Now intent on a career in the sciences, he returned home for a while before resuming his travels. He visited Rostock, Freiburg, and Basel. In 1568, he enrolled at the University of Basel with a view to moving to the town at a future date. Between 1569 and 1570, he resided in Augsburg, Germany. While there, he worked as an assistant astronomer to the town's mayor. He was put in charge of constructing a quadrant, built from oak, with a radius of 6 m (19 ft 8 in). It was large enough to be graduated in minutes of an arc and could be turned about a vertical axis. However, Tycho found that it was too heavy and clumsy to yield the expected measuring accuracy[7,1].

After the death of his father in 1571, Tycho went to live with another uncle, Steen Bille. Towards the end of that year, he met and fell in love with Kirsten Jørgensdatter, the daughter of a Lutherian minister. As she was a commoner, Tycho was unable to marry her without renouncing his noble privileges. Under Danish law, a nobleman and a common woman were permitted to live together openly as husband and wife for three years, and their relationship then became a legally binding marriage. However, the woman would not be ennobled, and any children would be considered commoners. Tycho's family disapproved of the relationship, but the Danish king, Frederick II (reigned 1559-1588) was sympathetic, having been prevented from marrying courtier Anne Hardenberg some years earlier. Tycho and Kirsten had eight children, of whom six lived to adulthood. Kirsten remained with Tycho until the latter's death.

During his stay, Tycho observed and wrote an account of the supernova of 1572 in

Cassiopeia (SN 1572), the first visible to the naked eye for nearly four hundred years and one of only eight in the historical record. The supernova was first reported on 6 November and by 11 November, when it was first reported by Tycho, it had attained a magnitude intermediate between those of Jupiter and Venus. Between 16 and 17 November, it was as bright as Venus, with an estimated magnitude of -3.2. By the middle of December, it had begun to wane; and it was fainter than Jupiter by the middle of January. Thereafter, it faded gradually, although it remained visible to the naked eye until February 1574. From the light curve, it is now believed that SN 1572 was a Type 1a supernova, reaching peak luminosity around 21 November 1572[11].

SN 1592 is often referred to as Tycho's supernova because of Tycho's account entitled *De Nova et nullius aevi memoria prius visa Stella* ('Concerning the Star, new and never before seen in the life or memory of anyone'), published in 1573. The work contained both Tycho's own observations and those by many other observers. Tycho made precise measurements of angular distance from other stars in Cassiopeia and demonstrated that the star lay beyond the Moon and was not an atmospheric phenomenon. The event had profound philosophical implications: medieval theologians interpreted Aristotelian cosmology to mean that the heavens were perfect. Perfection implies completeness, a view at odds with a new star appearing from nowhere. Tycho's account concluded with some astrological prognostications: that the star's influence would last from 1592 until 1632 and that religions full of pomp and splendour would disappear. The life of King Gustav II Adolph (1594-1632), one of Sweden's most influential rulers and a champion of Protestantism, was taken by some as a vindication of this prediction[7].

Following the publication of *De Nova Stella*, Tycho began to plan for his proposed move to Basel. In 1574, he lectured in Copenhagen, and the following year he visited William IV, Landgrave of Hesse-Kassel (1533-1592), who was himself an astronomer. The pair struck up a friendship and corresponded for many years, but Tycho's move to Basel never happened. In 1576, King Frederick II approached Tycho with a view to funding the construction of an observatory on the remote island of Hveen (now Swedish territory and known as Ven) in the Øresund[1].

Tycho agreed to the king's proposal. Work began in August that year on the observatory, which was built in a Renaissance Gothic style and named Uraniborg. Tycho was working there by the end of the year, although the observatory was not completed until 1580. In addition to an observatory, Uraniborg comprised a library, printing press, and a residence. Tycho developed various instruments for making his observations. The sextant (not to be confused with the navigational sextant) was a medium-sized quadrant cut down from a quarter to a sixth of a circle. It was intended for the measurement of visual angles in random planes. It was attached to a stand with a universal joint and could be turned in any direction. Quadrants were used for altitude measurement, and armillary spheres for the measurement of coordinates in relation to the ecliptic or the celestial equator. Tycho's instruments used traversals, or zigzag patterns of dotted lines to facilitate reading subdivisions on scales. He also improved the sights on his instruments. The backsight comprised four slits arranged in a square and the foresight comprised a square of the same size. The observer lines a star up with the upper edge of the foresight through the top slit of the backsight, and the other edges of the foresight through the corresponding slits of the backsight. For bright objects, accuracy could be improved by narrowing the slits. Overall, Tycho's instruments could achieve an accuracy of about one arcsec. All these facilities made Uraniborg one of the last great observatories of the pre-telescopic era[7,1].

Tycho obtained a value of 23°31'30" for the obliquity of the ecliptic (correct value for date 23°30') and a value of 365 days, 5 hrs, 48 min, 45 sec for the tropical year. Copernicus had obtained 23°28' (correct value for date 23°31') for the obliquity, so Tycho's value was far more accurate. His error was due to his use of values for the solar distance that were still far too low, and which introduced an unnecessary and erroneous correction of 3 arcmin for parallax. However, Tycho improved on Copernicus by correcting for the previously unrecognised effects of atmospheric refraction. His value for the tropical year was too low by just 1½ sec. Starting in 1582, Tycho began a series of observations of the Moon. He built up a large file of accurate observations for all points of its orbit, enabling him to derive a greatly improved lunar theory[7].

Tycho observed a total of seven comets, and described the first of these, the Great Comet of 1577, in his main astronomical work *De Mundi Aetherei Recentioribus Phaenomenis* ('The World's Recent Ethereal Phenomena'), published at Hveen in 1588. In it, he showed by comparing parallax that that comets move through the heavens at a much greater distance than the Moon, and like the supernova they were not atmospheric phenomena. This was a further blow for Aristotelian cosmology: how could a comet move through impermeable planetary spheres? Despite this, Tycho could not accept the Copernican view of a moving Earth, which conflicted with the Bible. The Tychonic system, put forward in *De Mundi Aetherei Recentioribus Phaenomenis*, was a compromise. It placed the Earth at the centre, with the Sun and Moon circling around it, but the five planets circled around the Sun rather than the Earth[1].

Uraniborg became an exemplary research institution where Tycho developed instruments, carried out large numbers of observations, and published his work in scientific journals. He also produced an annual almanac for the king, produced horoscopes, issued prescriptions, and prepared medications. In 1584, he had a second observatory, Stjerneborg, built on a hill outside Uraniborg. All this came at a cost: possibly as much as one or two percent of the crown's annual revenue. As feudal lord of the island, Tycho had obligations towards the king and the peasants, but he frequently did not fulfil them. He failed to maintain the island's lighthouse, generally overworked the peasants, and treated them badly. Frederick II died in 1588. His successor, Christian IV (reigned 1588-1648), was only eleven and a regency was set up until he was old enough to rule. He was crowned in 1596, by then aged 19. He was less tolerant than his predecessor of Tycho's treatment of the peasants, and of his general arrogance. When, following year, the young king decided to cut back on research grants, Tycho's funding was stopped, although he was allowed to keep his fiefdom on Hveen. However, Tycho left the island and moved, first to Copenhagen, then Rostock, and finally in October 1597, to Wandesburg Castle near Hamburg. Attempt to regain the favour of the king came to nothing, as the latter insisted on setting his own terms[1].

Early in 1598, Tycho published a small edition of a work entitled *Astronomiae Instauratae Mechanica* ('Instruments for the Restoration of Astronomy') which contained illustrated descriptions of his most important instruments, as well as a short survey of the theoretical results of his work. He also released a star catalogue from his incomplete *Astronomiae Instauratae Progymnasmata* ('Exercises for the Restoration of Astronomy'), a three-volume work that also included his revised solar and lunar theories, his observations of the 1572 supernova, and a critical review of other published works about the new star. The shortened work was entitled *Stellarum Inerrantium Restitution* ('Restoration of the Fixed Stars'). The two publications were circulated to colleagues and a number of princes in the hope of obtaining the level of funding he had enjoyed in Denmark. In 1599, Tycho received an invitation from the Rudolph II (1552-1612), Holy Roman Emperor and King of Bohemia. In June of that year, Tycho arrived in Prague, but it was not until late in 1600 that he managed to have his instruments shipped to him. In the meantime, he had recruited Johannes Kepler as an assistant, and the two worked together until Tycho's death in October 1601[1].

Tycho Brahe is commemorated by the highly prominent lunar crater Tycho, the 105 km (65 mile) diameter Martian crater Tycho Brahe, the asteroid 1677 Tycho Brahe, and the exoplanet 55 Cancri A c Brahe. He has also given his name to the Tycho Brahe Planetarium in Copenhagen and the palm genus *Brahea.*

Kepler

Johannes Kepler (1571-1630) was born in the Free Imperial City of Weil der Stad, Baden-Württemberg. His grandfather was an innkeeper and former Lord Mayor of the city. Johannes was the third of three children, but by that time the family fortunes were in decline. His father, Heinrich Kepler was a mercenary who left when Johannes was five, and he is believed to have been killed fighting in the Dutch Wars of Independence. His mother, Katharina Guldenmann, was the daughter of the Mayor of Eltingen, a village near Weil der Stadt.

Kepler became interested in astronomy at an early age after seeing the Great Comet of 1577. At eighteen, he gained a scholarship to the University of Tübingen, where he studied for five years. The University had a rigorous Protestant curriculum, having been influenced by Philip Melanchthon (although one of its more recent academics was Joseph Ratzinger, later Pope Benedict XVI). The curriculum helped Kepler to develop his understanding of the roles of astronomy and mathematics. While he was at Tübingen, Kepler was influenced by Professor of Mathematics Michael Mästlin (1550-1631), who became his lifelong mentor. Mästlin was a keen supporter of Copernicus, and he taught the Copernican heliocentric view alongside the traditional Ptolemaic geocentric view. Kepler initially sought a career in the church, but in 1594 a vacancy arose at the Protestant School in Graz for teacher in mathematics and astronomy. He was put forward as the strongest candidate for the position by his tutors and he accepted it in April that year, aged 23[1].

During his time at Gratz, Kepler attempted to describe the structure of the Solar System in geometrical terms in a work entitled *Mysterium Cosmographicum* ('The Cosmic Mystery'), which was published in 1596. The model determines both the number of planets and their distances from the Sun by nesting the five Platonic solids (tetrahedron, cube, octahedron, dodecahedron, and icosahedron) within the spheres encompassing the planetary orbits as follows:

1. Within the sphere representing the orbit of Saturn is a cube;
2. Within the cube is the sphere representing the orbit of Jupiter and within this is a tetrahedron;
3. Within the tetrahedron is the sphere representing the orbit of Mars and within this is

a dodecahedron;

4. Within the dodecahedron is the sphere representing the orbit of Earth and within this is an icosahedron;
5. Within the icosahedron is the sphere representing the orbit of Venus and within this is an octahedron;
6. Within the octahedron is the sphere representing the orbit of Mercury.

This model, beautifully depicted in Figure III of the *Mysterium Cosmographicum*, approximately represented the proportional distances of the planets from the Sun as they were then known. Although the model was hopelessly wrong, it was the first time an astronomer had sought explanations as opposed to simply descriptions. Kepler had asked, why are there six planets, and why are they spaced as they are? But what were the origins of his model? What led Kepler to propose a Solar System based on the Platonic solids? He claimed that he had first tried modelling orbits by nesting regular polygons, but that did not answer the question as to why there should only be "*six mobile orbs rather than twenty or one hundred*". The answer supposedly came to him when he realised that there are only five possible regular – and hence perfect – solids. To him, the five solids had been chosen by God as archetypes of a cosmic order. In September 1595, Kepler wrote to Mästlin convinced that his new geometrical solution explained the vast disparities between the planetary distances.

However, there may also be a connection to the artistic fashions of the time. Two-dimensional engravings and three-dimensional models of nested regular polyhedra were popular in Germany in the late sixteenth century. Kepler would almost certainly have seen, read, or learned about them from books and models – and the style of the nested polyhedra in Figure III of the *Mysterium Cosmographicum* is in the style of many of these polyhedral works. It therefore seems likely that they were the inspiration for Kepler's model of the Solar System. Also noteworthy is the technical and artistic workmanship of Figure III, which is of a much higher standard than any of the other diagrams in the *Mysterium Cosmographicum* or indeed in Kepler's later work. It has been suggested that Figure III was not so much Kepler's initiative as a project by the publisher, Georg Gruppenbach, who wanted to produce an artwork that would appeal to the contemporary fashion for polyhedral geometry and could be sold separately from Kepler's work[12,13].

In December 1595, Kepler was introduced to twice-widowed Barbara Müller, 23, with a view to marriage. Barbara had inherited her late husband's estate and was (as her name might suggest) the daughter of a successful mill owner, Jobst Müller. Due to Kepler's modest means, Jobst was initially dubious about a marriage, but he relented after he completed the *Mysterium Cosmographicum*. Even so, Kepler's frequent absences while he was attending to the details of publication put a strain on the relationship, but the couple eventually married on 27 April 1597.

Early in 1600, Tycho Brahe invited Kepler to join him in Prague, where he was still seeking to be reunited with his instruments from Denmark for his new observatory. Although Tycho was not a supporter of either the Copernican or Keplerian cosmologies, he was impressed by Kepler's theoretical ideas. Kepler was employed as one of several mathematical assistants at the court of Rudolph II. Through most of 1601, he was supported directly by Tycho, who assigned him to analysing planetary observations. Kepler also wrote a tract in support of Tycho against his rival Nicolaus Reimers Bär in a priority dispute over the Tychonic system.

When, later that year, Tycho died suddenly, Kepler was appointed Imperial Mathematician in his place, just short of his thirtieth birthday.

Kepler now enjoyed an international reputation, and began work on the *Tabulae Rudolphinae*, ('Rudolphine Tables'), a star catalogue and planetary tables that used some of Tycho Brahe's observational data. It was a lengthy project that would continue off and on until the last years of his life. In 1604, he produced a major work on optics in relation to astronomical observation, entitled *Astronomiae Pars Optica* ('The Optical aspect of Astronomy'). The work was the first to suggest that the image projected onto the retina by the lens of the eye is inverted, and that it must be corrected somewhere in the brain.

In October 1604, little over three decades after SN 1572, a supernova flared up in the constellation of Ophiuchus. Kepler documented the event in *De Stella Nova*, published in 1606, although clouds prevented him from observing it until eight days after the first sighting. SN 1604 attained a peak magnitude of -2.25, comparable to Jupiter. In November, it disappeared into the evening twilight, but it was still a first magnitude object when it reappeared in the dawn skies in January 1605. It remained visible to the naked eye for eighteen months. The light curve suggests that like its predecessor, SN 1004 was a Type 1a supernova. No conspicuous supernova has occurred since.

Since the time of his employment under Tycho Brahe, Kepler had been working on an analysis of the former's observations of Mars, attempting to produce a model that could fit Tycho's observations to the limits of their accuracy. He started from the premise that the Sun is the source of motive force in the Solar System, and he supposed that this force weakened with distance, causing a planet to speed up or slow down as planets move closer in or further away. In 1602, based on measurements of the aphelia and perihelia of Earth and Mars, he derived a geometrical relationship: a line connecting a planet to the Sun will sweep over equal areas in equal periods of time. This relationship is now known as Kepler's Second law of planetary motion – even though he devised it before the First law[14]. Armed with this, Kepler set about trying to model the entire orbit of Mars, but all attempts failed until early 1605, when he stumbled upon the solution of an elliptical orbit with the Sun at one of the foci of the ellipse. This is the rule now known as Kepler's First law of planetary motion[15]. The significance of this cannot be understated: Kepler had finally freed astronomy from the ancient Greek paradigm of uniform circular motion, and the attendant paraphernalia of deferents and epicycles with which even Copernicus had persisted. The two laws were outlined in *Astronomia Nova* ('A New Astronomy'), publication of which was delayed until 1609 over disputes with Tycho's heirs about the use of his Mars data.

In 1609, Rudolph II was deposed by his younger brother Matthias and held under house arrest in Prague until 1611, when he agreed to abdicate as King of Bohemia. He retained the title of Holy Roman Emperor until his death the following year. Kepler had tried without success to reconcile the two brothers, and he was now dubious about his future in the court of the new king. These were not the extent of his troubles. His marriage to Barbara had produced five children, but two had died in infancy. In 1611, Barbara came down with spotted fever, and as she was recovering all three surviving children fell ill with smallpox; Friedrich, aged six, died. Kepler, by now looking for a new job, travelled to Austria to arrange a position as teacher and district mathematician in Linz. On his return, Barbara relapsed into illness and died. Faced with this appalling turn of events, Kepler postponed the move to Linz until after Rudolph's death early in 1612.

Although Matthias wanted Kepler to stay on as Imperial Mathematician, he allowed him to leave for Linz. In Linz, Kepler continued to work on the *Tabulae Rudolphinae*; his day job consisted of teaching at the district school and providing astrological and astronomical services. He was financially secure and free from the religious and political shenanigans that had characterised his later years in Prague. He had already set about finding a new wife, and as his marriage to Barbara had never been entirely happy, he was distrustful of arranged marriages. Accordingly, he decided to make his own choices. He considered no fewer than eleven matches over a two-year period, taking into account such factors as the virtues and drawbacks of each candidate, dowry, negotiations with parents, natural hesitations, the advice of friends, etc. It was an impressive feat in an era before Tinder or Match.com. In October 1613, he married the fifth interviewee, the 23-year-old Susanna Reuttinger. He went against the advice of friends, who preferred candidate number four, and were concerned about Susanna's lack of high social status, wealth, parentage, and the lack of a dowry due to her being an orphan. It nevertheless turned out to be a wise decision. Kepler said that Susanna's education "*must take the place of a dowry*", and she provided the tranquil home life he needed. They had seven children, although only three survived to adulthood[16]. Unfortunately, Kepler's problems were far from over. The Lutherian Church barred him from Eucharist over his views, and in 1615 his mother Katharina was accused of witchcraft. Kepler took charge of her defence, and she was eventually cleared. But the case dragged on for six years, and Katharina died just six months later, aged 76.

Through these extremely challenging times, Kepler managed to continue with his scientific work, and in 1618 he put forward his Third law of planetary motion. This law states that the square of a planet's orbital period in years is equal to the cube of its mean distance from the Sun in astronomical units. More generally, the square of the orbital period of any orbiting body is proportional to the cube its mean distance from the body it orbits, assuming the latter is significantly more massive. The Third law was included in *Harmonice Mundi* ('The harmonics of the World'), published in Linz in 1619. Kepler never formalised his findings as laws of planetary motion, but they have been referred to as such since at least the time of Joseph de Lalande (1734-1807) in the late eighteenth century[1].

The Keplerian telescope, developed by Kepler in 1611, was an improvement on Galileo's design (see below) from an astronomical point of view. It substituted Galileo's concave eyepiece lens for a convex lens, enabling a much wider field of view at the expense of inverting the image. Much higher magnifications can be achieved with this design. However, all early lenses suffered from chromatic aberration. This is the failure of a lens to focus all wavelengths of light to the same point, and it was caused by dispersion, i.e., the variation with wavelength of the refractive index of the lens elements. Chromatic aberration manifests itself as coloured fringes along the boundaries separating light and dark parts of an image. Modern achromatic lenses use either low-dispersive glass or 'doublets', which are two-element lenses where the chromatic aberration of one is cancelled out by the other. In the seventeenth century, however, the only known solution was to use objective lenses with very long focal lengths. For example, Johannes Hevelius (1611-1686), Mayor of Danzig and a keen astronomer, built a Keplerian telescope with a focal length of 46 m (150 ft).

Kepler's final years were severely disrupted by the Thirty Years' War. In 1625, supporters of the Catholic Counter-Reformation placed most of Kepler's library under seal, and in 1626 Linz was besieged by rebel farmers. Kepler moved to Ulm, where the *Tabulae Rudolphinae* were finally published at his own expense in 1627. Providing the basis for the calculation of ephemerides of greatly increased accuracy, they represent the culmination of Kepler's astronomical work. In 1628, Kepler became an official advisor to General Albrecht von Wallenstein (1583-1634), supreme commander of the armies of the Habsburg Emperor Ferdinand II. Kepler provided astronomical calculations for the General's astrologers. He spent much of his time traveling between the imperial court in Prague, Linz, and Ulm.

In November 1630, Kepler fell ill and died while visiting Regensburg. He was buried in the Protestant St Peter's Cemetery, but the cemetery was destroyed, and the site of his grave lost in 1633 as a result of fighting during the Thirty Years' War. Shortly before his death, Kepler composed an epitaph: "*Mensus eram coelos, nunc terrae metior umbras; Mens coelestis erat, corporis umbra iacet*" ("*I measured the skies, now the shadows I measure; Skybound was the mind, earthbound the body rests*"). His final work, the *Somnium* ('The Dream'), was published four years after his death by his son Ludwig. *Somnium* is a fictional account of a visit to the Moon; it is a very early example of science fiction.

As befits such a pivotal figure, Kepler has been extensively commemorated. His honours include a 32 km (20 mile) diameter lunar crater, a 228 km (142 mile) diameter Martian crater, and the asteroid 1134 Kepler. The Kepler Space Telescope was an orbital observatory operated by NASA between 2009 and 2018, during which time it surveyed over half a million stars and detected 2,662 planets. The *Johannes Kepler* ATV was one of five crewless cargo spacecraft built to resupply the International Space Station. *Kepler* is an opera in two acts by the influential American composer Philip Glass. It premiered on 20 September 2009 at the Landestheater in Linz.

Galileo

Galileo di Vincenzo Bonaulti de Galilei (1564-1642), almost always known by his first name Galileo, or occasionally Galileo Galilei, was born in Pisa. He was the first of six children of Vincenzo Galilei, a lutenist, composer, and music theorist, and Giulia Ammannati, who had married in 1562. Despite his well-known problems with the Inquisition in Rome, Galileo remained a faithful Catholic all his life.

The Galilei family moved to Florence around 1570. Galileo attended the monastery school at Vallombrosa, near Florence, and then in 1581 he enrolled at the University of Pisa to study medicine. However, he soon realised that his real interest lay in mathematics. His father was reluctant for him to abandon his medical studies, as a physician earned more than a mathematician. Galileo studied the works of Euclid and Archimedes, alongside medicine, but in 1585 he left the University without completing his degree.

Galileo met the painter Cigoli (1559-1613) around this time, and they became lifelong friends. Galileo's biographer Viviani records that that Galileo busied himself "*with great delight and marvellous success in the art of drawing, in which he had such great genius and talent that he would later tell his friends that if he had possessed the power of choosing his own profession at that age, ... he would absolutely have chosen painting.*" Sketches attributed to Galileo confirm this view and suggest that he had mastered such techniques as perspective from an early age. Later work attests to his ability to render extremes of light and shadow accurately. Galileo's abilities as a draftsman would stand him in good stead in later life[17].

After leaving the University of Pisa, Galileo taught mathematics both privately in Florence, in Siena where he held an appointment, and at the Vallombrosa Benedictine Abbey. One of Galileo's first published works was a short tract entitled *La bilancetta* ('The Little Balance'), published in 1586. In it, he described Archimedes' method of weighing small quantities of precious metals in air and water to determine their purity. It was this work that brought him to the attention of the academic world. In 1588, he applied unsuccessfully for the chair of mathematics at the University of Bologna. However, he made an important contact when he met German astronomer Christopher Clavius (1538-1616). He also began to correspond with mathematician Guido Baldo del Monte (1545-1607), who had been impressed by his work on Archimedean balances. Galileo was also invited to give a lecture at the Academy in Florence. Thanks to his growing reputation and the influence of Clavius and del Monte, Galileo was appointed Professor of Mathematics at the University of Pisa in 1589.

Vincenzo Galilei died in 1591, leaving Galileo with financial responsibility for his family. His brother Michelangelo and two sisters had survived to adulthood. Michelangelo, who had followed his father in becoming a lutenist and composer, was unable to contribute his share of the promised dowries for the two sisters. Moreover, he also had to borrow funds from Galileo to support his musical career. In 1592, thanks to the patronage of del Monte, Galileo secured a better-paid role as the Professor of Mathematics at the University of Padua, where he remained until 1910.

In Padua, Galileo began a lengthy relationship with a Venetian woman named Marina Gamba. They never married, possibly because Galileo felt he was not wealthy enough. The couple nevertheless had two daughters and a son. All survived to adulthood, but with no marriage prospects the daughters entered a convent. The son, Vincenzo, followed his grandfather and his uncle in becoming a lutenist.

During his time at Pisa and Padua, Galileo researched the nature of motion. The story of his dropping two spheres of unequal mass from the Leaning Tower of Pisa may be apocryphal, but he did carry out experiments with inclined planes and pendulums. However, these results did not appear in print until 1638, when Galileo's final work *Discorsi e dimostrazioni matematiche intorno a due nuove scienze* ('Discourses and Mathematical Demonstrations Relating to Two New Sciences') was published.

Invention of the telescope

In October 1608, Dutch spectacle-maker Hans Lipperhey (1570-1619) filed a patent for the refracting telescope. He was probably not the first to build a telescope, but his is the earliest written record of it. Although he was not granted a patent, the Dutch government paid him well for copies of his design. Word of the invention reached Galileo in May 1609, and based on vague descriptions of the Lipperhey instrument, he made a 3x telescope with lenses procured from spectacle-makers' shops. Soon, he began grinding his own lenses and in August he presented an 8x instrument to the Venetian Senate. He was rewarded with a doubling of his salary.

Galileo's telescopes used a concave (diverging) lens as an eyepiece, which has the advantage of not inverting the image. This is an important consideration for a terrestrial telescope or spyglass, but the reduced field of view makes such instruments less suitable for astronomical work. For this reason, most modern astronomical refractors use the Keplerian design (see above), with a convex (converging) eyepiece lens. However, Galileo would certainly have been aware of the greater commercial possibilities of his design over an inverting telescope. By November, he had built a 20x telescope, and he turned it to the heavens.

Galileo began to study the Moon, and he soon realised that it was not the smooth, flawless sphere that Aristotle had envisaged. The terminator appeared jagged, with isolated patches of illumination. As he observed over the course of two or three hours, these patches merged into the fully illuminated zone. He realised that the Moon's surface was covered with craters and mountains, the peaks of which were in sunlight while the bases remained in darkness. But as the terminator moved westwards, so the bases were illuminated, causing the previously isolated peaks to merge into the now illuminated bases. From his later published work, it can be seen that Galileo recorded the Moon four times between 30 November and 2 December 1609, twice on 17-18 December, and once on 19 January 1610[17]. He also used the shadows to obtain an estimate of the heights of the Lunar Apennines, and although his results were too high (four miles against 4 km), they were of the right order of magnitude[18].

A letter dated 7 January 1610 contains Galileo's first mention of Jupiter's moons, although on this occasion he saw only three of them, and he believed them to be stars. But as he continued his observations, he observed a fourth on 13 January, and he saw that they were not fixed stars but were circling Jupiter. The significance of the discovery should not be underestimated. The four major moons of Jupiter were the first moons to be discovered apart from our own, and the first celestial objects that unquestionably were not circling the Earth. With a cosmology that placed Earth at the centre of the Solar System, it would not be expected for other planets to have bodies orbiting around them. On the geocentric cosmology, the Moon was simply the closest of nine bodies revolving around the Earth – but these new moons had revealed a Copernican system within a Copernican system. Galileo's telescope was also able to resolve the luminous clouds of the Milky Way into immeasurable numbers of stars, and to show that constellations contained far more stars than are visible to the naked eye; for example, 35 stars in Taurus rather than six, and 80 stars in Orion rather than nine.

In March 1610, Galileo published his observations in a short pamphlet entitled *Sidereus Nuncius* ('The Starry Messenger'). His lunar drawings, depicting the phases of the Moon, attest to his skill as a draftsman. The pamphlet reported the greatly increased numbers of stars visible through a telescope. In the final part of the work, Galileo reported the discovery of the four Jovian moons, which he named the Medicean Stars, in honour of his former pupil Cosimo II de Medici (1590-1621), who had become Grand Duke of Tuscany a year earlier.

We now know that Galileo was not the first to observe the Moon through a telescope. That distinction belongs to English astronomer Thomas Harriot (1560-1621), who observed and sketched the Moon through a telescope as early as July 1609, although his drawings contain far less detail. Art historian Horst Bredekamp[17] suggests that Harriot lacked Galileo's ability as a draftsman to interpret what he was seeing and transfer it to his sketchpad. Conversely, Sir Patrick Moore[18], who himself mapped the Moon, claimed that Harriot's map was much more accurate than Galileo's.

German astronomer Simon Marius claimed to have independently discovered the Jovian moons ahead of Galileo, but he did not publish until a year later, and then only in a local almanac. Not until 1614 did he publish his discovery in a major work, *Mundus Iovialis* ('The World of Jupiter'). Galileo hit back with accusations of plagiarism, and forcefully asserted his priority. The confusion arises from Marius's use of the Julian calendar (Old Style), which remained the standard in Protestant Germany; while Galileo in Padua was using the Gregorian calendar (New Style), which was adopted in the Catholic world in 1582. Protestant countries were slow to follow suit: Britain did not adopt the Gregorian calendar until 1752, by which time there was an 11-day discrepancy between the two systems. When this difference is taken into account, it is seen that Marius did not make his discovery until 8 January 1610, a day after Galileo. However, it was the names Marius proposed for the moons in *Mundus Iovialis* – Ganymede, Callisto, Europa, and Io – that were eventually adopted, although the usage was rare until the mid-twentieth century[19].

There have been many claims that the Galilean moons have been seen with the naked eye, and it has been suggested that one (probably Ganymede but possibly two of the moons close together) was seen in China by astronomer Gan De (see Chapter 15) as long ago as 365 BC – almost two thousand years before Galileo. The astronomer noted a reddish star "*in an alliance*" with Jupiter[20]. With a magnitude of +4.6, Ganymede is theoretically is quite easy to see with the naked eye, and the other three moons are above the threshold of naked-eye visibility. The problem is that Ganymede never gets more than 5.9 arcmin from Jupiter, which is 760 times brighter and drowns it out. Io and Europa are even closer to Jupiter. Callisto can reach a distance of 10 arcmin from Jupiter, but at magnitude +5.65 it is the faintest of the Galilean moons. Although Sir Patrick Moore[18] regarded the Gan De story as credible, the description of the object he saw as reddish casts doubt as Ganymede is too faint for its colour to be perceptible.

Galileo next began to observe Saturn, which moved into the night sky in July 1610. He noted that it appeared triple: a large central body flanked by two smaller ones that were almost touching and, unlike the Jovian moons, never changed their positions with respect to one another. But by 1612, the two smaller bodies had vanished. It just so happened that Earth was passing through the plane of Saturn's rings, hiding them from view. By 1616, the ring system as viewed from Earth was opening, and Galileo now saw the rings as half-ellipses. He was unable to explain his observations, and it was not until 1655 did Dutch physicist and astronomer Christiaan Huygens (1629-1695) proposed that Saturn was surrounded by "*a thin, flat ring, nowhere touching* [the planet]*, and inclined to the ecliptic*". Observing the planet with a 50x refracting telescope, he also discovered Saturn's largest moon, Titan. In 1676, Giovanni Cassini (1625-1712) discovered the gap in the rings now named the Cassini Division. Between 1671 and 1684, he also found the Saturnian moons Rhea, Iapetus, Tethys, and Dione.

By September 1610, Galileo was back in Florence, having been appointed court philosopher and mathematician to Grand Duke Cosimo II de Medici. The job did not require him to teach, and it paid him a stipend of 1,000 *scudi* per annum. While it is difficult to give a present-day equivalent of this sum, it was considerable. Galileo's salary was among the ten highest in the Grand Dutchy of Tuscany at the time[21].

Now largely free to concentrate on astronomy, Galileo turned his attention to Venus, which moved into the evening sky in early October. By 30 October, it would have begun to appear gibbous, becoming half by 11 December, and crescent by the end of December.

Galileo claims to have begun observing as Venus became visible and grew progressively brighter, but that he did not note any departure from circularity until mid-November. In all probability, this was due to the limitations of his telescope. Galileo's pupil Benedetto Castelli (1578-1643) wrote to him on 11 December, pointing out that if Copernican astronomy was correct, then Venus would exhibit phases. He asked if Galileo had observed such a phenomenon. On 30 December, Galileo wrote to Castelli and also to the German astronomer Christopher Clavius (1538-1616) confirming the discovery of Venus's phases. It has been suggested that it is that Galileo did not actually begin observing until after receiving Castelli's letter, and that he falsely claimed credit for the discovery. In fact, it is likely that he simply held off replying to Castelli until the crescent shape became clearly discernible. In order to refute the Ptolemaic system, it was necessary to observe Venus as it went from near-full to crescent. Under the Ptolemaic system, there are two possibilities. If the orbit of Venus lies closer to Earth than that of the Sun, then its phase can never exceed half. If, on the other hand, its orbit lies beyond that of the Sun, then its phase can never be less than gibbous. What it can never do under either possibility is go from near full to crescent. Galileo would have needed a full set of observations from October to December for verification; observations from 11 December only would not have been sufficient[22].

To these epochal discoveries, Galileo could have added the planet Neptune, which was not actually found until 1846. French astronomer Urbain Jean Le Verrier (1811-1877) and British astronomer John Couch Adams (1819-1892) independently predicted Neptune's existence and where it would be found on the basis of its gravitational influence on the orbit of Uranus (which itself was not sighted until 1781). The first actual observation of Neptune was made at the Berlin Observatory by German astronomer Johann Gottfried Galle (1812-1910) and his student Heinrich Louis d'Arrest (1822-1875), working from Le Verrier's calculations.

But in 1980, American astronomer Charles Kowal and Canadian science historian Stillman Drake found that during the course of his Jovian observations, Galileo had recorded Neptune as an eighth magnitude object on 28 December 1612 and again on 28 January 1613 when it is shown close to the seventh magnitude star SAO 119234. Accompanying the drawings is a note that suggests that Galileo had observed (but did not record) the pair the previous night, and noticed that they then had seemed further apart[23]. In 2009, the Australian physicist David Jamieson noted a possible further Galilean observation of Neptune. Galileo's notes for 6 January 1613 show an unlabelled black dot, which is in the right position to have been Neptune. Jamieson believes that it is possible that the dot was actually added on 28 January. He suggests that Galileo went back to his notes to record where he had previously seen Neptune, when it had then been even closer to Jupiter. At the time, he had ignored it, thinking it to be just another unremarkable star. The implication is that on 28 January, Galileo realised that one 'star' was moving with respect to the others, and that he had had it under observation since at least 6 January[24]. Did Galileo think, to paraphrase Obi Wan Kenobi, "*that's no star*"?

If so, why did he not follow it up? Kowal and Drake suggested that the lack of a suitable mount for his telescope made it impossible to keep track of Neptune once Jupiter had moved away. Jamieson suggests bad weather prevented further observations. However, he also notes that Galileo often sent cryptic anagrams to his correspondents to establish priority for his discoveries. Jamieson believes that Galileo's literature might include a coded reference to

Neptune, although as of a decade later it has still not come to light.

Discovery of sunspots

Sunspots were reported in ancient China in 28 BC and in Russia in the fourteenth century AD (see Chapter 15), but it is not clear who was the first to observe them telescopically. The year 1610 was close to a solar maximum, so sunspot numbers would have been high and could have been seen by any of the increasing numbers of telescopic observers. German astronomer Johan Fabricus (1587-1616) published the first report, in Holland, in 1611. His drawings are undated, but his observations were probably made in late 1610. Jesuit lecturer Christoph Scheiner (*c.*1575-1650) noted sunspots in March 1611 in Ingolstadt, but initially took no further action. Galileo probably first observed sunspots in November 1610. Benedetti Castelli was probably the first to describe the method of projecting the Sun's image onto a background in order to safely observe sunspots. Galileo described this method as the one that "*any sensible person will use*", which tends to refute the suggestion that he damaged his eyesight by looking directly through his telescope at the Sun[18].

Scheiner again noted sunspots in October 1611. He considered the possibility of a defect in either his eye or the lens, or else an atmospheric disturbance. That other observers had reported the same phenomenon ruled out the first possibility. The atmospheric disturbance he ruled out because, firstly, no cloud could possibly follow the diurnal motion of the Sun from sunrise to sunset; secondly, the spots showed no parallactic displacements during the course of the day; thirdly, their motion across the Sun was at a constant speed; and fourthly, they could be seen through thin clouds. In November, Scheiner wrote to his friend Mark Welser (1558-1614) in Augsburg reporting that the spots were a real phenomenon, "*either on the sun, or outside the sun in some celestial region*". He preferred the second possibility, i.e., that the spots were planetary bodies transiting the Sun and therefore were not a threat to the immutability and perfection of the heavens. In two further letters to Welser, Scheiner reiterated his transit theory and claimed that it was now possible to "*free the sun completely from the injury of spots*". If the spots were on the Sun, they would periodically return to the same positions, and he was reasonably certain that this was not the case[25].

Welser had Scheiner's letters printed and sent copies abroad, including to members of Accademia dei Lincei, a learned society in Rome of which Galileo was a member. The letters were published anonymously in case Scheiner was proved wrong, and hence brought the Society of Jesus into disrepute. Galileo responded with a series of three letters to Welser claiming priority for the discovery of sunspots and refuting Scheiner's transit theory. Scheiner had failed to recognise that the shape and size of the sunspots could change considerably even while they were visible (as indeed was clear from his own diagrams) and therefore they would not necessarily return periodically to the same positions. Galileo's own observations suggested that the sunspots were physically part of the Sun's surface and were carried around by its rotation. The spots appeared thinner when they were near the edges than when they were close to the centre of the Sun; they moved faster as they approached the centre and decreased as they receded toward the circumference; and they separated more and more as they moved toward the centre. All of these phenomena Galileo correctly interpreted as foreshortening effects on the Sun's spherical surface. In 1613, Federico Cesi (1585-1630), the founder of the Accademia dei Lincei, published Galileo's letters in a pamphlet entitled *Istoria e Dimostrazioni intorno alle Macchie Solari* ('History and Demonstrations regarding Sunspots')[25].

That the Sun rotated was by no means obvious, given that even the Earth's rotation was still controversial. Galileo suggested that once set in motion, the Sun would continue to rotate forever. He drew a significant analogy: "*A ship that had received some impetus through the tranquil sea, would move continually around our globe without ever stopping, and placed at rest, it would rest for ever, if in the first place, all extrinsic impediments could be removed, and, in the second place, no external cause of motion were added*". Galileo would return to this concept, which Newton later termed inertia, in *Two New Sciences*[25].

Eppur si muove : Galileo vs the Inquisition

Galileo's views had been of increasing concern to the Catholic Church since the publication of *Sidereus Nuncius* in 1610. He came increasingly under attack from religious conservatives, who argued that his views were heretical because a moving Earth contradicted the Bible. Between late 1613 and early 2015, Galileo wrote private letters to Benedetti Castelli and to the dowager Grand Duchess Christina of Tuscany refuting the biblical arguments against Copernicus. The letters circulated widely, to the annoyance of the conservatives. In February 2015, Dominican friar Niccolò Lorini forwarded a copy of the Castelli letter to Inquisition in Rome, together with a formal complaint. An investigation was launched, and a committee reported that the Copernican theory was "*absurd and false*" in natural philosophy and heretical in theology. Galileo himself was not summoned because the letters had not been published; and none of his published works either explicitly endorsed Copernicus or denied the scientific authority of the Bible. However, Galileo went to Rome of his own accord to defend his views. He had cordial meetings with senior Church officials, and this probably influenced the Inquisition in not issuing a formal condemnation of the Copernican model. However, in February 1616, Galileo was given a private warning by Cardinal Robert Bellarmine (1542-1621), who had been instructed by Pope Paul V to forbid him from holding or defending the truth of the Earth's motion. Galileo agreed to comply. As noted above, Copernicus's *De Revolutionibus* was censored[1].

For the next few years, Galileo kept away from the forbidden topics, but he did test the boundaries of what was permitted. *Discorso Sul Flusso E Il Reflusso Del Mare* ('Discourse on the Tides'), published in 1616, was an attempt to link tides to the movements of the Earth rather than to the long-apparent (but still unexplained) influence of the Moon. In 1619, he entered into a dispute with Jesuit astronomer Orazio Grassi (1583-1654) about the nature of comets, following the unprecedented appearance of three conspicuous comets in a matter of weeks during the autumn of 1618. Grassi was a supporter of Tycho's geocentric/heliocentric system. Galileo responded with *Discorso delle Comete* ('Discourse on Comets'), published under the name of his friend Mario Guiducci. He also used this publication to continue his feud with Scheiner, who by now had published a work entitled *De Maculis solaribus et stellis circa Jovem errantibus accuratior disquisitio* ('An Accurate Inquiry into Sunspots and the wandering stars around Jupiter').

Following the death of Paul V in 1623, Cardinal Maffeo Barberini (1568-1644) became

Pope Urban VIII. Barberini was a longstanding friend of Galileo, and the latter now felt less constrained by Bellarmine's warning. His *Il Saggiatore* ('The Assayer'), published the same year, was dedicated to the new pope. In it, Galileo sets out the idea that the natural world is best understood by the use of what became known as the scientific method: i.e., formulating hypotheses based upon observations of the real world, and the use of experimentation to test the predictions of these hypotheses. But Galileo also used the work for a further attack on Grassi, whose methods of enquiry were heavily biased by his religious beliefs. The work was widely acclaimed, and it is said to have particularly pleased Pope Urban VIII.

Galileo now decided to present the Copernican and Ptolemaic systems in a book as two possible viewpoints, ostensibly without taking sides. *Dialogo sopra i due massimi sistemi del mondo* ('Dialogue Concerning the two main World Systems') was published in 1632. It was written as a dialogue between two philosophers and a layperson taking place over four days. Philosopher Salviati argues for the Copernican model and is opposed by Simplicio, a supporter of Ptolemy and Aristotle, who argues instead for the Ptolemaic model. Sagredo is an intelligent layperson. Salviati and Sagredo are named after friends of Galileo; Simplicio is supposedly named after Simplicius of Cilicia (*c.*AD 490-560), who wrote about Aristotle. If this was genuinely so, it was an unfortunate choice. In Italian, *semplice* means 'simple-minded'. Some even took Simplicio to be Pope Urban VIII.

Galileo was careful to emphasise cosmological, astronomical, physical, and philosophical arguments, and avoid biblical and theological matters. The discussions cover Galileo's telescopic discoveries, his conclusions about motion, and his theories about tides. The book was intended as a critical examination of the arguments on both sides; it was a hypothetical discussion, because the Earth's motion was being presented as no more than a hypothesis to explain observed phenomena. However, the overall theme is that there are stronger arguments in favour of the Copernican model than the Ptolemaic model[1].

Galileo's enemies argued that the book did not treat the movement of the Earth as a hypothesis, but as an actual possibility. As such it violated the interdiction on promoting heliocentricity, and indeed a supposed injunction against discussing it at all. Galileo was summoned to Rome for trial, which began in April 1633. He insisted that the work discussed a purely hypothetical situation, and he strenuously denied ever having received a special injunction against discussing heliocentricity. After three weeks of discussion, Galileo agreed to a plea bargain whereby he would admit to an unintentional transgression of the warning not to defend Copernicus, and in return the more serious charge of violating the special injunction would be dropped[1].

On 22 June 1633, the trial ended, and Galileo was handed a harsher sentence than he had been led to expect. He was found guilty of 'vehement suspicion of heresy', the 'objectional beliefs' that the Earth moves, and that the Bible is not a scientific authority. *Dialogue* was banned, and Galileo was sentenced to life imprisonment, commuted to house arrest. He was forced to recant and abjure, but popular legend has it that he defiantly muttered "*Eppur si muove*" ('But it does move') immediately thereafter.

Final work : *Two New Sciences*

Galileo spent the rest of his life under house arrest in his villa in Arcetri, just south of

Florence, though after going blind in 1638 he was allowed to travel to Florence for medical advice. It was during this final period of his life that he returned to his earlier work on motion, and in 1638 he finally published his results in *Two New Sciences*. The 'two new sciences' are the science of materials and the motion of objects. The work was published in Holland, beyond the reach of the Inquisition. Simplicio, Sagredo, and Salviati return to discuss and debate the various questions Galileo is seeking to answer. This time, Simplicio has a more positive role to play as representing Galileo's early beliefs. Sagredo represents Galileo's middle period, and Salviati his most recent views and models. The dialogue again takes place over the course of four days.

Galileo rejected the Aristotelian view that a terrestrial body will gradually come to a halt in the absence of an external agency. Instead, he argued the opposite: that in the absence of an external agency, the body would continue to move indefinitely. Aristotle and those who came after had failed to realise that it was precisely because an external force was acting – friction – that terrestrial bodies came to a halt. This is the principle of inertia, later enunciated by Newton as the First law of motion. Galileo established the law of falling bodies: neglecting the effects of air resistance, the velocity of a falling body is proportional to the elapsed time and the distance fallen is proportional to the square of the elapsed time; and – contrary to the prevailing view – objects of unequal mass fall at the same rate. He also found that a projectile follows a parabolic trajectory.

Two New Sciences also describes the thought experiment of dropping a ball from the mast of a ship. According to Aristotelian physics, the ball would fall straight down and land astern of the mast. Galileo stated that in addition to falling towards the deck, the ball would share the forward motion of the ship and would therefore land at the foot of the mast. The same argument was used to refute suggestions that the Earth cannot be rotating because an arrow shot vertically into the air would land to the west of the archer instead of coming straight down.

After Galileo's death early in 1642, Cosimo II's successor, Grand Duke Ferdinando II, wished to bury him in the main body of the Basilica of Santa Croce in Florence and erect a marble mausoleum in his honour. The plan was vetoed by Pope Urban VIII. Galileo was eventually reburied in the main body of the basilica in 1737, almost a century after his death. The ban on publishing Galileo's work was gradually lifted from 1718, although *Dialogue* remained prohibited until 1741, and it remained subject to censorship even then. It was not until 1992 that Pope John Paul II expressed regret for the way in which the Galileo affair had been handled by the Catholic Church.

In his book *A Brief History of Time*, Stephen Hawking[26] states that Galileo "*perhaps more than any other single person, was responsible for the birth of modern science*" and that *Two New Sciences* "*was to be the genesis of modern physics*". Galileo is honoured with the lunar crater Galilaei, the large Martian crater Galilaei, the asteroid 697 Galilea, the Galileo Regio on the moon Ganymede, and the exoplanet 55 Cancri A b Galileo. Galileo is to date the only individual to have an asteroid, and exoplanet, and features on three different celestial bodies named for him. NASA's first Jupiter orbiter was named *Galileo*. The European Global Navigation Satellite System (GNSS) is named Galileo. International visitors to Tuscany fly into Galileo Galilei Airport, Pisa. Galileo, like Kepler (and Einstein), was the subject of a Philip Glass opera. The ten-scene opera, entitled *Galileo Galilei*, premiered in 2002 at the Goodman Theatre, Chicago. Galileo is namechecked in the operatic section of Queen's 1975 chart-topping *Bohemian*

Rhapsody. Finally, in the original series of *Star Trek*, the USS *Enterprise* carried a space shuttle named *Galileo*.

With Galileo's death, this is how things stood. After 1,500 years, the Ptolemaic geocentric model had finally been retired, although it had been a gradual process. First, Copernicus had replaced geocentricism with heliocentrism, but he had attempted to hold on to circular motions. Then, Kepler had banished circular motions, epicycles, and celestial orbs, and had put forward in their place empirical laws of planetary motion. Tycho Brahe had stressed the need for accurate observations of the heavens. Galileo provided confirmation of the geocentric model with his observations of the phases of Venus. Finally, Galileo's *Two New Sciences* had demolished Aristotelian physics. It remained for these strands to be drawn together into a single coherent theory – and this was now about to happen.

18
Standing on the shoulders of giants

Newton

Sir Isaac Newton (1642-1727) was born in Woolsthorpe-by-Colsterworth, Lincolnshire. His father, Isaac Newton, Snr, died three months before he was born. His mother, Hannah Ayscough, remarried when Isaac was still young, and left him in the care of his maternal grandmother Margery Ayscough. Stephen Hawking[1] is rather less complimentary about Newton than he is about Galileo. He claims that Newton was "*not a pleasant man*", noting that "*His relations with other academics were notorious, with most of his later life spent embroiled in heated disputes*". In his book *A Short History of Nearly Everything*, Bill Bryson[2] describes Newton as "*a decidedly odd figure*". It has been suggested that Newton had Asperger's Syndrome, but his abilities as an administrator at the Royal Society and the Royal Mint makes this unlikely, and the explanation for his personality and at times odd behaviour probably lies with his childhood abandonment. He may also have suffered from mercury poisoning arising from the alchemical experiments he performed between 1678 and 1696[3].

Regardless of his personal shortcomings, Newton was one of the most influential scientists of all time. He was educated at The King's School, Grantham, where he distinguished himself. In 1661, he was admitted to Trinity College, Cambridge, on the recommendation of his uncle. There, he supplemented the Aristotle-based curriculum with his own studies of the works of Descartes, Galileo, Kepler, and the English astronomer Thomas Street (1621-1689). He also kept a notebook he referred to as *Quaestiones quaedam philosophicae* ('*Certain philosophical questions*'), in which he wrote notes about scientific issues that interested him. Newton obtained his degree in August 1665, shortly before the University closed temporarily due to the Great Plague. Spending the next two years at home in Woolsthorpe-by-Colsterworth, he began to formulate early ideas about gravitation, optics, and calculus.

In April 1667, Newton returned to Cambridge, and in October was elected as a fellow of Trinity College. Two years later, he succeeded his former advisor Isaac Barrow (1630-1670) as Lucasian Professor of Mathematics. Barrow and Newton were the first two holders of what is now one of the most prestigious academic posts in the world. The post was founded in 1663 by Henry Lucas (1610-1663), who was Member of Parliament for the University of Cambridge from 1639 to 1640. It was officially established by King Charles II on 18 January 1664. Later incumbents include Astronomer Royal George Biddell Airy, engineer Charles Babbage, and Nobel Prize laureate Paul Dirac. After Newton, the best-known Lucasian Professor is probably Stephen Hawking, who held the post for thirty years from 1979 to 2009.

Rainbows and telescopes

During his enforced absence from Cambridge, Newton decided to "*try therewith the celebrated Phaenomena of Colour*" with a prism. He found that not only could white light be split into colours; the colours could be recombined into white light with a second prism. This was a radical discovery: previous theories had assumed that by default light was white and had sought to explain colour. For example, Descartes believed that light consisted of pulses transmitted through an aether of minute, rotating spherical particles. When a pulse of light impinges on aether particles, the friction imparts spin: if they rotate faster than usual, they cause the sensation of redness, and if they spin slower, they give rise to blueness. Newton's discovery reversed these theories by showing that colour is a basic property that can be used to explain white light[4].

The discovery also showed how raindrops refract sunlight to form coloured rainbows, thus providing the explanation for the phenomenon that had been sought since Aristotle's time. Newton famously divided the rainbow into seven named colours, although the number and choice of colours in a continuous spectrum is arbitrary. I, for one, have never been able to make out indigo as a separate colour, and I am not alone. The cover of Pink Floyd's 1973 album *Dark Side of the Moon* depicts a prism splitting white light into six colours *sans* indigo; Gay Pride's rainbow flag also omits indigo.

Newton expanded on the 'corpuscular' theory of light proposed by the French mathematician Pierre Gassendi (1592-1655), which proposed that it is light itself that consists of minute spherical particles or corpuscles. The theory explained the reflection of light in terms of elastic and frictionless collisions with smooth surfaces. In a medium such as water or glass, a light corpuscle is surrounded by equal numbers of particles of the medium, and there would be an attractive force between the corpuscle and the medium. Deep within the medium, these forces would cancel out, leaving no net force on the light corpuscle, which would remain in a uniform motion in a straight line within the medium. But near a boundary, for example, between air and water, the light corpuscle would experience more attracting matter particles on one side than the other, and hence a non-zero net force. During the transition between media, it would experience a net attractive force towards the medium with more matter particles. This would increase the vertical component of the corpuscle's velocity, thus deflecting it towards the surface's normal (a line perpendicular to the surface) and explaining refraction. The theory predicts that because the vertical forces speed up the corpuscles, their speed would be greater in a denser medium; Huygens's rival wave theory of light predicts the opposite. It was not until 1850 that the latter prediction was confirmed experimentally. The corpuscular theory also failed to explain diffraction, interference, and polarization, and it was finally abandoned[5].

Newton realised that the chromatic aberration that bedevilled refracting telescopes had the same cause as the splitting of white light by prisms. He believed (wrongly) that it was impossible to eliminate chromatic aberration from lenses and used a parabolic mirror instead of an objective lens. The primary mirror came to a focus on a small diagonal mirror, which reflected the image into an eyepiece mounted on the side of the telescope. Newton used a mirror made from speculum, a bronze alloy using one-third tin and two-thirds copper. Speculum can be polished to make a highly reflective surface, but it tarnishes rapidly and requires frequent re-polishing. Speculum was nevertheless used for making telescope mirrors until the middle of the nineteenth century, when it was replaced by silvered glass. Newton built his first telescope in 1669. The design, now known as the Newtonian reflector, remains popular with amateur astronomers to this day.

Word of Newton's telescope reached the Royal Society in 1671, and they asked for a demonstration. The telescope was enthusiastically received, and Newton was elected a Fellow of the Royal Society in January 1672. He then submitted a letter to the Society's journal *Philosophical Transactions of the Royal Society*, published as *A Letter of Mr. Isaac Newton, Professor of the Mathematicks in the University of Cambridge; containing his New Theory about Light and Colors* [sic], which detailed his research with prisms, his theories of light and colour, how his work explained chromatic aberration, the impossibility of eliminating it from refracting telescopes, and the use of parabolic mirrors as an alternative. Though well-received, the paper did have its critics, notably physicist Robert Hooke (1635-1703). Newton provided very few experimental details with his papers, and others were unable to replicate his results. Newton's unhelpful response was that they were using the wrong type of prism – without giving any indication as to what type of prism should be used. Although Newton's theories were widely accepted, criticism rumbled on for decades, especially in France. It has to be said that Newton was notoriously intolerant of criticism, and he was particularly hurt by Hooke's remarks. Indeed, at one point he threatened to resign his membership of the Royal Society, but he was talked out of it. The two men attempted to patch up their differences, and they exchanged a number of letters over the years. Newton's 1675 letter to Hook, in which he made his famous (but not original) claim that "*If I have seen further, it is by standing on the shoulders of giants*" was not, as is often supposed, a jibe at the latter's lack of height (Hooke, in any case, was not unusually diminutive in stature). Newton nevertheless withheld publication of a more complete version of his work with optics until after Hooke's death. Published in 1704, the book was entitled *Opticks*, and it gave the full experimental details that had been missing from the 1672 paper[4].

The *Principia*

Stephen Hawking[1] describes Newton's *Principia* as "*- surely the most influential book ever written in physics*", and it is not for me to question this verdict. The *Philosophae Naturalis Principia Mathematica* ('Mathematical Principles of Natural Philosophy'), to give it its full name, was first published in 1687, with further editions in 1713 and 1726. In it, Newton set out his three laws of motion and the law of universal gravitation. These laws form the basis of what is now known as classical mechanics.

1. A body will remain at rest or in uniform motion unless it is acted upon by an external force (the law of inertia);
2. The rate of change in momentum of a body (mass m times velocity v) is proportional to the force F acting on it (the rate of change of velocity, i.e., acceleration, is proportional to force F divided by mass m; hence $F = ma$);
3. For every action, there is an equal and opposite reaction (the conservation of linear momentum).

The Third law explains why a gun recoils when fired. The momentum of the discharged projectile is exactly balanced by the recoil in the opposite direction. Similarly, a jet engine discharges a fast-moving jet of air to produce thrust in the opposite direction.

The Law of universal gravitation states that between two bodies of masses m_1 and m_2 there is an attractive force F that is proportional to the product of the masses divided by the square of the distance r between them, i.e., $F = Gm_1m_2/r^2$, where G is a constant known as the gravitational constant. This law is an example of the inverse square law whereby the force falls off by the square of the distance. Kepler's Third law may be derived from it by applying the laws of motion.

The *Principia* had a lengthy gestation. Some of the groundwork was done during the Plague period of 1665 to 1666, while Newton was absent from Cambridge. Whether he was hit on the head by an apple during this period, as he later related to his biographer William Stukeley (1687-1765), we shall never know. It is not even certain that Newton genuinely was working on gravitation in 1666, as he never published anything on the subject during this period. Regardless, he was not the first to conceive of a gravitational force or an inverse square law. Between 1662 and 1666, Robert Hooke conducted experiments with a 'gravity balance' in the galleries of the old St Paul's Cathedral (destroyed in the Great Fire of London), Westminster Abbey, and in a deep well on Banstead Downs. He compared the acceleration due to gravity at these sites with the local value at ground level, hoping to measure the difference. The idea was sound, but the minute differences were beyond the resolution of his apparatus. Hooke's *Micrographia*, published in 1665, was predominantly about microscopes, but in it he discussed the possibility that the Moon had its own gravity, and he made references to an inverse square law. When Newton's *Principia* was presented to the Royal Society, Hooke claimed that he had given Newton the notion of the inverse square law, so reigniting the long-running feud between the pair. More would now be known about Hooke's theories concerning gravitation, and we might be able to resolve the controversy, but unfortunately many of his papers were lost after his death[6,7].

In January 1684, Hooke met Edmond Halley (1656-1742) and Sir Christopher Wren (1632-1723) at the Royal Society. Although chiefly remembered as an architect, Wren also had a long and distinguished scientific career, and was the Savilian Professor of Astronomy at Oxford from 1661 to 1672. He was also President of the Royal Society from 1680 to 1682. At 28, the young Halley was already making his mark as an astronomer. In 1676, he had travelled to the island of St Helena in the South Atlantic to catalogue the stars of the Southern Hemisphere and observe a transit of Mercury. In September 1682, he had observed (but not discovered) the comet later named for him, but it would not be until 1705 that he successfully predicted its return.

Halley was attempting to find a physical explanation for Kepler's laws of planetary motion, and he believed that the Third law could be explained in terms of an inverse square law. During the course of his meeting with Hooke and Wren, the latter offered 40 shillings (£2.00) in books if either of the other two could show that Kepler's First law (elliptical orbits) also followed from an inverse square law. In August 1685, Halley visited Newton in Cambridge after having been delayed for several months by the death of his father and the settling of his estate. As a result, Newton wrote a paper entitled as *De Motu* ('On the Motion'), which was presented to the Royal Society in December. He then decided to expand this into what became the *Principia*, which Halley edited. The new work was presented to the Royal Society in April 1696. Halley was asked to prepare a report, which he did early in June. The Royal Society agreed to publish it, but the previous year it had published at great expense a tome entitled *The History of Fish*, by Francis Willughby (1635-1672). This work had sold poorly, leaving the Society's finances badly strained. Eventually, Halley agreed to meet the cost of publication of the *Principia* himself. Although the initial outlay was considerable – probably at least £60 and possibly as much as £100 – it was no more than a quarter of the annual income from his late father's estate[8].

The *Principia* is very different to a modern physics textbook, even when translated from Latin. It follows the pattern of classical treatises, with definitions and axioms, followed by the statement of propositions and their demonstrations. The most obvious difference is a lack of equations. The Second law, for example, is familiar to present-day physicists as the equation $F = ma$, but this formulation was not used in the *Principia*, nor indeed anywhere else in Newton's lifetime. Similarly, the mathematical statement of the Law of universal gravitation, $F = Gm_1m_2/r^2$, does not appear in the text. Many of the definitions given in the *Principia* seem laboured, but it should be remembered that concepts such as 'mass', which seem elementary to us, were completely new at the time[9].

Legacy

Two further editions of the *Principia* appeared in Newton's lifetime, in 1713 and 1726. Newton became President of the Royal Society in 1703 and was knighted in 1705. He served two brief terms as Member of Parliament for the University of Cambridge (which returned two MPs until as late as 1950), from 1689 to 1690, and from 1701 to 1702. He also served as Warden of the Royal Mint from 1696 and Master of the Royal Mint from 1700 until his death. Although intended as sinecures, Newton took these roles seriously, and he used them to instigate currency reforms and crack down on forgery.

Newton did not mellow with age, and his priority dispute with Gottfried von Leibniz (1646-1716) over calculus raged from 1699 until the latter's death. It is now generally accepted that Newton and Leibniz independently discovered calculus (although forerunners of integral calculus go back to Mesopotamian times, see Chapter 10). However, it was Leibnitz's notation that was eventually adopted by mathematicians. Newton also became embroiled in a dispute with Astronomer Royal John Flamsteed (1646-1719) over data he required for the second edition of the *Principia*. Newton had used Flamsteed's data for many years, but he grew impatient with the latter's refusal to release unverified observations. Newton argued that the Astronomer Royal was a civil servant, and that therefore his work was public property, but Flamsteed was unmoved. Eventually, Newton used his position as President of the Royal Society to compel the release of the data he required. In this endeavour, he was assisted by Halley, who had fallen out with Flamsteed some years earlier. Following his death, Newton was buried in Westminster Abbey, the first scientist to be so honoured. He has since been joined by Charles Darwin and Stephen Hawking.

The poet Alexander Pope (1688-1744) composed an epitaph:

NATURE and Nature's Laws lay hid in Night:
God said, "Let Newton be!" and all was light.

Newton is commemorated by a lunar and a Martian crater, and the asteroid 8000 Isaac Newton. The newton is the SI unit of force. His quote "*Standing on the shoulders of giants*" appears on the rim of the British two-pound coin and he featured on the one-pound banknote from 1978 until the denomination was replaced by a coin in 1983. Halley has a lunar and a Martian crater, the asteroid 2688 Halley, and of course the comet P1/Halley, which returned as he had predicted in late 1758. A more tenuous connection is the differently-pronounced Bill Hayley and his rock-and-roll band Bill Hayley and the Comets, best known for the 1954 hit *Rock around the Clock* (Halley's name is thought to have been pronounced 'Haw-ley'). Hooke, likewise, has a lunar and a Martian crater, and the asteroid 3514 Hooke. Flamsteed has a lunar crater and the asteroid 4987 Flamsteed, but he has so far not been commemorated with a Martian crater.

It had taken just under one and a half centuries to shatter a cosmology that had endured for a one and a half millennia. The Ptolemaic cosmology had begun to unravel with *De revolutionibus*; now the *Principia* ushered in a Solar System governed by laws that will be familiar to anybody with a secondary-level schooling in the physical sciences. Modern science was beginning to emerge from the classical discipline of natural philosophy in terms of both theories and in the ways that science was practiced.

It is now just under three centuries since Newton's death, and astronomy has come a long way in that time; indeed, it has made astonishing progress in my lifetime. Thanks to robotic space probes, we now know more about Pluto than was known about the Moon in the 1950s. We now know that the Sun is just one of hundreds of billions of stars in our galaxy, which in turn is just one of around 2,000 billion galaxies in the observable universe. Since the first tentative discoveries in the 1990s, more than four thousand planets have been discovered circling other stars.

Rather unexpectedly, our Solar System appears to be something of an outlier. Most multi-planet systems are far more compact than the Solar System; the planets are more similar to one another in size; and their orbits are more regularly spaced. Is this a detection bias, or is the Solar System genuinely unusual? If so, why, and what implications does this have for our finding life on other planets? In the coming decades, we may well find answers to these and other questions, but more questions will surely arise in the meantime. The story that began when our distant ancestors first looked into the night skies still has a long way to run.

Appendix
The Celestial Clockwork

Even the most casual observer will be aware that the skies present a differing appearance from hour to hour, as the Sun, Moon, and stars march steadily across the heavens. They will realise, however, that the stars are fixed in relation to one another, often making distinctive patterns in the sky. They will also realise that the Moon changes its appearance from night to night, sometimes waxing sometimes waning. But why does the Sun rise and set in different places at different times of the year? Why is the Moon sometimes visible in broad daylight? Why do some stars remain close to the same point in the sky and never set? And what is to be made of the bright star-like objects that do not remain in a fixed position in relation to their neighbours, but move at differing speeds across the starry background, always keeping to the roughly same plane as the Moon in its wanderings.

Today, we know that the Moon goes around the Earth and the Earth and other planets go around the Sun. However, this does not even begin to tell the full picture, and the achievements of people such as the Maya and the ancient Babylonians – who were not even armed with these basic facts – cannot be overstated. The workings of the celestial clockwork make the achievements of the finest Swiss watchmaker pale into insignificance, yet from a modern perspective they are not difficult to understand; the object of this introduction is to give the reader just such an understanding.

The Celestial Sphere

Astronomers often use a model known as a **celestial sphere** to illustrate the movements of the Sun, Moon, planets, and stars as seen from Earth. This can be thought of as being similar to a geographer's globe, except it surrounds the Earth and we look at it from the inside. Another way of thinking of it is as a grid system projected onto the heavens to help us find our way around. To demonstrate such a grid would at one time have required the facilities of a planetarium, but there are now many software planetariums and 'augmented reality' smartphone apps available that can achieve almost the same effect.

Let us first consider the finer points of the celestial sphere itself. Like the geographer's globe, it will have two poles, an equator, and lines of latitude and longitude. The **celestial poles** (north and south) and the **celestial equator** (often referred to simply as the equator) are projections of their terrestrial counterparts onto the grid system. The coordinate system used to define a point on the grid system differs slightly to that used by geographers. The position of an object to the north or south of the equator is given by **declination** (Dec.). Like latitude, it is measured in degrees, but instead of suffixing the declination with an N or an S, northern declinations are given positive values and southern declinations negative values. The equivalent of longitude is **right ascension** (R.A.). It has a zero similar to the

Greenwich meridian, which passes through a point known as the **First point of Aries** (to be discussed shortly) or vernal point. Right ascension is measured eastwards from the First point of Aries. It can be measured in degrees but is usually measured in hours, minutes, and seconds. An hour corresponds to 15°, so 24 hours is equivalent to 360°.

From anywhere in the world, our field of view will be bounded by the horizon, which divides the celestial sphere exactly into two. The point directly above us is known as the **zenith**. Running through both poles and the zenith is a great circle known as the **meridian**. The angle of elevation of any object above the horizon is known as the **altitude**; the angular distance around the horizon, measured clockwise from due north is known as the **azimuth**.

Let us take a virtual trip to the (geographical) north pole. The north celestial pole is directly overhead, at the zenith, and the equator lies exactly on the horizon. Between the equator and the pole are a series of concentric circles, growing ever smaller the nearer they are to the pole. These represent differing declinations. A series of lines extend upwards from the equator, converging at the pole. These are the lines of right ascension. Note that because the declination circles are parallel to the horizon, we can see them in their entirety. However, we cannot see anything south of the equator at all.

If we observe the sky for a few hours, the grid and stars will appear to revolve clockwise around the north celestial pole. This is happening because the Earth is rotating on its axis, in a west to east direction, or anticlockwise as viewed from 'above' the north pole. This motion is known as **diurnal motion**. During the course of its diurnal motion, a star will reach its highest point in the sky when it is on the meridian. When a star reaches the meridian, it is said to **culminate**.

If we look close to the pole, we will see a bright star. This is Polaris, the Pole Star. It appears to be almost fixed while everything else wheels around it. The further a star is from the pole, the larger the circle in which it moves, with those near the equator moving in the largest circle of all. Note that no star ever rises or sets; we can see half of the stars the whole of the time.

Before we move on, how long do you think it takes the stars to make a complete circuit of the skies? The answer is of course the same length of time it takes the Earth to make a complete turn on its axis – but this is not 24 hours. The Earth makes a complete turn on its axis once every 23 hours 56 minutes and four seconds. This period of time is known as the **sidereal day**. However, we reckon time by the Sun rather than the stars, and as we shall see, the **solar day** is slightly longer.

Now we will travel to London, which is at latitude 51°30' N. The first thing we notice is that the grid now appears tipped over towards one side. The north celestial pole is no longer at the zenith; its altitude will be equal to the geographical latitude. The declination circles are now no longer parallel to the horizon, so only those close to the pole will be visible in their entirety and the others will be visible only as ever-decreasing arcs. However, we can now partially at least see declination circles that are south of the equator. Exactly half of the equator is visible, and it cuts the horizon at points due east and due west. The north celestial pole lies due north.

There is a declination circle whose southernmost point just touches the northern horizon, and only the stars within this circle now remain permanently above the horizon. Such stars are said to be **circumpolar** from that latitude (at either pole, as we have seen, all the stars are circumpolar). Stars further south do spend increasing amounts of time below the horizon, though those north of the equator are still up for more than half of a sidereal day. Stars lying directly on the equator are visible for exactly half of a sidereal day. Note that these stars rise due east and set due west. Stars located still further south are visible for less than half of a sidereal day. Finally, on the southern horizon, is the northernmost point of a declination circle that lies entirely below the horizon. Stars lying within this circle are permanently out of view and include those making up the Southern Cross and our nearest stellar neighbour, α Centauri.

Next, let us travel to the equator. The grid will now appear to be completely tipped onto its side, and the declination circles will now lie at right-angles to the horizon. Exactly half of each circle will be visible, but our observer can now see all of them. The north celestial pole lies exactly on the northern horizon and, 180° away, the south celestial pole has come into view. Every single star will be above the horizon for half of the sidereal day.

It will now be clear that were we to continue south, the north celestial pole would dip below the horizon and the process we have just witnessed would occur in reverse until upon reaching the geographical south pole, we would see the south celestial pole at the zenith.

The Sun and Seasons

Now let us observe the Sun and stars as a sidereal day goes by. At the end of one sidereal day, all the stars will be back where they started – but the Sun will be lagging behind. In fact, it will take approximately four minutes for the Sun to catch up. This is because while the Earth has been spinning on its axis, it has also been moving in its orbit around the Sun, completing a 1/365.24th of a circuit. Like the axial rotation, the orbital motion is west to east, or anticlockwise as viewed from a point to the north of the Solar System. (Astronomers refer to such motion as prograde or direct; clockwise motion, which is rare in the Solar System, is said to be retrograde.) The Sun, as viewed from Earth, appears to have changed its position slightly with respect to the stars. A solar day is defined as the time between successive crossings of the meridian by the Sun, but because the Earth's orbital speed varies slightly over the course of a year, the solar day is not constant in length. It is only over the course of a year that it averages out to the familiar 24 hours.

The result of the solar day being about four minutes longer than the sidereal day is that any given star will rise four minutes earlier each solar day. This is why the constellations visible at a given time vary over the course of the year. After a year has passed, the solar and sidereal days come back into step. Our star will rise at the same time that it did on the corresponding day a year ago.

If we follow the Sun's apparent movement over the course of a year, we will see that it will make a complete circuit of the celestial sphere. The path it traces out is known as the ecliptic. The ecliptic represents the plane of the Earth's orbit around the Sun and it is inclined to the equator at an angle of 23°26'. The reason for this is that the Earth's axis of rotation is not perpendicular to the plane of its orbit; it is inclined at an angle of 23°26'. This inclination is known as the **obliquity of the ecliptic** or axial tilt.

Constellations straddling the ecliptic are said to be zodiacal. These include the familiar twelve signs of the Zodiac, but to make matters confusing there is actually a thirteenth zodiacal constellation, Ophiuchus the Serpent Bearer, that has been ignored by astrologers and is not considered to be part of the Zodiac. The two points at which the ecliptic and the equator intersect are known as the **First point of Aries** (which we have already encountered) and the **First point of Libra**.

What effect will the Sun's movement along the ecliptic have over the course of a year? The ecliptic is inclined to the equator, so the Sun will spend half of the year north of the equator and the other half of the year south of it. Recall that for the Northern Hemisphere, if a star is north of the equator it will be up for more than half of the sidereal day, but if it is south of the equator it will be up for less than half of the sidereal day. The same applies to anything on the celestial sphere, and this includes the Sun.

Thus, for half of the year, day will be longer than night and for the other half it will be shorter. On the two days of the year known as the **equinoxes**, day and night will be of equal length; this will occur when the Sun is at the First point of Aries or the First point of Libra. The First point of Aries is known as the Sun's ascending node because this is where it crosses into the Northern Hemisphere from the south. Similarly, the First point of Libra is known as the descending node. Mid-way between the equinox points, the Sun will reach its most northerly and most southerly positions; these points are known as the **solstices**. In the Northern Hemisphere, the summer solstice occurs when the Sun is at the most northerly point on the celestial sphere and the winter solstice when it is at the most southerly point. This is the explanation for the seasons.

In lower latitudes, the Sun attains greater elevations above the horizon. When it is at either equinox, the Sun will be directly overhead at midday along the equator. At the summer solstice, it will be directly overhead at midday in the latitude defined by the Tropic of Cancer and at the winter solstice it will be directly overhead at midday in the latitude defined by the Tropic of Capricorn. This is why it gets rather hot in these parts of the world. Conversely, the Sun is circumpolar during the summer months within the Arctic Circle, but never rises at all during the winter months. This is the explanation for the famous 'Midnight Sun'. The situation is reversed for the Antarctic Circle.

Most people think of the Sun rising in the East and setting in the West. But the rising point of the Sun is actually only due East on two days of the year, those when it is at one or other of the equinoctial points. In the summer months, the azimuth of the rising point moves north, reaching its maximum extent at the summer solstice, before returning south. In the winter months, the azimuth of the rising point moves south reaching its maximum extent at the winter solstice, before returning north. Around the solstices, the rising appears to stand still for a few days; the word 'solstice' is derived from this phenomenon. The setting points move in the same manner, with the winter sunset limit lying opposite the summer sunrise limit, and the summer sunset limit lying opposite the winter sunrise limit. These seasonal variations in the rising and setting points vary with latitude, being more pronounced in higher latitudes.

The Earth's Orbit

Let us now consider the Earth's orbit around the Sun in a little more detail. The orbit is not circular but elliptical, meaning that the distance between the Earth and the Sun is not constant but varies over the course of the year, ranging from 147 million km (when Earth is said to be at **perihelion**) to 152 million km (**aphelion**). The Earth's orbital speed is at its greatest at perihelion and at its least at aphelion. This is because its movements are governed by Kepler's laws of planetary motion, which we will examine in more detail presently. The mean distance is 149.6 million km (93 million miles), and this distance is referred to as the astronomical unit (AU). The departure of the orbit from a perfect circle is known as the orbital eccentricity. Perihelion does not occur in the same place each year, but advances by 11.64 arcsec on each orbit. This is due to gravitational effects of other planets in the Solar System.

Precession, nutation, and other cycles

There are in addition a number of more gradual motions that are only significant over the long term. The most important of these is **precession**. In addition to rotating on its axis, the Earth also oscillates like a spinning top, each oscillation taking about 25,772 years and causing the Earth's spatial orientation to gradually change. This motion is due chiefly to the pull of the Sun and the Moon, though the other planets make a small contribution.

The observable effect is to make the nodes of the ecliptic gradually move westwards at 50.3 arcsec per year or one degree every 71.6 years. The stars will remain fixed in relation to the ecliptic, but as our celestial sphere grid system uses the Earth as its frame of reference, they will appear to move very slightly against it. This means that star maps have to be calibrated for a particular **epoch** which since 1984 has been Epoch 2000.0, the start of the year 2000. The effect, though small, is cumulative and amounts to two lunar diameters over a lifetime. Computerised amateur telescopes must take precession into account, or stars would begin to depart from their predicted positions in a matter of months if not weeks. More significant effects are experienced over longer periods – in antiquity, for example, the Southern Cross could be seen from Greece and the ancient Greeks included it in their star charts. 14,000 years from now, the Southern Cross will be visible all over Britain.

The precessional motion is not smooth but slightly wavy. This irregularity is known as **nutation** of the Earth's axis: a slight nodding of the Earth due to changes in tidal forces acting on the planet through variations in the distances and positions of the Sun and the Moon. The largest component of Earth's nutation corresponds to and has the same periodicity as the Moon's 18.61-year nodal cycle. This produces a variation of $\pm$ 9.2 arcsec in the obliquity of the ecliptic. Smaller components are due to the Earth's annual motion around the Sun and by the Moon's monthly motion around the Earth.

In addition to the effects of nutation, the obliquity of the ecliptic varies with time though the gravitational effects of other planets on the plane of Earth's orbit.

Finally, the eccentricity of Earth's orbit also varies, albeit very gradually. Again, this is due to the gravitational effects of other planets. The precessional cycle and the cyclical changes in the obliquity and orbital eccentricity are now known as the Milanković cycles. They are named for the Serbian mathematician Milutin Milanković, who proposed a link between them and cyclical changes in Earth's climate while interned in Budapest during World War I. Climatologists now accept that Milanković was right, but his views attracted little interest in his lifetime.

Four types of year

Up until now, we have used the term 'year' rather loosely. Most people think of a 'year' as being the time it takes the Earth to go once around the Sun. That is certainly a type of year, but not the only type. The Earth in fact has four different types of year. The first is the **sidereal year** of 365.256363 days (365 days 6 hrs 9 min 9.7632 sec) on average, and this is indeed the time it takes the Earth to go once around the Sun – but this is not the 'year' we base our calendar on.

The Gregorian calendar, which is now used throughout the Christian world, is actually based on the **tropical year**, which is defined as the time between the Sun making two passages through the First point of Aries. The First point of Aries is moving slowly in the opposite direction to the Sun along the ecliptic as a result of precession, so the Sun 'arrives' there about twenty minutes before it completes its circuit of the celestial sphere. Hence the tropical year is slightly shorter than the sidereal year. However, the progression of seasons is dictated by the former, so the calendar is based upon it. The mean value of the tropical year is 365.242189 days (365 days 5 hrs 48 min 46.6339 sec). The actual value can vary by up to around half an hour; for example, the interval between the spring equinox in 2015 and 2016 was 365 days, 5 hrs, 44 min, 56 sec; the corresponding 2016-17 interval was 365 days, 5 hrs, 58 min, 36 sec; and the 2017-18 interval was 365 days, 5 hrs, 46 min, 41 sec.

The third type of year is the **anomalistic year** of 365.259636 days (365 days 6 hrs 13 min 52.5504 sec) on average, which is defined as the time between the Earth making two returns to perihelion. As we have seen, the perihelion advances with each circuit, the Earth requires a bit longer to 'catch up'; hence the anomalistic year is fractionally longer than the sidereal year.

Finally, there is the **eclipse year** (or draconic year) of 346.620076 days (346 days 14 hrs 52 min 54.5664 sec) on average, which we shall encounter presently.

The slight variations in the lengths of these four 'years' are caused by the gravitational effects of other planets on Earth's orbit.

The Moon and its Orbit

Most people will think nothing if they happen to see the Moon in the night sky, but they are often surprised to see it in broad daylight. In fact, the Moon spends on average as much of its time above the horizon in daytime as it does in night-time. The most singular feature of the Moon is, in fact, something that most of us take completely for granted. This is that it appears almost exactly the same size as the Sun. The explanation is simple – the Sun is about four hundred times larger in diameter than the Moon, but it is also about four hundred times further away. The odds against this happening – if not quite astronomical – are pretty low.

The Moon's orbit around the Earth is inclined at an angle of 5° to the ecliptic. The Moon's apparent path around the celestial sphere intersects the ecliptic at two points again known as nodes; as with the intersections between the ecliptic and equator, there is an ascending node and a descending node. These nodes do not remain fixed; they move westwards on the celestial sphere at 19°20' per year due to perturbation by the Sun, taking 18.612958 years to complete a nodal cycle. This phenomenon is known as the **regression of nodes**.

The orbit itself is rather more elliptical than that of the Earth around the Sun. The Earth-Moon distance (centre to centre) varies from between 356,410 kilometres (minimum distance, or **perigee**) to 406,697 kilometres (maximum distance, or **apogee**). If the Moon is full when it is close to perigee, it will appear slightly larger and brighter than usual. In recent years, the media has begun referring to the phenomenon as a 'supermoon'.

The Moon's orbital speed increases at perigee, and it decreases at apogee. Like the Earth, this is due to Kepler's laws of planetary motion, which apply to all orbiting bodies. The eccentricity of the orbit is quite pronounced, so the effect is quite noticeable in terms of nightly movement on the celestial sphere, and this has been known since ancient times. In a manner similar to the Earth's perihelion, the Moon's perigee advances with each orbit, taking 8.85 years to complete a cycle.

Due to tidal interaction between the Moon and the Earth, the two are moving apart at about 38 mm (1.575 inches) per year. The Moon is, of course, responsible for the tides in Earth's oceans, but it also raises a 'tidal bulge' over the whole planet. As the Earth rotates, so this tidal bulge pulls on the Moon, causing it to speed up. At the same time, the Moon's pull on the tidal bulge has the effect of slowing the Earth's rotation, and the day gradually lengthening at a rate of about 1.7 milliseconds per century. These effects are small, but they add up over time and account for discrepancies amounting to several hours in the timing of eclipses observed in antiquity.

Phases of the Moon

The most noticeable feature of the Moon is that its appearance changes from night to night. These phases are due to differing portions of its day-lit side being presented to us as it moves around the Earth. At the start of the cycle, it cannot be seen because it lies in the same direction as the Sun, and its illuminated side faces away from us. A few days later it will have moved eastwards away from the Sun and will be seen as a slim crescent in the evening sky. As the days pass, the Moon is seen ever higher in the evening sky as it continues to grow or wax. After seven to eight days the Moon will be 90° east of the Sun in the sky; this point is known as the first quarter and the right half of the Earth-facing side is illuminated (in fact, because the Sun is not infinitely distant from the Earth and Moon, the angular separation is slightly less than 90°). At around fifteen days, the Moon's distance from the Sun reaches 180°. At this point, the Moon rises at sunset. The entire Earth-facing side is now illuminated, and we see a full Moon. Thereafter, the Moon begins to wane, going through its phases in reverse as its angular distance from the Sun begins to decrease once more. After about 22 days the Moon is 90° west of the Sun; this point is known as the third quarter and the left half of the Earth-facing side is illuminated. Subsequently, the Moon becomes an increasingly slim crescent, moving ever closer to the Sun and appearing only just before sunrise. Finally, after 29.530589 days on average, it disappears from view and the cycle begins again.

Five types of month

Most people think of this cycle of 29.530589 days as being a month, but they also think of the Moon going around the Earth once a month. In fact, the Moon takes only 27.321661 days (27 days 7 hrs 43 min 11.5104 sec) on average to go around the Earth. So, which 'month' is right? Well actually both are. It all depends on what is meant by a month. As we have seen, the Earth has four types of year; the Moon, not to be outdone, has five types of month.

The most obvious, perhaps, is the time the Moon takes to go once around the Earth. This is known as the **sidereal month** and as we have seen, it is 27.321661 days. But because the Earth is moving around the Sun at the same time the Moon is moving around the Earth, it takes the Moon a bit longer than a sidereal month to return to the same position with respect to the Sun and the Earth. As it is this which governs the phases, it takes more than a sidereal month to go through a complete cycle or **lunation**. The time for a lunation is known as the **synodic month** or lunar month. The mean value of the synodic month is 29.530589 days (29 days, 12 hrs, 44 min, 2.8896 sec), but it can vary by up to ± 7 hours. The maximum duration occurs when the New Moon is at apogee; the minimum when it is at perigee. The apogee duration is slightly shortened if it occurs when the Earth is at aphelion and hence moving more slowly than it does at perihelion. Conversely, the perigee duration is slightly lengthened if it occurs when the Earth is at perihelion. Overall, the duration cycles between minima and maxima and back over the course of around one year to 18 months.

The **tropical month** of 27.321582 days (27 days 7 hrs 43 min 4.6848 sec) on average is slightly shorter than the sidereal month. It is defined as the time from one lunar equinox to another. The lunar equinox occurs when the Moon crosses the equator; this takes slightly less than a sidereal month due to the effects of precession (*c.f.* tropical year).

Next is the **anomalistic month**, the time taken for the Moon to go from perigee to perigee. The perigee advances, so this is longer than the sidereal month and is 27.554550 days (27 days 13 hrs 18 min 33.1200 sec) on average.

Finally, there is the **draconic month** of 27.212220 days (27 days 5 hrs 5 min 35.8080 sec) on average. This is the time between successive passages by the Moon through the same node. The nodes are moving westwards, and the Moon is moving eastwards along the celestial sphere, so it takes less than a complete orbit for the Moon to return to the node, and thus the draconic month is shorter than the sidereal month. The word 'draconic' refers to a mythical dragon thought to devour the Sun and the Moon during solar and lunar eclipses; the eclipse year is also sometimes referred to as the draconic year for this reason.

The interrelationship of various types of month and year are of great importance when it comes to predicting eclipses, and these cycles may have been understood as far back as prehistoric times.

Lunar and Solar Calendars

In the Western world, we have long been accustomed to a year of 365 days, with a leap day inserted into February every fourth year. The Gregorian calendar, now the most widely used civil calendar in the world, does have exceptions to this leap year every four years rule, but the last such 'non-leap year' was in 1900 and the next will not occur until 2100. Most of us will live our lives without ever having been troubled by such details, but they are important.

The Gregorian calendar is an example of a solar calendar, or one based on the tropical year. Since this is not an exact number of days, a leap day must be intercalated (inserted) at intervals, and the Gregorian calendar provides for an extra day in February if the year is divisible by four. An exception to the rule is made if the year is divisible by 100 but not by 400, as is the case for 1900 and 2100, but not 2000. The Gregorian calendar was introduced in 1582 in the time of Pope Gregory XIII. It was a refinement to the earlier Julian calendar, which inserted the leap day every fourth year without exception. This gave a year of 365.25 days, which is slightly longer than the tropical year of 365.242189 days. The Julian calendar was introduced by Julius Caesar in 46 BC and the error, though small, had amounted to several days by the sixteenth century. The Gregorian calendar was not immediately adopted everywhere, due to resistance in Protestant countries to a Catholic innovation. In Britain, the changeover did not occur until 1752, by which time the correction amounted to eleven days and so Wednesday, 2 September was followed by Thursday, 14 September. The story that this led to riots by people demanding the return of their eleven days is probably apocryphal. In Russia, the new system was not adopted until early in the communist era, by which time thirteen days had to be dropped from the calendar. An ironic consequence was that the date of the Great October Socialist Revolution was shifted into November.

Solar calendars follow the seasons, but the months do not follow the phases of the Moon because there are not an exact number of synodic months in a tropical year. A lunar calendar is one based on the phases of the Moon and examples include the Islamic calendar, which comprises twelve synodic months and therefore lags the solar calendar by 11 to 12 days each (tropical) year. The Islamic calendar is the official calendar of Saudi Arabia, but elsewhere in the Islamic world, it is used mainly for religious purposes.

To cater for the problem of a lunar calendar fairly rapidly drifting out of step with the tropical year, some calendrical systems insert an intercalary month every so often, though various calendars use different systems for determining how and when these occur. Such systems are known as lunisolar; examples include the Hebrew and Chinese calendars.

Lunar and solar calendars generally come into line every 19 years. This is because 19 tropical years are almost exactly 235 synodic months; thus every 19 years the Moon will have the same phase on the same day of the year. This 19-year cycle is known as the Metonic cycle after the Greek philosopher Meton of Athens (ca 440 BC) who noticed it, though it was undoubtedly known earlier. The Metonic cycle formed the basis of the Greek calendar until 46 BC, when the Julian calendar was adopted.

Moonrise, Moonset, and lunar movements

Like the Sun, the Moon does not rise and set in exactly the same place every day. The azimuth of the rising and setting points varies cyclically over the course of a sidereal month between northern and southern limits and, as with the Sun, these variations are more pronounced in higher latitudes. Note that the 'month' in question here is the sidereal rather than synodic month, hence the Moon will not be at the same phase at two successive risings or settings at a particular point. Another way of looking at this is to consider only the azimuth of rising and setting of the full Moon, which will vary between the same limits over the course of a year.

However, these limits themselves open out and close up over the course of the 18.61-year nodal cycle. In the Northern Hemisphere, the variation reaches a maximum when the ascending node is coincident with the summer solstice; these are the **major standstill**

points. When the descending node reaches this point, the variation is at a minimum; these are the **minor standstill points**. Between these limits, the standstill points gradually close up and then re-open. The situation is reversed in the Southern Hemisphere.

In simpler terms, at the major standstill the Moon's 5-degree orbital inclination is added to the effect of the Earth's obliquity (23.5 + 5.0 = 28.5°); at the minor standstill it is subtracted (23.5 - 5 = 18.5°). Thus, the variation exceeds that of the Sun at major standstill, but it is less than it at the minor standstill.

As we have seen, the cycle is driven by the sidereal and not the synodic month, so different phases of the Moon will be best observed at different times of the year. The full Moon, for example, rides majestically high in the winter skies, but in summer its performance is decidedly lacklustre. It struggles into the sky, staggers wearily along the southern horizon for a few hours before giving up and disappearing again. The explanation is straightforward enough: when full the Moon is in the opposite part of the sky to the Sun, so in winter it behaves as the Sun in summer, and vice-versa. In spring, the waxing first quarter Moon is most favourably presented for observation, and in autumn it is the turn of the waning last quarter Moon. The waxing crescent is best seen in mid-spring; the waning crescent in mid-summer. These rules apply in both hemispheres, because the seasons are reversed in the Southern Hemisphere. As with the standstill points, these effects are accentuated and diminished over the course of the 18.61-year nodal cycle.

The Dark side of the Moon

When people refer to 'the dark side of the Moon' they really mean the side that cannot be seen from here on Earth. As is correctly pointed out in the eponymous Pink Floyd album, there is no dark side of the Moon and both sides experience equal portions of day and night. It is, however, true that the Moon's sidereal day is exactly one sidereal month, so in the main one side permanently faces the Earth. However, it is not strictly speaking true to say that we can only see one side from Earth.

The orbital speed is not constant, so the orbit and rotation get slightly out of step at times, which causes a slightly different face to be presented. This effect is known as libration in longitude. In addition, because the Moon's axis is inclined by 6.5° to its orbit, it appears to 'nod' back and forth over the course of a month – this is libration in latitude. Finally, **parallax** effects result in slightly different faces being presented to the observer at different times of the day; in total 59 percent of the Moon's surface may be seen from Earth (though of course no more than 50 percent at any one time).

The Wanderers

The word 'planet' comes from the Greek word *planetes*, meaning 'wanderer'. Long before the time of the Classical Greek civilisation, people would have been aware of bright star-like objects that did not remain fixed in relation to the stars but moved in roughly the same narrow band to which the Sun and Moon are constrained. Five planets (excluding the Earth) have been known since prehistoric times – Mercury, Venus, Mars, Jupiter, and Saturn. They fall into two groups, the **inferior planets**, whose orbits lie close to the sun that of the Earth

(Mercury and Venus) and the **superior planets** whose orbits whose orbits lie further away from the Sun (all the other planets, excluding Earth). The distance of each planet from the Sun is often given in astronomical units. Incidentally, the terms 'superior' and 'inferior' do not mean that the superior planets are 'better' planets.

The motion of each planet around the Sun is governed by Kepler's laws of planetary motion, which were formulated by the German mathematician Johannes Kepler between 1609 and 1618 and they apply not just to planets but all orbiting bodies, such as the Moon, satellites of other planets, and even artificial Earth satellites.

The First law states that the orbit of any planet around the Sun will be an ellipse, with the Sun at one focus. If you add the distances of any point on an ellipse from each of the two foci, you will always get the same result. By comparison, if you measure the distance of any point on a circle from the centre of that circle, you will always get the same result. In fact, these properties define circles and ellipses, which are both examples of what are termed conic sections by mathematicians.

The Second law states that the movement of any planet in its orbit is such that its radius vector (an imaginary line joining the planet to the Sun) sweeps out equal areas in equal times. This explains why the Earth and other planets move faster when they are close to perihelion and why the Moon moves faster when it is close to perigee. The sector swept out in, say, 24 hours, is shorter at these times, but because the Earth (or Moon) is moving faster, it is also 'fatter', and these two effects exactly cancel out.

The Third law states that the square of a planet's orbital period in years is equal to the cube of its mean distance from the Sun in astronomical units. More generally, the square of the orbital period of any orbiting body is proportional to the cube of its mean distance from the body it orbits, assuming that the latter is significantly more massive.

These laws arise naturally from Newton's Law of universal gravitation, which states that between any two objects, there exists an attractive force that is proportional to their masses multiplied together and divided by the square of their distance apart. Objects under consideration can be stars, planets, satellites, or even the apocryphal apple that is said to have given Newton the idea in the first place.

Aspects of the planets

As seen from the Earth, certain positions of the planets relative to the Sun are known as aspects. For superior planets, the two principal aspects are **opposition** and **conjunction**. At opposition, a planet is opposite to the Sun in the sky, i.e., they are 180° apart. It will be visible throughout the night and will reach the meridian at midnight. Opposition is the best time to observe a superior planet, because it is at its closest to the Earth. At conjunction, a superior planet is on the opposite side of the Sun to the Earth. It will not be visible from Earth at this time, being lost in the Sun's glare.

When a planet is at either opposition or conjunction (i.e., it, the Earth, and the Sun are in a straight line) it is said to be at **syzygy**. The Moon is at syzygy when it is both new and full. When a planet is at an angle of 90° from the Sun as seen from Earth, it is said to be at **quadrature**. We see a half-Moon when it is at quadrature.

Inferior planets cannot reach opposition or quadrature, but have two types of conjunction: inferior conjunction, when they lie between the Earth and the Sun; and superior conjunction, when they are on the far side of the Sun. When an inferior planet is at its greatest angular separation from the Sun, it is at greatest elongation. At its greatest western elongation, it will appear in the morning sky; at greatest eastern elongation it will appear in the evening sky. An inferior planet can never be seen throughout the night.

The inferior planets display phases like the Moon but when best seen (i.e., at elongation) they are crescent. They will be at full phase at superior conjunction and "new" at inferior conjunction, but they cannot be seen at these times. The superior planets show very little phase effect; only Mars shows a pronounced gibbous phase when it is at quadrature.

Movements of the planets

As seen from the Earth, the planets normally appear to move from west to east. However, around opposition a superior planet can appear to halt and then move briefly in an east to west direction before resuming its normal progress. This retrograde motion, so beloved of astrologers, occurs because the Earth, which is moving more rapidly, catches up and overtakes the planet in question. The points where the planet halts before changing direction are known as stationary points.

The planets all keep fairly close to the ecliptic, but all have orbits that are slightly inclined to it. Orbits are defined in terms of six elements or quantities. These are the semi-major axis (a) or mean distance from the Sun; the eccentricity (e); the inclination to the ecliptic (i); the longitude of the ascending node (Ω); the argument of perihelion (ω) which is angular displacement from **Ω**; and the time of perihelion passage (T).

A planet's 'year' is known as its **sidereal period**, corresponding to the Earth's sidereal year. The time taken for a planet to return to a particular aspect (such as opposition) as seen from Earth is known as the **synodic period** (*c.f.* the Moon's synodic month).

Eclipses

There is little doubt that a total eclipse of the Sun is one of the most awesome spectacles of Nature available anywhere in the Solar System. On no other planet is there such an exact match between the apparent size of the Sun and the apparent size of a satellite – despite some planets having as many as eighty to choose from, while we on Earth have to make do with just the one. Not quite as spectacular, perhaps, but still noteworthy is the sight of the Moon turning a deep blood red as it enters the Earth's shadow during a total lunar eclipse. In recent years, there has been an unfortunate tendency for the media to refer to a total lunar eclipse as a 'blood moon'.

The phenomena are related, but strictly speaking a solar eclipse is an occultation or hiding of a self-luminous body (in this case the Sun) by the Moon. In principle, there is no difference between this and the occultation of stars that occur throughout the month as the Moon pursues its course around the Earth. By contrast, a lunar eclipse entails the Moon being cut off from the source of its illumination as it enters the Earth's shadow.

Unlike point sources (such as a distant searchlight), extended luminous objects such as the Sun do not cast sharp shadows. A shadow will of course be cast when an object is interposed between the observer and the light source, but it will have two regions: the umbra in which the light source is wholly obscured and the penumbra in which it is only partially obscured.

For a disc such as the Moon, the Earth as seen from the Moon's surface, or a hot-air balloon drifting in front of the Sun as seen by an observer on the ground, the umbra will be cone-shaped, converging to a point; the umbra will be fan-shaped and diverging.

Types of Solar eclipse

The Moon's umbra under favourable conditions will just reach the Earth. It does not remain stationary but races across the Earth's surface as the Moon moves in its orbit. The path it follows is known as the track. Observers inside the umbra will see a total solar eclipse; those outside it but still within the penumbra will see a partial solar eclipse; those completely outside the Moon's shadow will see nothing.

The degree of obscuration of the Sun by the Moon or magnitude of the eclipse will increase the closer an observer is to the zone of totality. Magnitude ranges from 0 (no obscuration) to 1 (totality) and it refers to the solar diameter covered, not area. A 0.5 magnitude eclipse is one in which half the Sun's diameter is covered, but a little geometry will show that only 40 percent of the Sun's area will actually be hidden by such an eclipse.

The actual duration of totality for any eclipse varies and is dictated by three factors: the distance of the Earth from the Sun when the eclipse occurs; the distance of Moon from the Earth when the eclipse occurs; and the latitude at which the eclipse occurs.

If the Earth is at its maximum distance from the Sun its apparent diameter will be diminished, and if the Moon is at its minimum distance from Earth its apparent diameter will be increased; these factors favour long eclipses.

The Earth is rotating in the same direction as the Moon's shadow is moving, and this has the effect of prolonging the time the latter will linger over a particular region. The speed the Earth's surface is moving depends on latitude – at 40° north or south of the equator, the west to east motion is 1,270 kilometres per hour, but at the equator it is 1,670 kilometres per hour. The relative speed of the Moon's shadow is thus lower at lower latitudes and thus eclipses that take place in tropical latitudes tend to be of greater duration than those occurring in temperate latitudes.

If the Moon is at or close to its maximum distance from Earth, even if it passes directly in front of the Sun the umbra will not quite reach Earth and a ring of sunlight is left showing. Such eclipses are said to be annular. Total and annular eclipses are referred to as central eclipses, and annular eclipses are the slightly more frequent of the two types.

Occasionally, an eclipse is just total at mid-track, but due to the curvature of the Earth the umbra does not touch the endpoints. The result is a hybrid total/annular eclipse with observers at mid-track experiencing a total eclipse but those at either endpoint viewing only an annular eclipse.

Finally, in about one-third of all solar eclipses, only the penumbra reaches the Earth with the umbra missing it altogether. Such eclipses are partial only; nowhere on Earth is a total eclipse seen.

Stages of a Solar eclipse

The key events in a solar eclipse as viewed from a particular site are known as contacts. First Contact occurs when the Moon's western edge begins to slide across the Sun and is the point at which the penumbra first begins to move across the site. It is abbreviated to P1, for first penumbral contact. Second Contact occurs when the Moon's eastern edge touches the Sun's eastern edge. The marks the onset of totality or annularity, and for a total eclipse is the point at which the umbra begins to move across the site. It is abbreviated to U1 for first umbral contact (though strictly speaking this term is only appropriate for a total eclipse). Third Contact occurs when the Moon's western edge leaves the Sun's western edge. This marks the end of totality or annularity and is the point at which the umbra leaves the site. It is abbreviated to U2. Finally, Fourth Contact, abbreviated to P2, marks the departure of the penumbra from the site and the end of the eclipse. In a partial eclipse, only P1 and P2 occur.

Types of Lunar eclipse

Whereas the Moon's umbra will affect only a small portion of the Earth, the Earth's umbra is large enough to fully immerse the Moon. During a lunar eclipse, the Moon never entirely disappears from view but appears reddish. This is due to **refraction** or bending of sunlight by the Earth atmosphere into the umbra; red light is more easily refracted. There are three types of lunar eclipse; total, when the whole of the Moon enters the umbra; partial when only a portion does; and penumbral when the Moon just grazes the penumbra. The latter type generally results in only a slight dimming of a portion of the Moon and is often undetectable to the naked eye. Unlike a solar eclipse, a lunar eclipse may be viewed anywhere on Earth where the Moon is above the horizon.

Stages of a Lunar eclipse

As with solar eclipses, the key stages of a lunar eclipse are referred to as contacts though unlike a solar eclipse these are the same from any point on Earth. P1 occurs when the Moon begins to enter the Earth's penumbra. U1 is the point at which the Moon begins to enter the umbra; U2 is the point at which it is fully inside the umbra, marking the onset of totality. U3 is the point at which the Moon begins to leave the umbra, marking the end of totality; U4 is the point at which the Moon leaves the umbra altogether. P2 is the point at which the Moon leaves the penumbra and the eclipse ends. U2 and U3 do not occur in a partial eclipse. In a penumbral eclipse, U1 and U4 do not occur either.

When do solar eclipses occur?

As you might have inferred, a solar eclipse can only occur at new Moon – but why don't they occur at every new Moon, i.e., once every lunation? Recall that the Moon's orbit is inclined at about 5° to the ecliptic. Thus, the Moon usually 'misses' the Sun. Recall that there are two nodes where the Moon's path crosses the ecliptic. Only when the Sun is close to a node at new Moon can an eclipse occur, although it does not have to be exactly at a node; for the two discs to touch in a 'grazing' encounter will at minimum produce a partial eclipse.

The region the Sun has to occupy at new Moon to produce a solar eclipse is known as the **eclipse limit**. This varies, depending on the distance of the Moon from Earth at the time the new Moon occurs, and that of the Earth from the Sun. It ranges from 15°21' to 18°30' on each side of the node (a region of 30°42' to 37°00' in total). For a central eclipse to occur, the limit is less, ranging from 9°52' to 11°50' on each side of the node.

With the Sun moving along the celestial sphere at just under one degree per day, it will be apparent that it will take more than a synodic month of 29.530589 days to traverse even the minimum distance. In other words, the Sun will never be able to get through one of these 'danger zones' without the Moon catching up with it at some stage and causing an eclipse. Furthermore, if the Sun has only just entered the eclipse limit when the Moon comes around, the latter will have time to cause a second eclipse before the Sun can get out of the way.

The time period during which the Sun is within the eclipse limit is known as an **eclipse season**, and it ranges in duration from 31 days 8 hours to 37 days 13 hours. Eclipses can only occur during an eclipse season and as we have just seen, at least one must occur. How many eclipses will occur in a calendar year, given at least one must occur whenever the Sun approaches a node?

Recall the nodes are moving along the celestial sphere in the opposite direction to the Sun, completing a complete cycle every 18.612958 years. It will therefore take the Sun slightly less than a year to make successive passages through the same node. This is the 'fourth kind of year', the eclipse year mentioned earlier, of 346.620076 days. There will be two eclipse seasons in each eclipse year and a minimum of two solar eclipses and a maximum of four.

The calendar year is longer than an eclipse year and so the eclipse year will end at different times of the calendar year. Normally there will only be two eclipse seasons (and hence a minimum of two eclipses) in a calendar year, but if an eclipse year ends in December, a portion of a third eclipse season can be squeezed into that calendar year, meaning that a maximum of five solar eclipses could occur. However, you will have to wait until 2206 before this next happens.

When do lunar eclipses occur?

Just as solar eclipses can only occur at new Moon, so a lunar eclipse can only occur when the Moon is full. The condition for a lunar eclipse is for the Moon to pass through the opposite node to the one through which the Sun is passing during an eclipse season. As with a solar eclipse, this must happen at least once during an eclipse season, and can happen twice. For an eclipse to occur, the full Moon must be within 11°38' of a node. The maximum number of both types combined in an eclipse season is only three, because it would take 1 ½ lunations to produce two solar and two lunar eclipses, which is longer than the maximum length of an eclipse season. However, there will always be at least one of each. Note that this rule includes penumbral lunar eclipses, which many authorities omit from eclipse statistics.

The Saros

The word 'saros' is Greek, but it was originally applied to eclipse cycles by the seventeenth century English astronomer Edmond Halley. The saros results from a series of coincidences of nature: 223 **synodic months** (6585.3213 days or 18 years 11 days 07 hrs 43 min) is almost exactly the same length of time as 242 **draconic months** (6585.3572 days or 18 years 11 days 8 hrs 34 min) and also coincides with 239 **anomalistic months** (6585.5374 days = 6585 days 12 hrs 54 min). The net effect is that at the conclusion of 223 synodic months from the time of an eclipse, the Moon will again be both at **syzygy** and at a **node**, and it will also be at the same distance from Earth. This will not only give rise to another eclipse, but also for a solar eclipse the **eclipse limit** will be the same. This latter factor is equally important because were the Moon to be at a greater distance from Earth than previously, the eclipse limit would be smaller, and an eclipse might not occur at all. The period of 223 synodic months is referred to as a saros cycle. Eclipses do of course occur more frequently than every 18 years, so there are a number of saros series in operation at any one time. A saros series is a collection of eclipse seasons each of which is separated by 223 synodic months (i.e., one saros cycle) from the preceding eclipse season.

Saros series are numbered: those involving the Moon's descending node receive even numbers and those involving the ascending node receive odd numbers. Each saros series evolves and has a finite life. For a saros series involving the Moon's descending node, the series begins with an eclipse at the south pole. Each successive eclipse then has a track more northerly than the last, until a final eclipse at the north pole concludes the cycle. For a saros cycle involving the Moon's ascending node, the reverse happens, with the series beginning at the north pole and concluding at the south pole. When one saros series concludes at one pole another one will begin at the other pole. A complete series lasts from between 1,206 to 1,442 years; at any one time there are 43 saros series in operation.

~

Glossary

Achronycal rising
The last visible rising of a star or constellation in the eastern sky at sunset; thereafter it will have already risen by dusk. See *Cosmical setting*.

Altitude
The angular distance of an object above the horizon.

Anomalistic month
The time between the Moon making two returns to perigee (*q.v.*), 27.554550 days (27 days 13 hrs 18 min 33.1200 sec) on average. The perigee advances, so this is longer than the sidereal month (*q.v.*) of 27.321661 days.

Anomalistic year
The time between the Earth making two returns to perihelion, 365.259636 days (365 days 6 hrs 13 min 52.5504 sec) on average. As the perihelion advances with each circuit, the Earth requires a bit longer to 'catch up'; hence the anomalistic year is fractionally longer than the sidereal year (*q.v.*) of 365.256363 days (365 days 6 hrs 9 min 9.7632 sec).

Aphelion
The point in the orbit of a planet, asteroid, or comet when it reaches its maximum distance from the Sun.

Apogee
The point in the orbit of the Moon (or a satellite or spacecraft) when it reaches its maximum distance from the Earth; in a geocentric model (*q.v.*) of the Solar System the Sun may also have an apogee.

Asterism
Any grouping of stars; can be a recognised constellation, a portion of a constellation (for example, the Plough is an asterism within Ursa Major), or an 'unofficial' group of prominent stars (for example, the 'summer triangle' of Vega, Deneb, and Altair).

Azimuth
The direction of a celestial object from the observer, expressed as the angular distance from the north point of the horizon to the point at which a great circle passing through the object intersects the horizon.

Celestial equator
A great circle on the celestial sphere (*q.v.*) with a declination (*q.v.*) of zero, envisaged to be 'above' the terrestrial equator.

Celestial latitude
The latitude of an object on the celestial sphere (*q.v.*) in an ecliptic coordinate system, i.e., with the ecliptic (*q.v.*) as the datum plane; also known as ecliptic latitude.

Celestial longitude
The angular distance of an object east of the First point of Aries (*q.v.*) along the ecliptic (*q.v.*) in an ecliptic coordinate system; also known as ecliptic longitude.

Celestial pole
In an equatorial coordinate system, the points on the celestial sphere (*q.v.*) corresponding to declinations (*q.v.*) of +90° (north) and -90° (south) are the celestial poles. They are envisaged to be above the geographical poles.

Celestial sphere
An imaginary sphere concentric with the Earth upon which the stars, planets, Sun, and Moon are conceived as being projected.

Circumpolar
A star or constellation that never sets at the latitude from which it is being observed. For an observer at latitude θ, a star will be circumpolar if its declination $\delta + \theta$ is greater than +90° in the northern hemisphere, or $\delta - \theta$ is less than -90° in the southern hemisphere.

Conjunction
A planet is at conjunction when it is either on the opposite side of the Sun to Earth (superior conjunction) or directly between Earth and the Sun (inferior conjunction). Only an inferior planet (*q.v.*) can ever be at inferior conjunction. The Moon can never be at superior conjunction.

Cosmical setting
The first visible setting of a star or constellation before sunrise; previously it did not set until after sunrise. See *Achronycal rising*.

Culmination
The point at which an object reaches an observer's meridian (*q.v.*). This will happen twice a day: the upper culmination represents the highest point reached above the horizon and the lower culmination the lowest. However, for a non-circumpolar constellation the latter will occur when the object is below the horizon.

Declination
The latitude of an object on the celestial sphere (*q.v.*) in an equatorial coordinate system, i.e., with the celestial equator (*q.v.*) as the datum plane. Positive declinations (0 to +90°) correspond to the northern celestial hemisphere (*q.v.*); negative declinations (0 to -90°) to the

southern celestial hemisphere (*q.v.*).

Diurnal motion
The apparent motion of celestial bodies due to the rotation of the Earth.

Draconic month
The time between successive passages by the Moon through the same node (*q.v.*), 27.212220 days (27 days 5 hrs 5 min 35.8080 sec). The nodes are moving westwards, and the Moon is moving eastwards along the celestial sphere (*q.v.*), so it takes less than a complete orbit for the Moon to return to the node, and thus the draconic month is shorter than the sidereal month (*q.v.*) of 27.321661 days.

Eclipse limit
The region the Sun has to occupy at new Moon to produce a solar eclipse. It varies, depending on the distance of the Moon from Earth at the time the new Moon occurs, and that of the Earth from the Sun. It ranges from 15°21' to 18°30' on each side of the node (a region of 30°42' to 37°00' in total). For a central eclipse to occur, the limit is less, ranging from 9°52' to 11°50' on each side of the node. For a lunar eclipse, the full Moon must be within 11°38' of a node.

Eclipse season
The two periods during an eclipse year (*q.v.*) when the Sun is within the eclipse limit (*q.v.*) and an eclipse is possible. The duration of an eclipse season varies from 31 days 8 hours to 37 days 13 hours.

Eclipse year
The time between successive passages of the Sun through the same node (*q.v.*), 346.620076 days (346 days 14 hrs 52 min 54.5 sec) on average.

Ecliptic
The great circle on the celestial sphere (*q.v.*) representing the Sun's apparent path during the year; corresponds to the plane of the Earth's orbit around the Sun.

Epoch
A moment in time used as a reference point for an astronomical quantity that varies in time; for example, the coordinates of a star, which vary due (principally) to the effects of precession (*q.v.*)

Equatorial mean Sun
An imaginary Sun that crosses the meridian (*q.v.*) at local noon every 24 hours; not to be confused with mean Sun (*q.v.*).

Equation of centre
The difference between the celestial longitude (*q.v.*) of the Sun or a planet and its mean longitude (*q.v.*). For the Sun, it is also referred to as the solar equation.

Equation of time
The difference between the right ascension (*q.v.*) of the equatorial mean Sun (*q.v.*) and that of the actual Sun. It may be expressed as an angular distance or converted to units of time.

Equinox
The equinoxes occur twice a year when the Sun crosses the celestial equator (q.v.). The vernal or spring equinox occurs when the Sun moves into the northern celestial hemisphere (*q.v.*); the autumn equinox when it returns to the southern celestial hemisphere (*q.v.*). The day and night are of equal length in both northern and southern hemispheres.

Extinction
The disappearance of an astronomical object when close to the horizon due to its light having to pass a greater distance through the Earth's atmosphere. The extinction angle is the angle above the horizon below which an object ceases to be visible. It depends on the brightness of the object and prevailing atmospheric conditions, ranging typically between four to five degrees.

First point of Aries
The point on the celestial sphere (*q.v.*) where the Sun crosses into the northern skies at the spring equinox (*q.v.*) although due to precession (*q.v.*) it is currently in Pisces; also known as the vernal point.

First point of Libra
The point on the celestial sphere (*q.v.*) where the Sun crosses into the southern skies at the autumn equinox (*q.v.*), due to precession (*q.v.*) it is currently in Virgo.

Geocentric model
A model that puts the Earth at the centre of the Solar System.

Heliacal rising
The reappearance of a star or constellation in the eastern sky at dawn after a period of invisibility due to its proximity to the Sun in the sky. See *Heliacal setting*.

Heliacal setting
The temporary disappearance of a star or constellation in the western sky at dusk to its proximity to the Sun in the sky, following a period of visibility. See *Heliacal rising*.

Heliocentric model
A model that puts the Sun at the centre of the Solar System.

Inferior planet
A planet whose orbit is closer to the Sun than that of Earth (*c.f. Superior planet*).

Lunar anomaly
the variation of the Moon's speed along the ecliptic (*q.v.*) from a maximum at perigee (*q.v.*) to a minimum at apogee (*q.v.*).

Lunar theory
Any mathematical model used to describe or predict the movements of the Moon as seen from Earth.

Lunation
A full cycle of the Moon's phases; corresponds to the Moon's synodic period (*q.v.*) of 29.530589 days on average, i.e., a synodic or lunar month (*q.v.*).

Magnitude
A logarithmic measure of brightness of a celestial body, with five magnitudes corresponding to a one hundred-fold difference in brightness. Lower numbers correspond to higher magnitudes; for example, magnitude +1 is 100 times brighter than magnitude +6 and a negative magnitude is brighter than a positive magnitude.

Major standstill
The point in the Moon's nodal cycle (*q.v.*) when its range of declinations (*q.v.*) is at a maximum. See *standstill limits.*

Maximum solar equation
The maximum value of the equation of centre (*q.v.*) for the Sun; occurs when the Sun is 90° away from apogee (*q.v.*).

Mean longitude
The celestial longitude (*q.v.*) of an object if it is assumed to be moving at a uniform speed along the ecliptic (*q.v.*); the actual celestial longitude of the object will differ due to solar anomaly (*q.v.*) and zodiacal anomaly (*q.v.*).

Mean Sun
An imaginary Sun that moves along the ecliptic (*q.v.*) at a uniform speed with an angular separation from the real Sun equal to the equation of centre (*q.v.*); not to be confused with the equatorial mean Sun (*q.v.*).

Meridian
A great circle on the celestial sphere (*q.v.*) taking in an observer's zenith (*q.v.*) and the north and south celestial poles (*q.v.*); it is coplanar with the observer's terrestrial line of longitude.

Minor standstill
The point in the Moon's nodal cycle (*q.v.*) when its range of declinations (*q.v.*) is at a minimum. See *standstill limits.*

Nodal cycle of the Moon
The nodes (*q.v.*) of the Moon's orbit do not remain fixed; they move westwards on the celestial sphere at 19°20' per year due to perturbation by the Sun, taking 18.612958 years (6,798.383 days) on average to complete a full cycle; also referred to as the draconic cycle.

Nodes of the Moon's orbit
The two points on the celestial sphere (*q.v.*) where the Moon's orbit intersects the ecliptic (*q.v.*). The ascending node is where the Moon passes from the south of the ecliptic plane to the north; the descending node is where the Moon passes from the north of the ecliptic plane to the south.

Northern celestial hemisphere
The portion of the sky at a declination (*q.v.*) of greater than zero.

Nutation
The precessional motion of the Earth's axis is not smooth but slightly wavy. This irregularity is known as nutation: a slight nodding of the Earth due to changes in tidal forces acting on the planet through variations in the distances and positions of the Sun and the Moon. The largest component of Earth's nutation corresponds to and has the same periodicity as the Moon's 18.61-year nodal cycle (*q.v.*). This produces a variation of ± 9.2 arcsec in the obliquity of the ecliptic. Smaller components are due to the Earth's annual motion around the Sun and by the Moon's monthly motion around the Earth.

Obliquity of the ecliptic
The tilt of the Earth's axis of rotation, and hence the equator, with respect to the ecliptic (*q.v.*); it is currently 23°26' but is decreasing by 47 arcsec) per century due to perturbations of the plane of Earth's orbit by other planets.

Occam's Razor
A philosophical principle named for the Franciscan friar William of Occam; it states that if multiple explanations can be proposed for a phenomenon, the one involving the fewest assumptions is the one most likely to be correct.

Opposition
A planet is said to be at opposition when it is 180° away from the Sun on the celestial sphere (*q.v.*).

Parallax
The difference in the apparent position of an object when viewed along two different sightlines. The further away an object is, the smaller the parallax as viewed from the same pair of sightlines. Parallax can thus be used to determine distances.

Perigee
The point in the orbit of the Moon (or a satellite or spacecraft) when it reaches its minimum distance from the Earth; in a geocentric model (*q.v.*) of the Solar System the Sun may also have a perigee.

Perihelion
The point in the orbit of a planet, asteroid, or comet when it reaches its minimum distance from the Sun.

Poles of the ecliptic
In an ecliptic coordinate system, the poles of the ecliptic are analogous to the celestial poles (*q.v.*). They correspond to the points on the celestial sphere (*q.v.*) with celestial latitude (*q.v.*) +90° and -90°.

Precession of the equinoxes
The secondary motion of the Earth's axis, equivalent to the 'wobble' of a spinning top and causing a westward drift of the equinoxes along the ecliptic relative to the fixed stars. A precessional cycle takes around 25,772 years.

Proper motion
The actual physical movements of stars through space, resulting in changes to the appearance of constellations as viewed from Earth. Owing to the great distances involved, these changes are very gradual, manifesting themselves over tens of millennia.

Quadrature
When a planet is at an angle of 90° from the Sun as seen from Earth, it is said to be at quadrature.

Refraction
The change in direction of light passing from one medium to another or from a gradual change in the medium; for example, changes in air density. Since the pressure is lower at higher altitudes, light rays are refracted towards the surface when traveling long distances through the atmosphere. This shifts the apparent positions of stars slightly when they are close to the horizon and makes the sun visible before it geometrically rises above the horizon during a sunrise.

Retrogression
A period when a planet appears to move in a retrograde manner against the background stars due to Earth overtaking or being overtaken by the planet in question; the points where retrogression begins, and when it ends are known as stationary points or stations.

Right Ascension
The longitude of an object on the celestial sphere (*q.v.*) using an equatorial coordinate system, i.e., with the celestial equator (*q.v.*) as the datum plane. It is given in hours, minutes, and seconds east of the First point of Aries (*q.v.*) up to 24 hours; hence one hour corresponds to 15 degrees.

Seasonal hours
The division of the day from dawn to dusk into twelve 'hours'; the length of these hours will thus depend on the time of year being at their longest in summer and shortest in winter.

Sidereal day
The time taken for the Earth to complete a single turn on its axis, 23 hrs 56 min 4.091 sec.

Sidereal month
The time taken for the Moon to complete a single orbit of the Earth, 27.321661 days (27 days 7 hrs 43 min 11.5104 sec) on average.

Sidereal period
The time taken for a planet to complete a single orbit of the Sun.

Sidereal year
The time taken for the Earth to complete a single orbit of the Sun, 365.256363 days (365 days 6 hrs 9 min 9.7632 sec) on average.

Solar anomaly
In a geocentric model of the Solar System, the retrograde motion of a planet along the zodiac is referred to as the solar anomaly. In reality, it is a consequence of Earth 'overtaking' or being 'overtaken' by the planet in question in the course of both orbiting the Sun. Also referred to as inequality with respect to the Sun or second inequality. See *Zodiacal anomaly*.

Solar day
The time between two successive transits of the meridian by the Sun, averages 24 hours.

Solar theory
Any mathematical model used to describe or predict the apparent movements of the Sun as seen from Earth.

Solstice
The solstices occur twice a year when the Sun reaches its maximum (northern summer; southern winter) and minimum (northern winter; southern summer) declination (*q.v.*). The summer solstice corresponds to the longest day of the year; the winter solstice to the shortest.

Solstitial colure
A great circle on the celestial sphere passing through the celestial poles and the solstitial points.

Solstitial limits
The northernmost and southernmost rising and setting positions of the Sun, occurring at the summer and winter solstices respectively in the northern hemisphere; reversed for the southern hemisphere.

Southern celestial hemisphere
The portion of the sky at a declination (*q.v.*) of less than zero.

Standstill limits
The northernmost and southernmost rising and setting positions of the Moon when at major standstill (*q.v.*) or minor standstill (*q.v.*).

Superior planet
A planet whose orbit is further from the Sun than that of Earth (*c.f. Inferior planet*).

Synodic month
The synodic period (*q.v.*) of the Moon, averaging 29.530589 days (29 days, 12 hrs, 44 min, 2.8896 sec), or one lunation (*q.v.*); typically referred to as a lunar month.

Synodic period
The time taken for the Moon or a planet to make successive returns to the same point in the sky in relation to the Sun.

Syzygy
A straight-line configuration of three celestial bodies. When the Moon is at syzygy, it will be either new or full; i.e., the Sun, Moon, and Earth are in a straight line.

Tropics
Two circles on the celestial sphere marking the declinations (*q.v.*) of +23.5° and -23.5°; these are the maximum northerly and southerly declinations the Sun can reach, which it does at the solstices (*q.v.*). They correspond to the geographical Tropic of Cancer and Tropic of Capricorn, where the Sun is directly overhead at noon on the day of the summer solstice in the northern and southern hemispheres, respectively. The Tropic of Cancer was so named because at the time, the Sun was in Cancer during the summer solstice; similarly, it was in Capricornus during the winter solstice. Due to precession (*q.v.*), this is no longer the case.

Tropical month
The time for the Moon to make two successive passages through the First Point of Aries (*q.v.*), 27.321582 days (27 days 7 hrs 43 min 4.6848 sec) on average. This takes slightly less than a sidereal month (*q.v.*) of 27.321661 days due to the effects of precession (*q.v.*).

Tropical period
The average time for a planet, as seen from Earth, to make successive transits through the First point of Aries (*q.v.*). For a superior planet (*q.v.*), this will be slightly less than the sidereal period (*q.v.*); for example, for Saturn the tropical period is 10,746.94 days and the sidereal period is 10,759.22 days. For Mars, the difference between the two is just 10 min 5 sec. Inferior planets (*q.v.*) are 'tethered' to the Sun, and their transit times through the First point of Aries will average one tropical year (*q.v.*).

Tropical year
The time taken for the Sun to make two successive passages through the First Point of Aries (*q.v.*), averaging 365.242189 days (365 days 5 hrs 48 min 46.6339 sec); also known as the solar year. Due to the effects of precession (*q.v.*), the tropical year is slightly less than the sidereal year (*q.v.*) of 365.256363 days.

Vernal point
See First point of Aries.

Zenith
The point on the celestial sphere (*q.v.*) that is directly overhead from the point of view of an observer.

Zodiacal anomaly
In a geocentric model of the Solar System, the variable speed of the Sun or a planet along the ecliptic (*q.v.*) is referred to as the zodiacal anomaly. In reality, it is a consequence of the elliptical orbits of Earth and other planets around the Sun, going from a maximum at perihelion (*q.v.*) to a minimum at aphelion (*q.v.*). Also referred to as zodiacal inequality or first inequality. See *Solar anomaly*.

~

References

1: Astronomy in the Upper Palaeolithic

1. Richter, D. *et al.*, The age of the hominin fossils from Jebel Irhoud, Morocco, and the origins of the Middle Stone Age. *Nature* **546**, 293-296 (2017).
2. Neubauer, S., Hublin, J. & Gunz, P., The evolution of modern human brain shape. *Science Advances* **4** (1), eaao5961 (2018).
3. Hershkovitz, I. *et al.*, The earliest modern humans outside Africa. *Science* **359**, 456-459 (2018).
4. Liu, W. *et al.*, The earliest unequivocally modern humans in southern China. *Nature* **526**, 696-699 (2015).
5. Clarkson, C. *et al.*, Human occupation of northern Australia by 65,000 years ago. *Nature* **547**, 306-310 (2017).
6. Dillehay, T. *et al.*, New Archaeological Evidence for an Early Human Presence at Monte Verde, Chile. *PLoS One* **10** (11) (2015).
7. Carbonell, E. *et al.*, The first hominin of Europe. *Nature* **452**, 465-470 (2008).
8. Mellars, P., A new radiocarbon revolution and the dispersal of modern humans in Eurasia. *Nature* **493**, 931-935 (2006).
9. Higham, T. *et al.*, The timing and spatiotemporal patterning of Neanderthal disappearance. *Nature* **512**, 306-309 (2014).
10. Mellars, P., Neanderthals and the modern human colonization of Europe. *Nature* **432**, 461-465 (2004).
11. Mellars, P., Origins of the female image. *Nature* **439**, 176-177 (2009).
12. D'Errico, F. & Cacho, C., Notation versus Decoration in the Upper Palaeolithic: a Case-Study from Tossal de la Roca, Alicante, Spain. *Journal of Archaeological Science* **21**, 185-200 (1994).
13. D'Errico, F. *et al.*, Archaeological Evidence for the Emergence of Language, Symbolism, and Music—An Alternative Multidisciplinary Perspective. *Journal of World Prehistory* **17** (1), 1-70 (2003).
14. Robinson, J., Not counting on Marshack: A reassessment of the work of Alexander Marshack on notation in the Upper Palaeolithic. *Journal of Mediterranean Studies* **2** (1), 1-17 (1992).
15. Calude, A. & Verkerk, A., The typology and diachrony of higher numerals in Indo-European: a phylogenetic comparative study. *Journal of Language Evolution* **1** (2), 91-108 (2016).
16. Epps, P., Bowern, C. & Hill, J., On numeral complexity in hunter-gatherer languages. *Linguistic Typology* **16**, 41-109 (2012).
17. Wobst, M., Archaeology The Archaeo-Ethnology of Hunter-Gatherers or the Tyranny of the Ethnographic Record in Archaeology. *American Antiquity* **43** (2), 303-309 (1978).
18. Gould, R.. & Watson, P., A Dialogue on the Meaning and Use of Analogy in Ethnoarchaeological Reasoning. *Journal of Anthropological Archaeology* **1**, 355-381 (1982).
19. Hayden, B. & Villeneuve, S., Astronomy in the Upper Palaeolithic? *Cambridge Archaeological Journal* **21** (3), 331-355 (2011).
20. Barker, G., *The Agricultural Revolution in Prehistory: Why did Foragers become Famers* (Oxford University Press, Oxford, 2006).
21. Norris, R. & Hamacher, D., in *Handbook of Archaeoastronomy and Ethnoastronomy*, edited by Ruggles, C. (Springer, New York, NY, 2015), pp. 2216-2222.
22. Clarke, P., in *Handbook of Archaeoastronomy and Ethnoastronomy*, edited by Ruggles, C. (Springer, New York, NY, 2015), pp. 2223-2229.
23. Norris, R. & Hamacher, D., The Astronomy of Aboriginal Australia. *The Role of Astronomy in Society and Culture, Proceedings of the International Astronomical Union, IAU Symposium* **260**, 39-47 (2011).

2: Upper Palaeolithic lunar calendars

1. Gaffney, V. *et al.*, Time and a Place: A luni-solar 'time-reckoner' from 8th millennium BC Scotland. *Internet Archaeology* **34**, https://doi.org/10.11141/ia.34.1 (2013).
2. Marshack, A., *The Roots of Civilization* (McGraw-Hill, New York, 1972).
3. McBrearty, S. & Brooks, A., The revolution that wasn't: a new interpretation of the origin of modern human behaviour. *Journal of Human Evolution* **39** (5), 453-563 (2000).
4. Diamond, J., *The Third Chimpanzee* (Random, London, 1991).
5. Marshack, A., Some implications of the Paleolithic symbolic evidence for the origin of language. *Current Anthropology* **17**, 274-82 (1976).
6. Marshack, A., Upper Paleolithic Symbol Systems of the Russian Plain: Cognitive and Comparative Analysis. *Current Anthropology* **20** (2), 271-311 (1979).
7. Marshack, A., Evolution of the Human Capacity. *Yearbook of Physical Anthropology* **32**, 1-34 (1989a).
8. Marshack, A., The Berekhat Ram figurine: a late Acheulian carving from the Middle East. *Antiquity* **71**, 327-338 (1997).
9. Marshack, A., A Lunar-Solar Year Calendar Stick from North America. *American Antiquity* **50** (1), 27-51 (1985).
10. Marashack, A., The female image - a time-factored symbol. *Proceedings of the Prehistoric Society* **57**, 17-31 (1991a).
11. Clottes, J., *Cave Art* (Phaidon, New York, 2008).
12. Marshack, A., Lunar notation on Upper Paleolithic remains. *Science* **146**, 743-745 (1964).
13. Robinson, J., Not counting on Marshack: A reassessment of the work of Alexander Marshack on notation in the Upper Palaeolithic. *Journal of Mediterranean Studies* **2** (1), 1-17 (1992).
14. Bourrillon, R. *et al.*, A new Aurignacian engraving from Abri Blanchard, France: Implications for understanding Aurignacian graphic expression in Western and Central Europe. *Quaternary International* **491**, 46-64 (2018).
15. Marshack, A., The La Marche Antler Revisited. *Cambridge Archaeological Journal* **6** (1), 99-111 (1996).
16. Marshack, A., The Taï plaque and calendrical notation in the Upper Paleolithic. *Cambridge Archaeological Journal* **1** (1), 25-61 (1991b).
17. Zeilik, M., Historic lunar calendars in the Pueblo Southwest: Examples from Zuñi and Hopi. *Astronomy Quarterly* **8** (2), 89-107 (1991).
18. Lewis-Williams, D. & Pearce, D., *Inside the Neolithic Mind* (Thames & Hudson, London, 2005).
19. Hudson, T. & Underhay, E., *Crystals in the Sky: an Intellectual Odyssey Involving Chumash Astronomy, Cosmology and Rock Art* (Ballena Press, Socorro, NM, 1978).
20. Movius, H., Comment on: Cognitive Aspects of Upper Paleolithic Engraving. *Current Anthropology,* **13** (3/4), 467-469 (1972).
21. Hayden, B. & Villeneuve, S., Astronomy in the Upper Palaeolithic? *Cambridge Archaeological Journal* **21** (3), 331-355 (2011).
22. Elkins, On the Impossibility of Close Reading: The Case of Alexander Marshack. *Current Anthropology* **37** (2), 185-226 (1996).
23. D'Errico, F. & Cacho, C., Notation versus Decoration in the Upper Palaeolithic: a Case-Study from Tossal de la Roca, Alicante, Spain. *Journal of Archaeological Science* **21**, 185-200 (1994).
24. D'Errico, F., A New Model and its Implications for the Origin of Writing: The La Marche Antler Revisited. *Cambridge Archaeological Journal* **5** (2), 163-206 (1995).
25. D'Errico, F., Marshack's Approach: Poor Technology, Biased Science. *Cambridge Archaeological Journal* **6** (1), 111-117 (1996).
26. D'Errico, F. *et al.*, Archaeological Evidence for the Emergence of Language, Symbolism, and Music—An Alternative Multidisciplinary Perspective. *Journal of World Prehistory* **17** (1), 1-70 (2003).
27. D'Errico, F., Palaeolithic Lunar Calendars: A Case of Wishful Thinking? *Current Anthropology* **30** (1), 117-18 (1989a).
28. D'Errico, F., A reply to On Wishful Thinking and Lunar "Calendars". *Current Anthropology* **30** (4), 491-500 (1989b).
29. Marshack, A., On Wishful Thinking and Lunar "Calendars". *Current Anthropology* **30** (4), 491-493

(1989b).
30. Mellars, P., in *Prehistoric Europe*, edited by Cunliffe, B. (Oxford University Press, Oxford, 1994), pp. 42-78.

3: Ice age star maps

1. Mellars, P., in *Prehistoric Europe*, edited by Cunliffe, B. (Oxford University Press, Oxford, 1994), pp. 42-78.
2. Lewis-Williams, D., *The Mind in the Cave:Consciousness And The Origins Of Art* (Thames & Hudson, London, 2002).
3. Clottes, J., *Cave Art* (Phaidon, New York, 2008).
4. Hayden, B. & Villeneuve, S., Astronomy in the Upper Palaeolithic? *Cambridge Archaeological Journal* **21** (3), 331-355 (2011).
5. Congregado, L., 1991.
6. Edge, F., 1995.
7. Rappenglück, M., A Palaeolithic planetarium underground - the caves of Lascaux Part 1. *Migration & Diffusion* **5** (18), 93-119 (2004a).
8. Matossian, M., Symbols of Seasons and the Passage of Time: Barley and Bees in the New Stone Age. *Griffith Observer* **44** (11), 9-17 (1980).
9. Genty, D. *et al.*, Dating the Lascaux Cave gour formation. *Radiocarbon* **53** (3), 479-500 (2011).
10. Rappenglück, M., in *Earth-Moon Relationships*, edited by C., B. & F., R. (Springer, Dordrecht, 2001), Vol. 85–86, pp. 391-404.
11. Marshack, A., *The Roots of Civilization* (McGraw-Hill, New York, 1972).
12. Rappenglück, M., A Palaeolithic planetarium underground - the caves of Lascaux Part 2. *Migration & Diffusion* **5** (19), 6-46 (2004b).
13. Rappenglück, M., Ice Age People Find their Ways by the Stars: A Rock Picture in the Cueva de El Castillo (Spain) May Represent the Circumpolar Constellation of the Northern Crown (CrB). *Migration and Diffusion* **1** (2), 15-28 (2000).
14. Rappenglück, M., The anthropoid in the sky: Does a 32,000-year old ivory plate show the constellation Orion combined with a pregnancy calendar? *Uppsala Astronomical Observatory Reports* **59**, 51-55 (2003).
15. Higham, T. *et al.*, Testing models for the beginnings of the Aurignacian and the advent of figurative art and music: The radiocarbon chronology of Geißenklösterle. *Journal of Human Evolution* **62** (6), 664-676 (2012).

4: Göbekli Tepe and the ancient peril

1. Dietrich, O. & Schmidt, K., A radiocarbon date from the wall plaster of Enclosure D of Göbekli Tepe. *Neo-Lithics* **2** (10), 82-83 (2010).
2. Notroff, J. *et al.*, More than a vulture. *Mediterranean Archaeology and Archaeometry* **17** (2), 57-74 (2017).
3. Schmidt, K., Investigations in the early Meospotamian Neolithic: Göbekli Tepe and Gürcütepe. *Neo-Lithics* (2/95), 9-10 (1995).
4. Schmidt, K., Beyond Daily Bread: Evidence of Early Neolithic Ritual from Göbekli Tepe. *Neo-Lithics* (2/98), 1-5 (1998).
5. Schmidt, K., Göbekli Tepe, Southeastern Turkey A Preliminary Report on the 1995-1999 Excavations. *Paléorient* **26** (1), 45-54 (2000).
6. Schmidt, K., The 2003 Campaign at Göbekli Tepe (Southeastern Turkey). *Neo-Lithics* (2/03), 3-8 (2003).
7. Peters, J. & Schmidt, K., Animals in the symbolic world of Pre-Pottery Neolithic Göbekli Tepe, south-eastern Turkey: a preliminary assessment. *Anthropozoologica* **39** (1), 179-218 (2004).
8. Lewis-Williams, D. & Pearce, D., *Inside the Neolithic Mind* (Thames & Hudson, London, 2005).
9. De Laubenfels, M., Dinosaur Extinction: One More Hypothesis. *Journal of Paleontology* **30** (1), 207-212

(1956).
10. Velikovsky, I., *Worlds in Collision* (Macmillan, New York, NY, 1950).
11. Morrison, D., Velikovsky at Fifty: Cultures in Collision on the Fringes of Science. *Skeptic* **9** (1), 62-76 (2001).
12. Steel, D., *Rogue asteroids and doomsday comets* (John Wiley & Sons, Inc., New York, 1993).
13. Alvarez, L., Alvarez, W., Asaro, F. & Michel, H., Extraterrestrial Cause for the Cretaceous-Tertiary Extinction. *Science* **208**, 1095-1108 (1980).
14. Hildebrand, A. *et al.*, Chicxulub Crater: A possible Cretaceous/Tertiary boundary impact crater on the Yucatán Peninsula, Mexico. *Geology* **19** (9), 867-871 (1991).
15. Renne, P. *et al.*, Time Scales of Critical Events Around the Cretaceous-Paleogene Boundary. *Science* **339**, 684-687 (2013).
16. Bottke, W., Vokrouhlicky, D. & Nesvorny, D., An asteroid breakup 160 Myr ago as the probable source of the K/T impactor. *Nature* **449**, 48-53 (2007).
17. Napier, W. & Clube, V., A theory of terrestrial catastrophism. *Nature* **282**, 455-459 (1979).
18. Clube, V. & Napier, W., The role of episodic bombardment in geophysics. *Earth and Planetary Science Letters* **57**, 251-262 (1982).
19. Clube, V. & Napier, W., Spiral Arms, Comets and Terrestrial Catastrophism. *Quarterly Journal of the Royal Astronomical Society* **23**, 45-66 (1983).
20. Clube, V. & Napier, W., Comet capture from molecular clouds: a dynamical constraint on star and planet formation. *Monthly Notices of the Royal Astronomical Society* **208**, 575-588 (1984a).
21. Clube, V. & Napier, W., The microstructure of terrestrial catastrophism. *Monthly Notices of the Royal Astronomical Society* **211** (4), 953-968 (1984b).
22. Brown, P., The Leonid Meteor Shower: Historical Visual Observations. *Icarus* **138**, 287-308 (1999).
23. Napier, W. & Clube, V., Our cometary environment. *Reports on Progress in Physics* **60**, 293-343 (1997).
24. Steel, D. & Asher, D., The orbital dispersion of the macroscopic Taurid objects. *Monthly Notices of the Royal Astronomical Seciety* **280**, 806-822 (1996).
25. Spurný, P., Borovička, J., Mucke, H. & Svoreň, J., Discovery of a new branch of the Taurid meteoroid stream as a real source of potentially hazardous bodies. *Astronomy & Astrophysics* **605**, A68 (2017).
26. Kresák, L., The Tunguska object - A fragment of Comet Encke. *Bulletin of the Astronomical Institutes of Czechoslovakia* **29** (3), 129-134 (1978).
27. Longo, G., in *Comet/Asteroid Impacts and Human Society: an Interdisciplinary Approach*, edited by Bobrowsky, P. & Rickman, H. (Springer-Verlag, New York, NY, 2007), pp. 303–330.
28. Lewis, L., *Rain of Iron and Ice: the very real threat of comet and asteroid bombardment* (Helix Books Addison-Wesley Publishing Company, Reading, MA, 1996).
29. Hartung, J., Was the Formation of a 20-km Diameter Impact Crater on the Moon Observed on June 18, 1178? *Meteoritics* **11** (3), 187-194 (1976).
30. Withers, P., Meteor storm evidence against the recent formation. *Meteoritics & Planetary Science* **36**, 525-529 (2001).
31. Morota, T. *et al.*, Formation age of the lunar crater Giordano Bruno. *Meteoritics & Planetary Science* **44** (8), 1115-1120 (2009).
32. Clube, V., The Nature of Punctuational Crises and the Spenglerian Model of Civilization. *Vistas in Astronomy* **39** (4), 673-698 (1995).
33. Napier, W. & Asher, D., The Tunguska impact event and beyond. *Astronomy & Geophysics* **50**, 1.18-1.26 (2009).
34. Shoemaker, E., Asteroid and comet bombardment of the Earth. *Annual Review of Earth and Planetary Sciences* **11**, 461-494 (1983).
35. Chapman, C. & Morrison, D., Impacts on the Earth by asteroids and comets: assessing the hazard. *Nature* **367**, 33-40 (1994).
36. Chapman, C., "Rogue Asteroids and Doomsday Comets" by Duncan Steel (John Wiley & Sons, New York, 1995, 308 pp, $24.95). Book Review by Clark R. Chapman. *Meteoritics & Planet Science* **31**, 313-314 (1996).
37. Schmitz, B., Farley, K., Goderis, S., Heck, P. & Bergström, S., An extraterrestrial trigger for the mid-Ordovician ice age: Dust from the breakup of the L-chondrite parent body. *Science Advances* **5**, eaax4184 (2019).

38. Firestone, R. *et al.*, Evidence for an extraterrestrial impact 12,900 years ago that contributed to the megafaunal extinctions and the Younger Dryas cooling. *PNAS* **104** (41), 16016-16021 (2007).
39. Kennett, D. *et al.*, Nanodiamonds in the Younger Dryas Boundary Sediment Layer. *Science* **323**, 94 (2009).
40. Fayek, M., Anovitz, L., Allard, L. & Hull, S., Framboidal iron oxide: Chondrite-like material from the black mat, Murray Springs, Arizona. *Earth and Planetary Science Letters* **319**, 251-258 (2012).
41. Kurbatov, A. *et al.*, Discovery of a nanodiamond-rich layer in the Greenland ice sheet. *Journal of Glaciology* **56** (199), 747-757 (2010).
42. Bunch, T. *et al.*, Very high-temperature impact melt products as evidence for cosmic airbursts and impacts 12,900 years ago. *PNAS* **109** (28), E1903-E1912 (2012).
43. Israde-Alcántara, I. *et al.*, Evidence from central Mexico supporting the Younger Dryas extraterrestrial impact hypothesis. *PNAS* **109** (13), E738-E747 (2012).
44. Wittke, J. *et al.*, Evidence for deposition of 10 million tonnes of impact spherules across four continents 12,800 y ago. *PNAS* **110** (33), E2088-E2097 (2013a).
45. Kinzie, C. *et al.*, Nanodiamond-rich layer across three continents consistent with major cosmic impact at 12,800 cal BP.. *Journal of Geology* **122** (5), 475-506 (2014).
46. Turner, R., Roberts, N., Eastwood, W., Jenkins, E. & Rosen, A., Fire, climate and the origins of agriculture: Micro-charcoal records of biomass burning during the last glacial-interglacial transition in Southwest Asia. *Journal of Quaternary Science* **25** (3), 371-386 (2010).
47. Wolbach, W. *et al.*, Extraordinary Biomass-Burning Episode and Impact Winter Triggered by the Younger Dryas Cosmic Impact ~12,800 Years Ago. 1. Ice Cores and Glaciers. *The Journal of Geology* **126** (2), 165-184 (2018a).
48. Wolbach, W. *et al.*, Extraordinary Biomass-Burning Episode and Impact Winter Triggered by the Younger Dryas Cosmic Impact ~12,800 Years Ago. 2. Lake, Marine, and Terrestrial Sediments. *Journal of Geology* **126** (2) (2018b).
49. Petaev, M., Huang, S., Jacobsen, S. & Zindler, A., Large Pt anomaly in the Greenland ice core points to a cataclysm at the onset of Younger Dryas. *PNAS* **110** (32), 12917-12920 (2013).
50. Moore, C. *et al.*, Widespread platinum anomaly documented at the Younger Dryas onset in North American sedimentary sequences. *Scientific Reports* **7**, 44031 (2017).
51. Kennett, J. *et al.*, Bayesian chronological analyses consistent with synchronous age of 12,835–12,735 Cal B.P. for Younger Dryas boundary on four continents. *PNAS* **112** (32), E4344-E4353 (2015).
52. Hagstrum, J., Firestone, R., West, A., Weaver, J. & Bunch, T., Impact-related microspherules in Late Pleistocene Alaskan and Yukon "muck" deposits signify recurrent episodes of catastrophic emplacement. *Scientific Reports* **7**, 16620 (2017).
53. Kjær, K., Larsen, N., Binder, T., Bjørk, A. & Eisen, O., A large impact crater beneath Hiawatha Glacier in northwest Greenland. *Science Advances* **4**, eaar8173 (2018).
54. Taylor, K. *et al.*, The Holocene–Younger Dryas Transition Recorded at Summit, Greenland. *Science* **278**, 825-827 (1997).
55. Severinghaus, J., Sowers, T., Brook, E., Alley, R. & Bender, M., Timing of abrupt climate change at the end of the Younger Dryas interval from thermally fractionated gases in polar ice. *Nature* **391**, 141-146 (1998).
56. Broecker, W., Was the Younger Dryas Triggered by a Flood? *Science* **312**, 1146-1148 (2006).
57. Carlson, A., What Caused the Younger Dryas Cold Event? *Geology* **38** (4), 383-384 (2010).
58. Moreno, P., Jacobson, G., Lowell, T. & Denton, G., Interhemispheric climate links revealed by a late-glacial cooling episode in southern Chile. *Nature* **409** , 804-808 (2001).
59. Murton, J., Bateman, M., Dallimore, S., Teller, J. & Yang, Z., Identification of Younger Dryas outburst flood path from Lake Agassiz to the Arctic Ocean. *Nature* **464**, 740-743 (2010).
60. Renssen, H. *et al.*, Multiple causes of the Younger Dryas cold period. *Nature Geoscience* **8**, 946-950 (2015).
61. Tian, H., Schryvers, D. & Claeys, P., Nanodiamonds do not provide unique evidence for a Younger Dryas impact. *PNAS* **108** (1), 40-44 (2011).
62. Daulton, T., Suspect cubic diamond "impact" proxy and a suspect lonsdaleite identification. *PNAS* **109** (34), E2242 (2012).
63. Pigati, J. *et al.*, Accumulation of impact markers in desert wetlands and implications for the Younger

Dryas impact hypothesis. *PNAS* **109** (19), 7208-7212 (2012).
64. Gill, J. *et al.*, Paleoecological changes at Lake Cuitzeo were not consistent with an extraterrestrial impact. *PNAS* **109** (34), E2243 (2012).
65. Holliday, V., Surovell, T., Meltzer, D., Grayson, D. & Boslough, M., The Younger Dryas impact hypothesis: A cosmic catastrophe. *Journal of Quaternary Science* **29** (6), 515-530 (2014).
66. Marlon, J. *et al.*, Wildfire responses to abrupt climate change in North America. *PNAS* **106** (8), 2519-2524 (2009).
67. Thy, P., Willcox, G., Barfod, G. & Fuller, D., Anthropogenic origin of siliceous scoria droplets from Pleistocene and Holocene archaeological sites in northern Syria. *Journal of Archaeological Science* **54**, 193-209 (2015).
68. Meltzer, D., Holliday, V., Cannon, M. & Miller, S., Chronological evidence fails to support claim of an isochronous widespread layer of cosmic impact indicators dated to 12,800 years ago. *PNAS* **111** (21), E2162-E2171 (2014).
69. Holliday, V., Problematic dating of claimed Younger Dryas boundary impact proxies. *PNAS* **112** (49), E6721 (2015).
70. Wittke, J. *et al.*, Reply to van Hoesel et al.: Impact-related Younger Dryas boundary nanodiamonds from The Netherlands. *PNAS* **110** (41), E3897-E3898 (2013b).
71. Boslough, M., Harris, A., Chapman, C. & Morrison, D., Younger Dryas impact model confuses comet facts, defies airburst physics. *PNAS* **110** (45), E4170 (2013).
72. Sweatman, M. & Dimitrios Tsikritsis, D., Decoding Göbekli Tepe with archaeoastronomy: What does the Fox say? *Mediterranean Archaeology and Archaeometry* **17** (1), 233-250 (2017).
73. Sweatman, M., *Prehistory Decoded* (Matador, Kibworth, 2019).
74. Sweatman, M. & Coombs, A., Decoding European Palaeolithic Art:Extremely Ancient knowledge of Precession of the Equinoxes. *Athens Journal of History* **X** (Y), 1-30 (2018).
75. Asher, D. & Clube, V., Towards a dynamical history of proto-Encke. *Celestial Mechanics and Dynamical Astronomy* **69** (1), 149-170 (1998).
76. Rogers, J., Origins of the ancient constellations: I. The Mesopotamian traditions. *Journal of the British Astronomical Association* **108** (1), 9-28 (1998a).
77. Thurston, H., *Early Astronomy* (Springer, New York, NY, 1994).

5: Stonehenge

1. Renfrew, C., *Before Civilization* (Jonathon Cape, London, 1973).
2. Pitts, M., *Hengeworld* (Century, London, 2000).
3. Darvill, T., Marshall, P., Parker Pearson, M. & Wainwright, G., Stonehenge remodelled. *Antiquity* **86**, 1021-1040 (2012).
4. Parker Pearson, M. *et al.*, Who was buried at Stonehenge? *Antiquity* **83**, 23-39 (2009).
5. Parker Pearson, M. *et al.*, Craig Rhos-y-felin: a Welsh bluestone megalith quarry for Stonehenge. *Antiquity* **89**, 1331-1352 (2015).
6. Parker Pearson, M., Pollard, J., Richards, C. & Welham, K. ., The origins of Stonehenge: on the track of the bluestones. *Archaeology International* **20**, 52-57 (2017).
7. Snoeck, C. *et al.*, Strontium isotope analysis on cremated human remains from Stonehenge support links with west Wales. *Scientific Reports* **8**, 10790 (2018).
8. Parker Pearson, M., Researching Stonehenge: Theories Past and Present. *Archaeology International* **16**, 72-83 (2013).
9. Steel, D., *Rogue asteroids and doomsday comets* (John Wiley & Sons, Inc., New York, 1993).
10. Craig, O. *et al.*, Feeding Stonehenge: cuisine and consumption at the Late Neolithic site of Durrington Walls. *Antiquity* **89**, 1096-1109 (2015).
11. Wright, E., Viner-Daniels, S., Parker Pearson, M. & Albarella, A., Age and season of pig slaughter at Late Neolithic Durrington Walls (Wiltshire, UK) as detected through a new system for recording tooth wear. *Journal of Archaeological Science* **52**, 497-514 (2014).
12. Viner, C., Evans, J., Albarella, U. & Parker Pearson, M., Cattle mobility in prehistoric Britain:

strontium isotope analysis of cattle teeth from Durrington Walls (Wiltshire, Britain). *Journal of Archaeological Science* **37**, 2812-2820 (2010).
13. Madgwick, R. *et al.*, Multi-isotope analysis reveals that feasts in the Stonehenge environs and across Wessex drew people and animals from throughout Britain. *Science Advances* **5** (3), eaau6078 (2019).
14. Parker Pearson, M. *et al.*, Materializing Stonehenge: The Stonehenge Riverside Project and New Discoveries. *Journal of Material Culture* **11**, 227-261 (2006).
15. Ruggles, C., *Astronomy in Prehistoric Britain and Ireland* (Yale University Press, New Haven, CT, 1999).
16. Ruggles, C., Astronomy and Stonehenge. *Proceedings of the British Academy* **92**, 203-229 (1997).
17. Lockyer, N., *Stonehenge and Other British Stone Monuments Astronomically Considered* (Macmillan, London, 1906).
18. Hawkins, G., Stonehenge Decoded. *Nature* **200**, 306-308 (1963).
19. Hawkins, G., Stonehenge: a Neolithic computer. *Nature* **202**, 1258-1261 (1964).
20. Hawkins, G., Sun, Moon, Men, and Stones. *American Scientist* **53** (4), 391-408 (1965).
21. Hawkins, G. & White, J., *Stonehenge Decoded* (Doubleday, New York, NY, 1965).
22. Newham, C., Stonehenge - a Neolithic observatory. *Nature* **211**, 456-458 (1966).
23. Newham, C., *The Astronomical significance of Stonehenge* (Coates & Parker Ltd, Warminster, Wilts., 1972).
24. Hoyle, F., *On Stonehenge* (W. H. Freeman and Co., San Francisco, CA, 1977).
25. Atkinson, R., Moonshine on Stonehenge. *Antiquity* **40**, 212-216 (1966).
26. Sadler, D., Prediction of eclipses. *Nature* **211**, 1119-1121 (1966).
27. Colton, R. & Martin, R., Eclipse Cycles and Eclipses at Stonehenge. *Nature* **213**, 476-478 (1967).

6: Megalithic yards, standardised stone circles, and lunar observatories

1. Thom, A. S., in *Records in Stone: Papers in memory of Alexander Thom*, edited by Ruggles, C. (Cambridge University Press, Cambridge, 1988), pp. 3-13.
2. Ferguson, L., in *Records in stone: Papers in memory of Alexander Thom*, edited by Ruggles, C. (Cambridge University Press, Cambridge, 1988), pp. 31-131.
3. Sixsmith, E., The megalithic story of Professor Alexander Thom. *Significance* **6** (2), 94-96 (2009).
4. Thom, A., A Statistical Examination of the Megalithic Sites in Britain. *Journal of the Royal Statistical Society. Series A (General)* **118** (3), 275-295 (1955).
5. Thom, A., The Geometry of Megalithic Man. *The Mathematical Gazette* **45** (352), 83-93 (1961).
6. Thom, A., The Megalithic Unit of Length. *Journal of the Royal Statistical Society. Series A (General)* **125** (2), 243-251 (1962).
7. Thom, A., *Megalithic sites in Britain* (Clarendon Press, Oxford, 1967).
8. Thom, A., *Megalithic lunar observatories* (Clarendon Press, Oxford, 1971).
9. Thom, A. & Thom, A. S., in *Records in Stone: Papers in memory of Alexander Thom*, edited by Ruggles, C. (Cambridge University Press, Cambridge, 1988), pp. 132-151.
10. Ruggles, C., *Astronomy in Prehistoric Britain and Ireland* (Yale University Press, New Haven, CT, 1999).
11. Thom, A. & Thom, A. S., *Megalithic Remains in Britain and Brittany* (Clarendon Press, Oxford, 1978b).
12. Thom, A., The Lunar observatories of Megalithic Man. *Vistas in Astronomy* **11**, 1-29 (1969).
13. Thom, A. & Thom, A. S., Megalithic Astronomy. *The Journal of Navigation* **30** (1), 1-14 (1977a).
14. Thom, A. & Thom, A. S., A reconsideration of the lunar sites. *Journal for the History of Astronomy* **9**, 170-179 (1978a).
15. Thom, A. & Thom, A. S., A new study of all megalithic lunar lines. *Journal for the History of Astronomy* **11**, S78 (1980).
16. Thom, A., Thom, A. S. & Thom, A. S., Stonehenge as a possible lunar observatory. *Journal for the History of Astronomy* **6** (1), 19-30 (1975).
17. Thom, A. & Thom, A. S., The Standing Stones in Argyllshire. *Glasgow Archaeological Journal* **6** (6), 5-10 (1979).
18. Thom, A. & Thom, A. S., A megalithic lunar observatory in Orkney: the Ring of Brogar and its cairns. *Journal for the History of Astronomy* **4**, 111-123 (1973a).

19. Thom, A. & Thom, A. S., A Fourth Lunar Foresight for the Brogar Ring. *Journal for the History of Astronomy* **8**, 54-55 (1977b).
20. Thom, A. & Thom, A. S., The astronomical significance of the large Carnac menhirs. *Journal for the History of Astronomy* **2**, 147-160 (1971).
21. Thom, A. & Thom, A. S., The Carnac Alignments. *Journal for the History of Astronomy* **3**, 11-26 (1972a).
22. Thom, A. & Thom, A. S., The uses of the alignments at Le Menec Carnac. *Journal for the History of Astronomy* **3**, 151-164 (1972b).
23. Thom, A. & Thom, A. S., The Keriescan cromlechs. *Journal for the History of Astronomy* **4**, 168-173 (1973b).
24. Thom, A. & Thom, A. S., The Kermario alignments. *Journal for the History of Astronomy* **5**, 30-47 (1974).
25. Thom, A., Thom, A. S. & Gorrie, J., The two megalithic lunar observatories at Carnac. *Journal for the History of Astronomy* **7**, 11-26 (1976).
26. Atkinson, R., Alexander Thom: Megalithic Sites in Britain. Oxford: Clarendon Press, 1967. 174 pp.,77 figs. 63s. *Antiquity* **42** (165), 77-78 (1968).
27. Freeman, P., A Bayesian Analysis of the Megalithic Yard. *Journal of the Royal Statistical Society. Series A (General)* **139** (1), 20-55 (1976).
28. Madgwick, R. *et al.*, Multi-isotope analysis reveals that feasts in the Stonehenge environs and across Wessex drew people and animals from throughout Britain. *Science Advances* **5** (3), eaau6078 (2019).
29. Davis, A., in *Records in Stone: Papers in memory of Alexander Thom*, edited by Ruggles, C. (Cambridge University Press, Cambridge, 1988), pp. 392-422.
30. Ponting, M., in *Records in Stone: Papers in memory of Alexander Thom*, edited by Ruggles, C. (Cambridge University Press, Cambridge, 1988), pp. 423-441.
31. Chamberlain, A. & Parker Pearson, M., in *From Stonehenge to the Baltic. Living with Cultural Diversity in the Third Millennium BC: British Archaeological Reports International Series 1692*, edited by Larsson, M. & Parker Pearson, M. (Archaeopress, Oxford, 2007), pp. 169-174.
32. Kenny, J. & Teather, A., New insights into the Neolithic chalk drums from Folkton (North Yorkshire) and Lavant (West Sussex). *PAST* (83), 5-6 (2016).
33. Teather, A., Chamberlain, A. & Parker Pearson, M., The chalk drums from Folkton and Lavant: Measuring devices from the time of Stonehenge. *British Journal for the History of Mathematics* **34** (1), 1-11 (2019).
34. Curtis, R., in *Records in Stone: Papers in memory of Alexander Thom*, edited by Ruggles, C. (Cambridge University Press, Cambridge, 1988), pp. 351-371.
35. Cowan, T., in *Records in Stone: Papers in memory of Alexander Thom*, edited by Ruggles, C. (Cambridge University Press, Cambridge, 1988), pp. 378-389.
36. MacKie, E., in *Records in Stone: Papers in memory of Alexander Thom*, edited by Ruggles, C. (Cambridge University Press, Cambridge, 1988), pp. 206-231.
37. Burl, A., in *Records in Stone: Papers in memory of Alexander Thom*, edited by Ruggles, C. (Cambridge University Press, Cambridge, 1988), pp. 177-205.
38. Burl, A., Intimations of Numeracy in the Neolithic and Bronze Age Societies of the British Isles (c. 3200–1200 B.C.). *Archaeological Journal* **133** (1), 9-32 (1976).
39. Ruggles, C., A reassessment of the high precision lunar sightlines, 1: backsights, indicators and the archaeological status of the sightlines. *Journal for the History of Astronomy* **13** (Archaeoastronomy, No. 4), S21-S40 (1982).
40. Ruggles, C., A reassessment of the high precision lunar sightlines, 2: foresights and the problem of selection. *Journal for the History of Astronomy* **14** (Archaeoastronomy, No. 5), S1-S36 (1983).
41. Cooke, J., Few, R., Morgan, J. & Ruggles, C., Indicated declinations at the Callanish megalithic sites. *Journal for the History of Astronomy* **8**, 113-133 (1977).
42. Norris, R., in *Records in Stone: Papers in memory of Alexander Thom*, edited by Ruggles, C. (Cambridge University Press, Cambridge, 1988), pp. 262-276.

7: Drawing down the Moon

1. Ruggles, C., A New Study of the Aberdeenshire Recumbent Stone Circles, 1 Site data. *Archaeoastronomy: Supplement to the Journal for the History of Astronomy* **15** (6), S55-79 (1984).
2. Burl, A., The Recumbent Stone Circles of North-East Scotland. *Proceedings of the Society of Antiquaries of Scotland* **102**, 56-81 (1970).
3. Ruggles, C. & Burl, A., A New Study of the Aberdeenshire Recumbent Stone Circles, 2 Interpretation. *Archaeoastronomy: Supplement to the Journal for the History of Astronomy* **16** (8), S25-S60 (1985).
4. Ruggles, C., *Astronomy in Prehistoric Britain and Ireland* (Yale University Press, New Haven, CT, 1999).
5. Ruggles, C., in *Handbook of Archaeoastronomy and Ethnoastronomy*, edited by Ruggles, C. (Springer Science, New York, NY, 2014a), pp. 1277-1285.
6. Cope, J., *The modern antiquarian* (Thorsons, 1998).
7. Bradley, R., *The Moon and the Bonfire: An Investigation of Three Stone Circles in Aberdeenshire* (Society of Antiquaries of Scotland, Edinburgh, 2005).
8. Burl, A., Science or symbolism: problems of archaeo-astronomy. *Antiquity* **54**, 191-200 (1980).
9. Henty, E., The Archaeoastronomy of Tomnaverie Recumbent Stone Circle: A Comparison of Methodologies. *Papers from the Institute of Archaeology,* **24** (1) (2014).
10. Burl, A., in *Astronomy and Society in Britain during the Period 4000 - 1500 BC*, edited by Ruggles, C. & Whittle, A. (British Archaeological Reports, Oxford, 1981), pp. 243-274.
11. Ruggles, C. & Prendergast, F., in *Proceedings of the Second SEAC Conference, August 29th-31st, 1994*, edited by Schlosser, W. (Rhur-Universität, Bochum, 1996), pp. 5-15.
12. Ruggles, C., The linear settings of Argyll and Mull. *Journal for the History of Astronomy* **16** (Archaeoastronomy, No. 9), S105-S132 (1985).
13. Ruggles, C., in *Records in stone: Papers in memory of Alexander Thom*, edited by Ruggles, C. (Cambridge University Press, Cambridge, 1988), pp. 232-250.
14. Ruggles, C., in *Handbook of Archaeoastronomy and Ethnoastronomy*, edited by Ruggles, C. (Springer, New York, NY, 2014b), pp. 1287-1296.
15. Ruggles, C. & Burl, A., Astronomical Influences on Prehistoric Ritual Architecture in North-Western Europe: The Case of the Stone Rows. *Vistas in Astronomy* **39**, 517-528 (1995).
16. Martlew, R. & Ruggles, C., The North Mull project (4): excavations at Ardnacross 1989-91. *Journal for the History of Astronomy* **24** (Archaeoastronomy, No. 18), S55-S64 (1993).
17. Martlew, R. & Ruggles, C., Ritual and Landscape on the West Coast of Scotland: an Investigation of the Stone Rows of Northern Mull. *Proceedings of the Prehistoric Society* **62**, 17-131 (1996).
18. Ruggles, C., The stone rows of south-west Ireland: a first reconnaissance. *Archaeoastronomy 19 (Journal for the History of Astronomy)* **25**, S1-S20 (1994).
19. Ruggles, C., Stone rows of three or more stones in south-west Ireland. *Archaeoastronomy 21 (Journal for the History of Astronomy)* **27**, S55-S71 (1996).
20. Silva, F. & Pimenta, F., The crossover of the Sun and the Moon. *Journal for the History of Astronomy* **43** (2), 191-208 (2012).
21. Harding, J., in *Handbook of Archaeoastronomy and Ethnoastronomy*, edited by Ruggles, C. (Springer, New York, NY, 2014), pp. 1239-1247.

8: Spherical Earth

1. Engels, D., The Length of Eratosthenes' Stade. *The American Journal of Philology* **86** (3), 298-311 (1985).
2. Newton, R., The Sources of Eratosthenes Measurement of the Earth. *Quarterly Journal of the Royal Astronomical Society* **21**, 379-387 (1980).
3. Thurston, H., *Early Astronomy* (Springer, New York, NY, 1994).
4. Rawlins, D., The Eratosthenes-Strabo Nile Map. Is it the earliest surviving instance of spherical cartography? Did it supply the 5000 stades arc for Eratosthenes' experiment? *Archive for History of Exact Sciences* **26** (3), 211-219 (1982).
5. MacPherson, H., The Development of Cosmological Ideas. *Popular Astronomy* **21**, 78-84 (1913).

6. MacPherson, H., The cosmological ideas of the Greeks. *Popular Astronomy* **24**, 358-369 (1916).
7. Dicks, D., *Early Greek astronomy to Aristotle* (Cornell University Press, Ithaca, NY, 1970).
8. Evans, J., *The History and Practice of Ancient Astronomy* (Oxford University Press, Inc., New York, NY, 1998).
9. Lynch, D., Visually discerning the curvature of the Earth. *Applied Optics* **34** (34), H39-H43 (2008).
10. Alcock, S. & Cherry, J., in *The Human Past*, edited by Scarre, C. (Thames & Hudson, London, 2013), pp. 472-517.
11. Fitton, L., *Minoans* (The British Museum Press, London, 2002).
12. Hadjidaki, E., Underwater Excavations of a Late Fifth Century Merchant Ship at Alonnesos, Greece : the 1991-1993 Seasons. *Persée* **120** (2), 561-593 (1996).
13. Kristiansen, K. & Larsson, T., *The rise of Bronze Age society* (Cambridge University Press, Cambridge, 2005).
14. Beck, C., Criteria for "amber trade": The evidence in the eastern European Neolithic. *Journal of Baltic Studies* **16** (3), 200-209 (1985).
15. Todd, J., Baltic amber in the ancient near east: A preliminary investigation. *Journal of Baltic Studies* **16** (3), 292-301 (1985).
16. Shaw, I., in *The Oxford history of Ancient Egypt*, edited by Shaw, I. (Oxford University Press, Oxford, 2000), pp. 314-329.
17. Lewis-Williams, D., *The Mind in the Cave:Consciousness And The Origins Of Art* (Thames & Hudson, London, 2002).

9: The origin of the ancient constellations

1. Frank, R., in *Handbook of Archaeoastronomy and Ethnoastronomy*, edited by Ruggles, C. (Springer, New York, NY, 2015), pp. 147-163.
2. Rogers, J., Origins of the ancient constellations: I. The Mesopotamian traditions. *Journal of the British Astronomical Association* **108** (1), 9-28 (1998a).
3. Gibbon, W., Asiatic Parallels in North American Star Lore: Ursa Major. *The Journal of American Folklore* **77** (305), 236-250 (1964).
4. Dillehay, T. *et al.*, New Archaeological Evidence for an Early Human Presence at Monte Verde, Chile. *PLoS One* **10** (11) (2015).
5. Moreno-Mayar, V. *et al.*, Terminal Pleistocene Alaskan genome reveals first founding population of Native Americans. *Nature* **553**, 203-207 (2018).
6. Schaefer, B., The Latitude and the Epoch for the Formation of the Southern Greek Constellations. *Journal for the History of Astronomy* **33**, 313-350 (2002).
7. Mellars, P., in *Prehistoric Europe*, edited by Cunliffe, B. (Oxford University Press, Oxford, 1994), pp. 42-78.
8. Clottes, J., *Cave Art* (Phaidon, New York, 2008).
9. Congregado, L., 1991.
10. Frank, R. & Arregi Bengoa, J., in *Astronomy, Cosmology, and Landscape*, edited by Ruggles, C., Prendergast, F. & Ray, T. (Ocarina Books, Bognor Regis, 2001), pp. 15-43.
11. Rogers, J., Origins of the ancient constellations: II. the Mediterranean traditions. *Journal of the British Astronomical Association* **108** (2), 79-89 (1998b).
12. Altschuler, E., Calude, A., Meade, A. & Page, M., Linguistic evidence supports date for Homeric epics. *Bioessays* **35** (5), 417-420 (2013).
13. Evans, J., *The History and Practice of Ancient Astronomy* (Oxford University Press, Inc., New York, NY, 1998).
14. Maunder, E., *The Astronomy of the Bible* (Hodder & Stoughton, London, 1909).
15. Maunder, E., Origin of the Constellations. *The Observatory* **36**, 329-334 (1913).
16. Crommelin, A., in *Splendour of the heavens*, edited by Phillips, T. & Steavenson, W. (Hutchinson, London, 1923), pp. 640-669.
17. Ovenden, M., The origin of the constellations. *The Philosophical Journal* **1** (3), 1-18 (1966).

18. Schaefer, B., The latitude and epoch for the origin of the astronomical lore of Eudoxus. *Journal for the History of Astronomy* **35**, 161-223 (2004).
19. Roy, The origin of the constellations. *Vistas in Astronomy* **27**, 171-197 (1984).
20. Dicks, D., *Early Greek astronomy to Aristotle* (Cornell University Press, Ithaca, NY, 1970).

10: Astronomy in Ancient Mesopotamia

1. Rogers, J., Origins of the ancient constellations: I. The Mesopotamian traditions. *Journal of the British Astronomical Association* **108** (1), 9-28 (1998a).
2. Friberg, J., A Remarkable Collection of Babylonian Mathematical Texts. *Notices of the AMS* **55** (9), 1076-1086 (2008).
3. Evans, J., *The History and Practice of Ancient Astronomy* (Oxford University Press, Inc., New York, NY, 1998).
4. Neugebauer, O., The Alleged Babylonian Discovery of the Precession of the Equinoxes. *Journal of the American Oriental Society* **70** (1), 1-8 (1950).
5. McIntosh, J., *Ancient Mesopotamia: New perspectives* (ABC-CLIO, Santa Barbara, CA, 2005).
6. Powell, B., *Writing: Theory and History of the Technology of Civilization* (Wiley-Blackwell, Chichester/Malden, MA, 2009).
7. Nissen, H., Damerow, P. & Englund, R., *Archaic Bookkeeping: Early Writing and Techniques of Economic Administration in the Ancient Near East* (University of Chicago Press, Chicago, IL, 1993).
8. Dicks, D., Solstices, Equinoxes, &the Presocratics. *The Journal of Hellenic Studies* **86**, 26-40 (1966).
9. Rochberg-Halton, F., Babylonian Seasonal Hours. *Centaurus* **32** (2), 146-170 (1989).
10. Stephenson, R. & Fatoohi, L., The Babylonian unit of time. *Journal of the History of Astronomy* **25**, 99-110 (1994).
11. Steele, J., Stephenson, F. & Morrison, L., The accuracy of eclipse times measured by the Babylonians. *Journal of the History of Astronomy* **28**, 337-345 (1997).
12. Thurston, H., *Early Astronomy* (Springer, New York, NY, 1994).
13. Hunger, H. & Reiner, E., A scheme for intercalary months from Babylonia. *Wiener Zeitschrift für die Kunde des Morgenlandes* **67**, 21-28 (1975).
14. Ratzon, E., Early Mesopotamian intercalation schemes and the sidereal month. *Mediterranean Archaeology and Archaeometry* **16** (4), 143-151 (2016).
15. Stern, S., The Babylonian Month and the New Moon: Sighting and Prediction. *Journal for the History of Astronomy* **39**, 19-42 (2008).
16. Englund, R., Administrative Timekeeping in Ancient Mesopotamia. *Journal of the Economic and Social History of the Orient* **31** (2), 121-185 (1988).
17. Steele, J., in *Handbook of Archaeoastronomy and Ethnoastronomy*, edited by Ruggles, C. (Springer, New York, NY, 2014), pp. 1841-1845.
18. Sachs, A., Babylonian observational astronomy. *Philosophical Transactions of the Royal Seciety of London A* **276**, 43-50 (1974).
19. Thurston, H., The Babylonian Theory of the Planets. *DIO* **13** (2), 3-9 (2006).
20. Steele, J. & Stephenson, F., Lunar Eclipse Times Predicted by the Babylonians. *Journal for the History of Astronomy* **28**, 119-131 (1997).
21. Nickiforov, M., On the discovery of the saros. *Bulgarian Astronomical Journal* **16**, 73-90 (2010).
22. Gray, J. & Steel, J., Studies on Babylonian goal-year astronomy I: a comparison between planetary data in Goal-Year Texts, Almanacs and Normal Star Almanacs. *Archive for History of Exact Sciences* **63**, 553-600 (2008).
23. Offord, J., The Deity of the Crescent Venus in Ancient Western Asia. *The Journal of the Royal Asiatic Society of Great Britain and Ireland*, 197-203 (1915).
24. Campbell, W., Is The Crescent Form of Venus Visible to the Naked Eye? *Publications of the Astronomical Society of the Pacific* **28** (162), 85 (1916).
25. Moore, P., *The Planet Venus*, 3rd ed. (Faber and Faber Limited, London, 1961).
26. Gray, J. & Steele, J., Studies on Babylonian goal-year astronomy II: the Babylonian calendar and goal-

year methods of prediction. *Archive for History of Exact Sciences* **63**, 611-633 (2009).
27. Ossendrijver, M., Ancient Babylonian astronomers calculated Jupiter's position from the area under a time-velocity graph. *Science* **351**, 482-484 (2016).
28. Ossendrijver, M., in *Handbook of Archaeoastronomy and Ethnoastronomy*, edited by Ruggles, C. (Springer-Verlag, New York, NY, 2015), pp. 1863-1870.
29. Pingree, D., Astronomy and Astrology in India and Iran. *ISIS* **54** (2), 229-246 (1963).
30. Neugebauer, O., Babylonian Planetary Theory. *Proceedings of the American Philosophical Society* **98**, 60-89 (1954).
31. Neugebauer, O., Problems and Methods in Babylonian Mathematical Astronomy Henry Norris Russell Lecture, 1967. *The Astronomical Journal* **72** (8), 964-972 (1967).
32. Aaboe, A., Scientific astronomy in antiquity. *Philosophical Transactions of the Royal Society A* **246**, 21-42 (1974).
33. Cohen, R., Ancient Babylonians took first steps to calculus. *Science* **351**, 435 (2016).
34. Ossendrijver, M., New Results on a Babylonian Scheme for Jupiter's Motion along the Zodiac. *Journal of Near Eastern Studies* **7** (2), 231-247 (2017).
35. Steele, J., Eclipse Prediction in Mesopotamia. *Archive for History of Exact Science* **54**, 421-454 (2000).
36. Brack-Bernsen, L. & Steele, J., Eclipse Prediction and the Length of the Saros in Babylonian Astronomy. *Centaurus* **47** (3), 181-206 (2005).
37. Steele, J., in *Under One Sky: Astronomy and Mathematics in the Ancient Near East*, edited by Steele, J. & Imhausen, A. (Ugarit-Verlag, Münster, 2002), pp. 405-420.
38. Aaboe, A., Remarks on the Theoretical Treatment of Eclipses in Antiquity. *Journal for the History of Astronomy* **3**, 115-118 (1972).
39. Steele, J., Solar Eclipse Times Predicted by the Babylonians. *Journal for the History of Astronomy* **28**, 133-139 (1997).
40. Goldstein, B., On Babylonian discovery of the periods of lunar motion. *Journal for the History of Astronomy* **33**, 1-13 (2002).
41. Goldstein, B., Ancient and Medieval Values for the Mean Synodic Month. *Journal of the History of Astronomy* **34** (114), 65-74 (2003).
42. Brack-Bernsen, L. & Schmidt, O., On the Foundations of the Babylonian Column φ: Astronomical Significance of Partial Sums of the Lunar Four. *Centaurus* **37**, 183-209 (1994).
43. Dicks, D., *Early Greek astronomy to Aristotle* (Cornell University Press, Ithaca, NY, 1970).
44. Jones, A., The Adaptation of Babylonian Methods in Greek Numerical Astronomy. *ISIS* **82**, 441-453 (1991b).

11: Calendars and Clocks of Ancient Egypt

1. Shaw, I., *The Oxford History of Ancient Egypt* (Oxford University Press, Oxford, 2000).
2. Toomer, G., in *The Legacy of Egypt*, edited by Harris, J. (Oxford University Press, Oxford, 1971), pp. 27-53.
3. Parker, R., Ancient Egyptian astronomy. *Philosophical Transactions of the Royal Society of London A* **276**, 51-65 (1974).
4. Thurston, H., *Early Astronomy* (Springer, New York, NY, 1994).
5. Parker, R., *The Calendars of Ancient Egypt* (University of Chicago Press, Chicago, IL, 1950).
6. Parker, R., in *The Legacy of Egypt*, edited by Harris, J. (Oxford University Press, Oxford, 1971), pp. 13-26.
7. Clagett, M., *Ancient Egyptian Science: A Sourcebook. Volume 2: Calendars, Clocks, and Astronomy* (American Philosophical Society, Philadelphia, PA, 1995).
8. Schaefer, B., The heliacal rise of Sirius and ancient Egyptian chronology. *Journal for the History of Astronomy* **31** (2), 149-155 (2000).
9. DeYoung, G., in *Astronomy Across Cultures. Science Across Cultures: The History of Non-Western Science, vol 1.*, edited by Selin, H. & Xiaochun, S. (Springer, Dordrecht, 2000), pp. 475-508.

12: Greek philosophers and astronomers

1. Dicks, D., *Early Greek astronomy to Aristotle* (Cornell University Press, Ithaca, NY, 1970).
2. Thurston, H., *Early Astronomy* (Springer, New York, NY, 1994).
3. Evans, J., *The History and Practice of Ancient Astronomy* (Oxford University Press, Inc., New York, NY, 1998).
4. Jones, A., The Adaptation of Babylonian Methods in Greek Numerical Astronomy. *ISIS* **82**, 441-453 (1991b).
5. Kahn, C., *Pythagoras and the Pythagoreans* (Hackett Publishing, Cambridge, MA, 2001).
6. Dicks, D., Thales. *The Classical Quarterly* **9** (2), 294-309 (1959).
7. Panchenko, D., Thales's Prediction of a Solar Eclipse. *Journal for the History of Astronomy* **25**, 275-288 (1994).
8. Dicks, D., Solstices, Equinoxes, &the Presocratics. *The Journal of Hellenic Studies* **86**, 26-40 (1966).
9. Swerdlow, N., Hipparchus's Determination of the Length of the Tropical Year and the Rate of Precession. *Archive for History of Exact Sciences* **21** (4), 291-309 (1980).
10. Hartner, W., The Rôle of Observations in Ancient and Medieval Astronomy. *Journal for the History of Astronomy* **8**, 1-11 (1977).
11. Thurston, H., Early Greek solstices and equinoxes. *Journal for the History of Astronomy* **32** (107), 154-156 (2001).
12. Thoren, V., Anaxagoras, Eudoxus, and the Regression of the Lunar Nodes. *Journal for the History of Astronomy* **2**, 23-28 (1971).
13. Russell, B., *History of Western Philosophy* (George Allen & Unwin Ltd, London, 1946).
14. Toomer, G., Hipparchus on the Distances of the Sun and Moon. *Archive for History of Exact Sciences* **14** (2), 126-142 (1974b).
15. Swerdlow, N., Hipparchus on the Distance of the Sun. *Centaurus* **14** (1), 287-305 (1969).
16. Pannekoek, A., The Planetary Theory of Ptolemy. *Popular Astronomy* **55** (9), 459-475 (1947).
17. Swerdlow, N., in *Wrong for the Right Reasons. Archimedes (New Studies in the History and Philosophy of Science and Technology), vol 11*, edited by J., B. & A., F. (Springer, Dordrech, 2005), pp. 41-71.
18. Toomer, G., The Chord Table of Hipparchus and the Early History of Greek Trigonometry. *Centaurus* **18** (1), 6-28 (1974a).
19. Jones, A., Ptolemy's First Commentator. *Transactions of the American Philosophical Society* **80** (7), i-iv,1-61 (1990).
20. Jones, A., Hipparchus's Computations of Solar Longitudes. *Journal for the History of Astronomy* **22**, 101-125 (1991a).
21. Neugebauer, O., Astronomical Fragments in Galen's Treatise on Seven-Month Children. *Rivista degli studi orientali* **24**, 92-94 (1949).
22. Krisciunas, K., Determining the Eccentricity of the Moon's Orbit without a Telescope, and Some Comments on "Proof" in Empirical Science. *American Journal of Physics* **78** (8), 834-838 (2010).
23. Goldoni, G., Copernicus decoded. *The Mathematical Intelligencer* **27** (3), 12-30 (2005).
24. Murschel, A., The Structure and Function of Ptolemy's Physical Hypotheses of Planetary Motion. *Journal for the History of Astronomy* **24**, 33-44 (1995).
25. Saliba, G., Early Arabic Critique of Ptolemaic Cosmology: A Ninth-Century Text on the Motion of the Celestial Spheres. *Journal for the History of Astronomy* **25**, 115-141 (1994b).

13: India

1. Farmer, S., Sproat, R. & Witzel, M., The collapse of the Indus-script thesis: The myth of a literate Harappan civilization. *Electronic Journal of Vedic Studies* **11** (2), 19-56 (2004).
2. Rao, R. *et al.*, Entropy, the Indus Script, and Language: A Reply to R. Sproat. *Computational Linguistics* **36** (4), 796-805 (2010).
3. Rao, R. *et al.*, Entropic Evidence for Linguistic Structure in the Indus Script. *Science* **325**, 1165 (2009).
4. Witzel, M., in *Proceedings of the conference on the Indus civilization*, edited by Kenoyer, J. (Madison, WI, 2000).

5. Silva, M. *et al.*, A genetic chronology for the Indian Subcontinent points to heavily sex-biased dispersals. *BMC Evolutionary Biology* **17** (88), 1-18 (2017).
6. Coningham, R., in *The human past*, edited by Scarre, C. (Thames & Hudson, London, 2013), pp. 518-551.
7. Pingree, D., Astronomy and Astrology in India and Iran. *ISIS* **54** (2), 229-246 (1963).
8. Ashfaque, S., Astronomy in the Indus Valley Civilization: A Survey of the Problems and Possibilities of the Ancient Indian Astronomy and Cosmology in the Light of Indus Script Decipherment by the Finnish Scholars. *Centaurus* **21** (2), 149-193 (1977).
9. Evans, J., *The History and Practice of Ancient Astronomy* (Oxford University Press, Inc., New York, NY, 1998).
10. Vahia, M. & Menon, S., in *Mapping the Oriental Sky. Proceedings of the Seventh International Conference on Oriental Astronomy*, edited by Nakamura, T., Orchiston, W., Sôma, M. & Strom, R. (National Astronomical Observatory of Japan., Tokyo, 2011), pp. 1-11.
11. Vahia, M. & Menon, S., A possible Harappan Astronomical Observatory at Dholavira. *Journal of Astronomical History and Heritage* **16** (3), 1-8 (2013).
12. Danino, M., New Insights into Harappan Town-Planning, Proportions and Units, with Special Reference to Dholavira. *Man and Environment* **33** (1), 66-79 (2008).
13. Witzel, M., in *Inside the Texts, Beyond the Texts: New Approaches to the Study of the Vedas*, edited by Witzel, M. (Harvard University Press, Cambridge, MA, 1997), pp. 269-270.
14. Anthony, D., *The Horse the Wheel and Language: how Bronze-Age riders from the Eurasian steppes shaped the modern world* (Princeton University Press, Princeton, NJ, 2007).
15. Kak, S., in *Astronomy across cultures: the history of non-Western astronomy*, edited by Selin, H. (Springer, Dordrecht, 2000), pp. 303-340.
16. Pingree, D., The Mesopotamian Origin of Early Indian Mathematical Astronomy. *Journal for the History of Astronomy* **4** (1), 1-12 (1973).
17. Holay, P., in *Treasures of ancient Indian astronomy*, edited by Abhyankar, K. & Sidharth, B. (Ajanta Publications, Delhi, 1993), pp. 13-34.
18. Sarma, N., Measures of time in ancient India. *Endeavour* **15** (4), 185-188 (1991).
19. Iyengar, R., *Ancient Indian Astronomy in Vedic Texts*, Pune, India, 2016 (unpublished).
20. Ohashi, Y., in *History of Oriental Astronomy*, edited by Ansari, R. (Klewer Academic Publishers, Dortrecht, 2002), pp. 75-82.
21. Abraham, G., in *Treasures of ancient Indian astronomy*, edited by Abhyankar, K. & Sidharth, B. (Ajanta, Delhi, 1993), pp. 85-91.
22. Van Wijk, W., *Decimal Tables for the Reduction of Hindu Dates from the Data of the Sūrya-Siddhānta* (Springer, Dordrecht, 1938).
23. Pant, M., in *Treasures of ancient Indian astronomy*, edited by Abhyankar, K. & Sidharth, B. (Ajanta, Delhi, 1993), pp. 42-46.
24. Iyengar, R., Eclipse period number 3339 in the Rigveda. *Indian Journal of History of Science* **40** (2), 139-152 (2005).
25. Yano, M., in *Sources and Studies in the History of Mathematics and Physical Sciences*, edited by Ramasubramanian, K., Hayashi, T. & Montelle, C. (Hindustan Book Agency, Singapore, 2019), pp. 232-245.
26. Parpola, A., Beginnings of Indian Astronomy, with Reference to a Parallel Development in China. *History of Science in South Asia* **1**, 21-78 (2013).
27. Weinstock, S., Lunar Mansions and Early Calendars. *The Journal of Hellenic Studies* **69**, 48-69 (1949).
28. Pingree, D. & Morrissey, P., On the Identification of the Yogataras of the Indian Naksatras. *Journal for the History of Astronomy* **20** (2), 99-119 (1989).
29. Velikovsky, I., *Worlds in Collision* (Macmillan, New York, NY, 1950).
30. Vartak, P., in *Treasures of ancient Indian astronomy*, edited by Abhyankar, K. & Sidharth, B. (Ajanta Publications, Delhi, 1993), pp. 114-117.
31. Thurston, H., *Early Astronomy* (Springer, New York, NY, 1994).
32. Selin, H., *Encyclopaedia of the History of Science, Technology, and Medicine in Non-Western Cultures* (Springer, New York, NY, 2008).
33. Anseri, R., Aryabhatta I. His Life and his Contributions. *Bulletin of the Astronomical Soceity of India* **5**, 10-18 (1977).

34. Clark, W., *The Aryabhatiya of Aryabhata: an ancient Indian work on Mathematics and Astronomy* (University of Chicago Press, Chicago, IL, 1930).
35. Chakravarty, A., in *Treasures of ancient Indian astronomy*, edited by Abhyankar, K. & Sidharth, B. (Ajanta, Delhi, 1993), pp. 78-84.
36. Neugebauer, O., On some aspects of early Greek astronomy. *Proceedings of the American Philosophical Society* **116** (3), 243-251 (1972).
37. Duke, D., The Equant in India: The Mathematical Basis of Ancient Indian Planetary Models. *Archive for History of Exact Sciences* **59** (6), 563-576 (2005).
38. Hockey, T. *et al.*, *Biographical Encyclopedia of Astronomers* (Springer, New York, NY, 2014).
39. Sidharth, B., in *Treasures of ancient Indian astronomy*, edited by Abhyankar, K. & Sidharth, B. (Ajanta, Delhi, 1993), pp. 98-106.
40. Sidharth, B., *The celestial key to the Vedas* (Inner Traditions International, Rochester, VT, 1999).
41. Dutt, N., in *Treasures of ancient Indian astronomy*, edited by Abhyankar, K. & Sidharth, B. (Ajanta, Delhi, 1993), pp. 118-130.
42. Van der Waerden, B., The conjunction of 3102 BC. *Centaurus* **24** (1), 117-131 (1980).
43. Acharya, E., Evidences of Hierarchy of Brahmi Numeral System. *Journal of the Institute of Engineering* **14** (1), 136-142 (2018).
44. Gupta, R., Spread and triumph of Indian numerals. *Indian Journal of History of Science* **18** (1), 23-38 (1983).
45. Al-Khalili, J., *The House of Wisdom: How Arabic Science Saved Ancient Knowledge and Gave Us the Renaissance* (Penguin, New York, NY, 2011).
46. Cajori, F., The Controversy on the Origin of Our Numerals. *The Scientific Monthly* **9** (5), 458-464 (1919).
47. Plofker, K., Keller, A., Hayashi, T., Montelle, C. & Wujastyk, D., The Bakhshālī Manuscript: A Response to the Bodleian Library's Radiocarbon Dating. *History of Science in South Asia* **5** (1), 134-15 (2017).

14: The Islamic Golden Age

1. Thurston, H., *Early Astronomy* (Springer, New York, NY, 1994).
2. Al-Khalili, J., *The House of Wisdom: How Arabic Science Saved Ancient Knowledge and Gave Us the Renaissance* (Penguin, New York, NY, 2011).
3. Goldstein, B., The Arabic Version of Ptolemy's Planetary Hypotheses. *Transactions of the American Philosophical Society* **57** (4), 3-55 (1967).
4. Gupta, R., Spread and triumph of Indian numerals. *Indian Journal of History of Science* **18** (1), 23-38 (1983).
5. King, D., A Survey of Medieval Islamic Shadow Schemes for Simple Time-Reckoning. *Oriens* **32**, 191-249 (1990).
6. Saliba, G., in *The Role of Astronomy in Society and Culture*, edited by Valls-Gabaud, D. & Boksenberg, A. (Cambridge University Press: , Cambridge, 2011), pp. 149-165.
7. Ragep, J. & al-Qūshjī, A., Freeing Astronomy from Philosophy: An Aspect of Islamic Influence on Science. *Osiris* **16**, 49-70 (2001).
8. Ni, L., Abdullah, S. & Yean, W., The World Of Science: Muslims' Philanthropy Astronomy Contribution. *International Journal of Advanced Computing, Engineering and Application* **2** (4), 57-60 (2013).
9. Kennedy, E., A Survey of Islamic Astronomical Tables. *Transactions of the American Philosophical Society* **46** (2), 123-177 (1956).
10. Pingree, D., The Fragments of the Works of Al-Fazārī. *Journal of Near Eastern Studies* **27** (2), 103-123 (1970).
11. Selin, H., *Encyclopaedia of the History of Science, Technology, and Medicine in Non-Western Cultures* (Springer, New York, NY, 2008).
12. Hockey, T. *et al.*, *Biographical Encyclopedia of Astronomers* (Springer, New York, NY, 2014).
13. Algeriani, A. & Mohadi, M., The House of Wisdom (Bayt al-Hikmah) and Its Civilizational Impact on Islamic libraries: A Historical Perspective. *Mediterranean Journal of Social Sciences* **8** (5), 179-187 (2017).
14. Abdukhalimov, B., Ahmad al-Farghani and his compendium of astronomy. *Journal of Islamic Studies* **10** (2), 142-158 (1999).
15. Goldstein, B., On the Theory of Trepidation. *Centaurus* **10**, 232-247 (1964).

16. Evans, J., *The History and Practice of Ancient Astronomy* (Oxford University Press, Inc., New York, NY, 1998).
17. Swerdlow, N., Al-Battani's Determination of the Solar Distance. *Centaurus* **17** (2), 97-105 (1972).
18. Swerdlow, N., Montucla's Legacy: The History of the Exact Sciences. *Journal of the History of Ideas* **54** (2), 299-328 (1993).
19. Saliba, G., Early Arabic Critique of Ptolemaic Cosmology: A Ninth-Century Text on the Motion of the Celestial Spheres. *Journal for the History of Astronomy* **25**, 115-141 (1994b).
20. Saliba, G., *A history of Arabic astronomy: planetary theories during the Golden Age of Islam* (New York University Press, New York, NY, 1994a).
21. Saliba, G., Theory and Observation in Islamic Astronomy: The Work of IBN AL-SHĀTIR of Damascus. *Journal for the History of Astronomy* **18**, 35-43 (1987).
22. Swerdlow, N., The Derivation and First Draft of Copernicus's Planetary Theory: A Translation of the Commentariolus with Commentary. *Proceedings of the American Philosophical Society* **113** (6), 423-512 (1973).
23. Saliba, G., Greek astronomy and the medieval Arabic tradition. *American Scientist* **90** (4), 360-367 (2002).

15: China

1. Sun, X., in *Handbook of Archaeoastronomy and Ethnoastronomy*, edited by Ruggles, C. (Springer, New York, NY, 2015c), pp. 2060-2068.
2. Cullen, C., *Astronomy and mathematics in ancient China: the Zhou bi suan jing* (Cambridge University Press, Cambridge, 1996).
3. Shi, Y., in *Handbook of Archaeoastronomy and Ethnoastronomy*, edited by Ruggles, C. (Springer, New York, NY, 2015), pp. 2031-2042.
4. Thurston, H., *Early Astronomy* (Springer, New York, NY, 1994).
5. He, N., Taosi: An archaeological example of urbanization as a political center in prehistoric China. *Archaeological Research in Asia* **14**, 20-32 (2018).
6. Pankenier, D., Liu, C. & de Mais, S., The Xiangfen, Taosi site: A Chinese Neolithic 'observatory'? *Archaeologica Baltica* **10**, 141-148 (2008).
7. Sun, X., in *Handbook of Archaeoastronomy and Ethnoastronomy*, edited by Ruggles, C. (Springer, New York, NY, 2015d), pp. 2105-2110.
8. Li, G., in *Handbook of Archaeoastronomy and Ethnoastronomy*, edited by Ruggles, C. (Springer, New York, NY, 2015), pp. 2095-2104.
9. Xu, F., in *Handbook of Archaeoastronomy and Ethnoastronomy*, edited by Ruggles, C. (Springer, Dordrecht, 2015), pp. 2111-2116.
10. Richards, E., in *Explanatory Supplement to the Astronomical Almanac*, edited by Urban, S. & Seidelmann, P. (University Science Books, Mill Valley, CA, 2013), pp. 610-615.
11. Pankenier, D., in *Handbook of Archaeoastronomy and Ethnoastronomy*, edited by Ruggles, C. (Springer, New York, NY, 2015), pp. 2069-2078.
12. Morgan, D., *Astral Sciences in Early Imperial China* (Cambridge University Press, Cambridge, 2017).
13. Sun, X., in *Handbook of Archaeoastronomy and Ethnoastronomy*, edited by Ruggles, C. (Springer, New York, NY, 2015b), pp. 2043-2049.
14. Sun, X., in *Handbook of Archaeoastronomy and Ethnoastronomy*, edited by Ruggles, C. (Springer, New York, NY, 2015e), pp. 2133-2140.
15. Dicks, D., Solstices, Equinoxes, & the Presocratics. *The Journal of Hellenic Studies* **86**, 26-40 (1966).
16. Dicks, D., *Early Greek astronomy to Aristotle* (Cornell University Press, Ithaca, NY, 1970).
17. Deng, K., in *Handbook of Archaeoastronomy and Ethnoastronomy*, edited by Ruggles, C. (Springer, Dortrecht, 2015), p. 2117.
18. Sun, X., in *Handbook of Archaeoastronomy and Ethnoastronomy*, edited by Ruggles, C. (Springer, Dortrecht, 2015f), pp. 2127-2132.
19. Sun, X., in *Handbook of Archaeoastronomy and Ethnoastronomy*, edited by Ruggles, C. (Springer, New York, NY, 2015a), pp. 2043-2049.

20. Pang, K., Extraordinary floods in early Chinese history and their absolute dates. *Journal of Hydrology* **96**, 139-155 (1987).
21. Guan, Y., in *Handbook of Archaeoastronomy and Ethnoastronomy*, edited by Ruggles, C. (Springer, New York, NY, 2015), pp. 2079-2084.
22. Lee, Y., Building the Chronology of Early Chinese History. *Asian Perspectives* **41** (1), 15-42 (2002).
23. Higham, C., in *The human past*, edited by Scarre, C. (Thames & Hudson, London, 2013b), pp. 552-593.
24. Qibin, L. & Meidong, C., in *History of Oriental Astronomy*, edited by Ansari, R. (Kluwer Academic Press, Dordrecht, 2002), pp. 227-235.
25. Chen, K., in *History of Oriental Astronomy*, edited by Ansari, R. (Kluwer Academic Publishers, Dordrecht, 2002), pp. 67-73.
26. Pang, K., Yau, K. & Chou, H., in *History of Oriental Astronomy*, edited by Ansari, R. (Kluwer Academic Publications, Dordrecht, 2002), pp. 95-119.

16: The Maya

1. Diamond, J., *Guns, Germs and Steel* (Chatto & Windus, London, 1997).
2. Diamond, J., Evolution, consequences and future of plant and animal domestication. *Nature* **418**, 700-707 (2002).
3. Webster, D. & Evans, S., in *The human past*, edited by Scarre, C. (Thames & Hudson, London, 2013), pp. 594-639.
4. Coe, M. & Koontz, R., *Mexico*, 7th ed. (Thames & Hudson, New York, NY, 2013).
5. Rickard, T., The Use of Meteoric Iron. *The Journal of the Royal Anthropological Institute of Great Britain and Ireland* **71** (1/2), 55-66 (1941).
6. Burger, R. & Gordon, R., Early Central Andean Metalworking from Mina Perdida, Peru. *Science* **282**, 1108-1111 (1998).
7. Aldenderfer, M., Craig, N., Speakman, R. & Popelka-Filcoff, R., Four-thousand-year-old gold artifacts from the Lake Titicaca basin, southern Peru. *PNAS* **105** (13), 5002-5005 (2008).
8. Cooke, C., Abbott, M. & Wolfe, A., in *Encyclopedia of the History of Science, Technology, and Medicine in Non-Western Cultures Vol. 2*, edited by Seline, H. (Kluwer Science, Dordrecht, 2008), pp. 1658-1662.
9. Hosler, D. & Macfarlane, A., Copper Sources, Metal Production, and Metals Trade in Late Postclassic Mesoamerica. *Science* **273**, 1819-1824 (1996).
10. Lewis, P., Ethnologue: Languages of the World - Summary by language family, Available at http://www.ethnologue.com/ethno_docs/distribution.asp?by=family (2009).
11. Pohl, M. *et al.*, Early Agriculture in the Maya Lowlands. *Latin American Antiquity* **7** (4), 355-372 (1996).
12. Coe, M., *The Maya*, 8th ed. (Thames & Hudson, London, 2011).
13. Pohl, M., Pope, K. & von Nagy, C., Olmec Origins of Mesoamerican Writing. *Science* **298**, 1984-1987 (2002).
14. Aveni, A., *Skywatchers* (University of Texas Press, Austin, TX, 2001).
15. Vail, G., in *Handbook of Archaeoastronomy and Ethnoastronomy*, edited by Ruggles, C. (Springer, Dordrecht, 2015), pp. 696-708.
16. Wells, A., Forgotten Chapters of Yucatán's Past: Nineteenth-Century Politics in Historiographical Perspective. *Mexican Studies/Estudios Mexicanos* **12** (2), 195-229 (1996).
17. Thompson, J., Maya astronomy. *Philosophical Transactions of the Royal Society of London A* **276**, 83-98 (1973).
18. Webster, D. & Evans, S., in *The human past*, edited by Scarre, C. (Thames & Hudson, London, 2013), pp. 594-639.
19. Thompson, E., Contributions to American Archaeology. *Maya chronology: the correlation question* **14**, 55-104 (1935).
20. Milbrath, S., *Star Gods of the Maya: Astronomy in Art, Folklore, and Calendars* (University of Texas Press, Austin, TX, 1999).
21. Malmstrom, V., Origin of the Mesoamerican 260-Day Calendar. *Science* **181**, 939-941 (1973).
22. Henderson, J., Origin of the 260-Day Cycle in Mesoamerica. *Science* **185**, 542-543 (1974).

23. Stokstad, E., Oldest New World Writing Suggests Olmec Innovation. *Science* **298**, 1873-1874 (2002).
24. Teeple, J., *Maya Astronomy* (Carnegie Institution of Washington Publications, Washington, DC, 1931).
25. Saturno, W., Stuart, D., Aveni, A. & Rossi, F., Ancient Maya Astronomical Tables from Xultun, Guatemala. *Science* **336**, 714-717 (2012).
26. Bricker, H., Aventi, A. & Bricker, V., Ancient Maya documents concerning the movements of Mars. *PNAS* **98** (4), 2107-2110 (2001).
27. Iwaniszewski, I., in *Handbook of Archaeoastronomy and Ethnoastronomy*, edited by Ruggles, C. (Springer, Dordrecht, 2015), pp. 709-715.
28. Linden, J., Glyph X of the Maya Lunar Series: An Eighteen-Month Lunar Synodic Calendar. *American Antiquity* **51**, 122-136 (1986).
29. Bricker, H. & Bricker, V., More on the Mars Table in the Dresden Codex. *Latin American Antiquity* **8** (4), 384-397 (1997).
30. Bricker, H. & Bricker, V., Classic Maya Prediction of Solar Eclipses. *Current Anthropology* **24** (1), 1-23 (1983).
31. Thurston, H., *Early Astronomy* (Springer, New York, NY, 1994).
32. Thompson, E., *A commentary on the Dresden Codex, a Maya hieroglyphic book* (American Philosophical Society Memoirs 93, Philadelphia, PA, 1972).
33. Aveni, A., Archaeoastronomy in the Maya region: a review of the past decade. *Journal for the History of Astronomy - Archaeoastronomy Supplement No. 3* **12**, S1-S16 (1981).
34. Closs, M., Comment on Bricker, H. & Bricker, V.: Classic Maya Prediction of Solar Eclipses. *Current Anthropology* **24** (1), 1-23 (1983).
35. Justeson, J., A cyclic-time model for eclipse prediction in Mesoamerica and the structure of the eclipse table in the Dresden codex. *Ancient Mesoamerica* **28**, 507-541 (2017).
36. Martin, F., A Dresden Codex Eclipse Sequence: Projections for the Years 1970-1992. *Latin American Antiquity* **41** (1), 74-93 (1993).
37. Martin, F., Venus and the Dresden Codex Eclipse Table. *Journal for the History of Astronomy* **26**, S57-S73 (1995).
38. Aveni, A., Comment on Bricker, H. & Bricker, V.: Classic Maya Prediction of Solar Eclipses. *Current Anthropology* **23** (1), 1-24 (1983).
39. Teeple, J., Maya Inscriptions: The Venus Calendar and Another Correlation. *American Anthropologist* **28**, 402-408 (1926).
40. Thompson, E., A correlation of the Mayan and European calendars. *Field Museum of Natural History Publication* **241** (1), 3-22 (1927).
41. Lounsbury, F., in *Dictionary of Scientific Biography*, edited by Gillispie, C. (Scribner's, New York, NY, 1978), pp. 759-818.
42. Bricker, V., in *New Directions in American archaeoastronomy*, edited by Aveni, A. (British Archaeological Reports International Series 454, Oxford, 1988), pp. 81-103.
43. Aylesworth, G., in *Handbook of Archaeoastronomy and Ethnoastronomy*, edited by Ruggles, C. (Springer, Dordrecht, 2015), pp. 784-791.
44. Aveni, A., Dowd, A. & Vining, B., Maya Calendar Reform? Evidence from Orientations of Specialized Architectural Assemblages. *Latin American Antiquity* **14** (2), 159-178 (2003).
45. Thompson, E., A survey of the northern Maya area. *American Antiquity* **11**, 2-241 (1945).
46. Aveni, A., Gibbs, S. & Hartung, H., The Caracol Tower at Chichen Itza: An Ancient Astronomical Observatory. *Science* **188**, 977-985 (1975).

17: The Copernican revolution

1. Hockey, T. *et al.*, *Biographical Encyclopedia of Astronomers* (Springer, New York, NY, 2014).
2. Rabin, S., Nicolaus Copernicus, Available at https://plato.stanford.edu/archives/fall2019/entries/copernicus/ (2019).
3. Lynn, W., Copernicus and Mercury. *The Observatory* **191**, 321-322 (1892).
4. Moore, P., *New guide to the planets* (Sidgwick & Jackson, London, 1993).

5. Evans, J., *The History and Practice of Ancient Astronomy* (Oxford University Press, Inc., New York, NY, 1998).
6. Gingerich, O., *The Eye of Heaven: Ptolemy, Copernicus, Kepler* (American Institute of Physics, New York, NY, 1993).
7. Thurston, H., *Early Astronomy* (Springer, New York, NY, 1994).
8. Kennedy, E., Late Medieval Planetary Theory. *Isis* **57** (3), 365-378 (1966).
9. Saliba, G., Greek astronomy and the medieval Arabic tradition. *American Scientist* **90** (4), 360-367 (2002).
10. Gingerich, O., Did Copernicus Owe a Debt to Aristarchus. *Journal for the History of Astronomy* **16**, 37-42 (1985).
11. Ruiz-Lapuente, P., Tycho Brahe's Supernova: Light from Centuries Past. *The Astrophysical Journal* **612**, 357-363 (2004).
12. Brecher, K., in *Proceedings of Bridges 2011: Mathematics, Music, Art, Architecture, Culture*, edited by Sarhangi, R. & Séquin, C. (Tessellations Publishing, Melbourne, 2011), pp. 379-386.
13. Andrews, N., Tabula III: Kepler's Mysterious Polyhedral Model. *Journal for the History of Astronomy* **48** (3), 281-311 (2017).
14. Barker, P. & Goldstein, B., Distance and Velocity in Kepler's Astronomy. *Annals of Science* **51**, 59-73 (1994).
15. Aiton, E., How Kepler Discovered the Elliptical Orbit. *The Mathematical Gazette* **59** (1975).
16. Ferguson, T., Who Solved the Secretary Problem? *Statistical Science* **4** (3), 282-289 (1989).
17. Bredekamp, H., Gazing Hands and Blind Spots: Galileo as Draftsman. *Science in Context* **13** (3-4), 423-463 (2000).
18. Moore, P., *The Guinness Book of Astronomy*, 5th ed. (Guinness Publishing Ltd, Enfield, 1995).
19. Pasachoff, J., Simon Marius's Mundus Iovialis: 400th Anniversary in Galileo's Shadow. *Journal for the History of Astronomy* **46** (2), 218-234 (2015).
20. Selin, H., *Encyclopaedia of the History of Science, Technology, and Medicine in Non-Western Cultures* (Springer, New York, NY, 2008).
21. Biagioli, M., Galileo the emblem maker. *Isis* **81** (2), 230-258 (1990).
22. Palmieri, P., Galileo and the Discovery of the Phases of Venus. *Journal for the History of Astronomy* **32**, 109-129 (2001).
23. Kowal, C. & Drake, S., Galileo's observations of Neptune. *Nature* **287**, 311-313 (1980).
24. Jamieson, D., Galileo's miraculous year 1609 and the revolutionary telescope. *Australian Physics* **46** (3), 72-76 (2009).
25. Shea, W., Galileo, Scheiner, and the Interpretation of Sunspots. *Isis* **61** (4), 498-519 (1970).
26. Hawking, S., *A Brief History of Time* (Bantam Press, London, 1988).

18: Standing on the shoulders of giants

1. Hawking, S., *A Brief History of Time* (Bantam Press, London, 1988).
2. Bryson, B., *A Short History of Nearly Everything* (Doubleday, London, 2003).
3. Keynes, M., Balancing Newton's mind: his singular behaviour and his madness of 1692–93. *Notes & Records of the Royal Society* **62**, 289-300 (2008).
4. Fara, P., Newton shows the light: a commentary on Newton (1672) 'A letter... containing his new theory about light and colours...'. *Philosophical Transactions of the Royal Society A* **373**, 20140213 (2015).
5. Nyambuya, G., Dube, A. & Musosi, G., 2017.
6. Patterson, L., Hooke's Gravitation Theory and Its Influence on Newton. I: Hooke's Gravitation Theory. *Isis* **40** (4), 327-341 (1949).
7. Patterson, L., Hooke's Gravitation Theory and Its Influence on Newton. II: The Insufficiency of the Traditional Estimate. *Isis* **41** (1), 32-45 (1950).
8. Cook, A., Edmond Halley and Newton's Principia. *Notes and Records of the Royal Society of London* **45** (2), 129-138 (1991).
9. Cohen, I., in *The Cambridge Companion to Newton*, edited by Iliffe, R. & Smith, G. (Cambridge University Press, Cambridge, 2016), pp. 61-92.

~

Index

~

By the same author

Humans: from the beginning

In just a few years, our understanding of the human past has changed beyond recognition as new discoveries and advances in genetic techniques overturn long-held beliefs and make international news.

Drawing upon expert literature and the latest research, *Humans: from the beginning* is a rigorous but accessible guide to the human story, presenting an even-handed account of events from the first apes to the rise of the first cities and civilisations. Along the way, we learn about the emergence of modern human behaviour, prehistoric art, early modern human migrations from Africa, the peopling of the world, and how farming and agriculture replaced hunter-gathering.

Humans: from the beginning is written for the non-specialist, but it is sufficiently comprehensive and well-referenced to serve as an ideal 'one-stop' text not only for undergraduate students, but also for postgraduates, researchers, and other academics seeking to broaden their knowledge.

ISBN: 978-0-9927620-7-0 (paperback) 978-0-9927620-8-7 (Kindle)

Prehistoric Investigations

How do you infer the existence of a hitherto-unknown human species from a fragmentary finger bone? Why do we walk on two legs? Were Neanderthals really dimwitted? How did a small, solitary predator become the world's most popular pet? What was the ancient link between languages spoken in places as far apart as Iceland and India?

These are just some of the questions faced by those seeking to unravel the secrets of the vast period of time that predates the last six thousand years of 'recorded history'. In addition to fieldwork and traditional methods, paleoanthropologists and archaeologists now draw upon genetics and other cutting-edge scientific techniques.

In fifty chapters, *Prehistoric Investigations* tells the story of the many thought-provoking discoveries that have transformed our understanding of the distant past.

ISBN: 978-0-9927620-6-3 (paperback) 978-0-9927620-5-6 (Kindle